Density-Functional Theory for Electrons

Hideaki Takahashi

Density-Functional Theory for Electrons

Basic Theory and New Formalism

Springer

Hideaki Takahashi
Department of Chemistry, Graduate School
of Science
Tohoku University
Sendai, Japan

ISBN 978-981-95-1228-7 ISBN 978-981-95-1229-4 (eBook)
https://doi.org/10.1007/978-981-95-1229-4

This Springer imprint is published by the registered company Springer Nature Singapore Pte Ltd.
The registered company address is: 152 Beach Road, #21-01/04 Gateway East, Singapore 189721, Singapore

Dedicated to Prof. Dr. Manabu Takahashi

Preface

The mean-field approximation is the basis of what we use to analyze the properties of an interacting many-particle system. The mean field is a potential clothed in the effect due to the complex correlations among the particles besides the bare external potential applied to the particles. The independent electron model associated with the mean field provides an intuitive physical picture to elucidate and to understand the mechanism underlying the chemical event of interest. In the wave-function theory, referred to as the Hartree-Fock (HF) method, the mean field is obtained through the minimization of the expectation value of the anti-symmetrized wave function of the composite system comprised of electrons in the molecule of interest. In the Kohn-Sham density-functional theory (KS-DFT), on the other hand, the mean field is obtained by the functional differentiation of the exchange-correlation functional with respect to the electron density. Then, the mean field and the resultant density on the potential are determined through a self-consistent manner. The first half of Chap. 1 is devoted to describe in detail how to construct the mean fields in the theories. The equations provided in these common textbook contents will be used for later references.

The optimized effective potential (OEP) method offers an implicit route to construct the *local* potential regarded as the mean field of the KS-DFT. The HF-OEP potential is designed so that the KS wave functions optimized on the potential reproduces the HF exchange energy given as a functional of the KS wave functions. The second half of Chap. 1 provides the outline of the HF-OEP method and the numerical problems associated with it. The OEP approach is quite important because it can be considered as a 'proof of principle' for the KS-DFT. It is of crucial importance for the justification of the KS-DFT to examine the possibility that the effect of the electron correlations can be cast onto a local potential. The inverse Kohn-Sham (inv-KS) is a method to determine the potential that uniquely corresponds to a given density. The inv-KS approach has a deep relevance to the OEP method. The relationship between the HF-OEP and the inv-KS will be discussed at the end of Chap. 1. It is shown in the numerical calculations that the inv-KS for the HF ground state density is capable of producing the local potential comparable to or even better than that obtained by the HF-OEP method.

A hydrogen molecule H_2 with an infinite bond length is known as a typical 'entangled' system. The system is considered to be entangled because of the fact that observing the electron spin state on a hydrogen atom simultaneously determines the spin state on the other atomic site. Such a system exhibiting the strong nonlocality of the electron correlation is a representative of the EPR (Einstein-Podolsky-Rosen) state. The EPR state is a characteristic manifestation of the quantum mechanical phenomena. The entangled hydrogen molecule can be properly described with a multi-configuration (MC) wave function of the electrons. The energy difference between the states described with the multi- and single-configuration wave functions is often referred to as the static correlation (SC). In the theoretical framework of DFT, *in principle*, the SC energy can be incorporated into the exchange-correlation functional. In practice, however, it is quite difficult to develop a SC functional by referring only to the local or semi-local properties of the electron density, as a common approximate functional does. In Chap. 2, an approach by A. D. Becke to develop a SC functional within the conventional theoretical framework of the DFT will be introduced. A brief review will also be given for the method to construct the local OEP potential that minimizes the ground state energy of the MC wave function. Through the MC-OEP approach, it is investigated whether the SC can be realized using the MC wave functions built from the one-electron wave functions optimized on the KS local potential.

Chapter 3 describes in detail the novel framework of the DFT for electrons, which constitutes the central subject of the book. The point of the theory is to project the electron density onto the energy coordinate ϵ to construct the energy electron density $n^e(\epsilon)$ which serves as the fundamental variable in the DFT. The energy coordinate ϵ is defined by the external potential that specifies the system under consideration. In parallel to the Hohenberg-Kohn (HK) theorem in the conventional DFT, it can be proved that the one-to-one correspondence is also established between the subset of the energy electron densities and the corresponding subset of the external potentials. It is, thus, shown that the HK functional for the energy electron density can also be formulated. This justifies constructing the mean field for the electrons on the energy coordinate. The approximate exchange functional for $n^e(\epsilon)$ will then be constructed using the conventional local density approximation (LDA) functional as a template. Its extension to more sophisticated functionals that include the derivatives of the density or the kinetic energy density as ingredients will also be investigated.

One may wonder about the advantage of using $n^e(\epsilon)$ instead of the conventional electron density. The notable property of $n^e(\epsilon)$ is that it captures the electron densities of the entangled atoms into the same energy coordinate ϵ no matter how long the atoms are separated. Therefore, we can say that the energy electron density has the spatially nonlocal information about the electron distribution. It is somewhat surprising that the nonlocality is included in the distribution $n^e(\epsilon)$, although the information content of the electron density is certainly degraded through the projection onto the energy coordinate. By taking advantage of this property of the energy electron density $n^e(\epsilon)$, Chap. 4 will be devoted to the development of the static correlation functional in terms of $n^e(\epsilon)$. The functional will be applied to calculate the dissociations of the single, double, and triple bonds in small molecules.

The Thomas-Fermi (TF) model is the kinetic energy functional derived for the homogeneous electron gas. It is well known that the TF model is not capable of realizing the formation of chemical bonds. In Chap. 5, it is shown that the source of the problem in the TF model is the same as that of the static correlation (SC) error in the LDA from the numerical point of view. Thus, it is possible to solve the TF problem in the same way as the method to include the SC in the LDA functional discussed in Chap. 4. On the basis of this idea, in Chap. 5, the kinetic energy functional is developed using the energy electron density to describe chemical bonds in small molecules. The nonlocal kinetic energy functional is also developed, where the response function of the molecule under consideration is also represented on the energy coordinate.

The major drawback of the KSDFT lies in the way to construct the approximate functionals, and not in the principle itself. In the systematic improvement of the functional, the zeroth step of it is usually the LDA approximation which refers only to the local property of the system, i.e. the density at the given position. It seems quite unlikely that the LDA and its modified version can describe the quantum entanglement in a stretched chemical bond. The new formalism of DFT developed in Chap. 3 offers a simple and unique route to describe the nonlocality of the entangled state with the mean field on the energy coordinate. Furthermore, the novel DFT also opens a new possibility for the development of the kinetic energy functional known as a toughest issue in the computational physics and chemistry. The author hopes that the book will serve as an introduction to the new DFT using the energy electron density as the fundamental variable.

Sendai, Japan
April 2025

Hideaki Takahashi

Acknowledgements The author is grateful to Prof. Emeritus Kizashi Yamaguchi of Osaka University for his continuous encouragement and fruitful suggestions in the field of theoretical chemistry. The author is also grateful to Prof. Nobuyuki Matubayasi of Osaka University for private communications with and without scientific issues. The theory of solutions developed by N. M. gave me a hint to construct the new density-functional theory for electrons which constitutes the motivation for writing this book.

Competing Interests The author has no competing interests to declare that are relevant to the content of this manuscript.

Contents

Chapter 1
Wave-Function Theory and Density-Functional Theory

1.1 Mean-Field Approximation

Mean-field approximation (MFA) is a method to describe and analyze the properties of many interacting particles. The approximation is effective and even essential especially when it is applied to the quantum mechanical objects such as electrons. Actually, MFA constitutes the basis of the methods used in both the wave-function theory [1–4] and in the density-functional theory [3, 5–7] for electrons. Note, however, that the physical meanings of MFA are somewhat different in each of these theories. In the present section, we briefly review how the approximations are used in the theories to clarify the difference between them. The discussions will reveal the major problems in the MFA. Then, at the end of this subsection, we will address the main subject of the present book, that is, to establish a new formalism of the DFT for solving the problems. This section serves as a guide to the main theme underlying the content described in Sect. 1.1 through Chap. 4 of this book.

Wave-Function Theory

The MFA in the wave-function theory [1–4] is known as the Hartree-Fock (HF) approximation. It is important to remind that the HF wave function Ψ_{HF} is constructed from an anti-symmetrized product of the one-electron wave functions $\{\varphi_i\}$ for *independent* electrons. The variational minimization of the expectation value of the exact electronic Hamiltonian with respect to $\{\varphi_i\}$ leads to the formation of the mean field felt by each independent electron. Thus, the mean-field approximation is a consequence of the independent particle model. Then, the wave function with the complex correlations among electrons can be obtained by means of a mathematically straightforward process. In the mathematical notion, the HF wave function Ψ_{HF} can be viewed as an element in a tensor-product space constructed from the vector space spanned by the functions $\{\varphi_i\}$. And the wave function for the correlating electrons

H. Takahashi, *Density-Functional Theory for Electrons*,
https://doi.org/10.1007/978-981-95-1229-4_1

can be constructed simply by a linear combination of the elements in the space. An element in the space is often referred to as an electronic configuration. The coefficient for each element in the correlated wave function is determined on the basis of the variational principle. This approach, termed as configuration interaction (CI) method [2–4], allows in principle to build the wave function that describes fully correlated electrons with a given basis set. Other approaches categorized in the wave-function theory mostly take the method more or less along this line. A cumbersome problem, however, arises from the fact that the number of the electronic configurations to be considered increases drastically with the increase in the number N of the electrons in the system. Furthermore, a huge number of electronic configurations are necessitated in general to describe even a subtle electron correlation. In fact, in the practical applications, the computational cost becomes so large that the calculation cannot be carried out in a common computing environment. This has long been a major obstacle for the wave-function theory to be applied to the systems with realistic sizes. Substantial efforts have been devoted to develop the methods to reduce the number of electronic configurations without loss of accuracy for the sake of saving the computational costs [2–4].

Density-Functional Theory

We next consider the mean field in the context of the density-functional theory (DFT) [3, 5–7]. In general, the term DFT often refers to the method formulated by Kohn and Sham (KS-DFT) [8] in 1965, because it is widely recognized as the method of choice in the DFT. It is surprising enough that the mean field provided by the KS-DFT can be formally exact in principle [5]. This means explicitly that the ground state wave function for the *independent* particles bound to the mean field of the KS-DFT allows to generate the exact ground state electron density of the real system of interest. The system of the independent particles is regarded as a *reference* for the real system. The wave function of the reference system differs from that of the real system with the interacting electrons. It is thus important to note that the non-interacting reference system only serves to provide the exact density of the real system through the mean field acting on the particles. Importantly, the existence of such a potential is guaranteed by the Hohenberg-Kohn (HK) theorem [9] for a set of υ-representable electron densities $n_{\upsilon\text{-rep}}(\boldsymbol{r})$ with $\boldsymbol{r}$ being the coordinate of the electron. Note that $n_{\upsilon\text{-rep}}$ is defined as a density derived from the ground state wave function for an N-electron system under a certain external potential $\upsilon_{\text{ext}}(\boldsymbol{r})$. Note, however, that the explicit conditions are not yet known to distinguish whether a given density is υ-representable or not [5]. Anyway, the HK theorem states that there exists one-to-one correspondence between the set of the potentials υ_{ext} and the set of the densities $n_{\upsilon\text{-rep}}(\boldsymbol{r})$. As a direct consequence of the HK theorem, it is possible to mention that any property of a system in its ground state is a functional of the corresponding density. Therefore, the exchange and correlation energy E_{xc} of the electrons in the system can also be described as a functional of the density, that is, $E_{xc}[n_{\upsilon\text{-rep}}]$. Fortunately, it was turned out later by the formalism of Levy's constraint search [10] that the domain of definition of the functional can be extended from the set of the v-rep. densities to that of N-rep. densities $n_{N\text{-rep}}$. Note that $n_{N\text{-rep}}$ is defined as a density derived from an anti-symmetric wave function of N particles, and the necessary and sufficient conditions to be fulfilled by an N-rep. density are known explicitly [5].

The functional derivative of E_{xc} with respect to the electron density, i.e. $v_{xc}[n](\boldsymbol{r}) = \frac{\delta E_{xc}[n]}{\delta n(\boldsymbol{r})}$ plays a decisive role in constructing the mean field in KS-DFT. Actually, the exact electronic energy as well as the electron density of the ground state of the system can be obtained provided that the exact functional for $E_{xc}[n_{N\text{-rep}}]$ is given. In this sense, the mean field created in DFT is not merely an approximate object, but a quantity that allows to deliver the exact ground state electronic energy and the density of the system of interest. Thus, the DFT for electrons offered a revolution in the method of electronic structure calculations. Along with the efforts to establish the basic formalism of the KS-DFT, a lot of E_{xc} functionals have been developed by many theoretical physicists and chemists, making the KS-DFT a practical computational tool that achieves almost the same accuracy as the advanced wave-function theories at a much less computational cost [6]. Nowadays, KS-DFT method is well established as an essential tool in the modern theoretical physics and material science [7].

In the previous paragraph, we introduced the exchange-correlation potential $v_{xc}[n](\boldsymbol{r})$ given by the functional derivative of E_{xc} with respect to the density n. It should be noticed that the potential v_{xc} is merely a *local* operator in the real space, regardless of the fact that in principle it allows to generate the exact ground state density. One might think that this implies some inconsistency in the formalism of KS-DFT since the correlation among the electrons is inherently nonlocal. Actually, it seems impossible to incorporate the quantum mechanical effect into the local potential. As will be shown in the next subsection, even the exchange potential of the mean field created by the HF method is found to be *nonlocal* in the real space. Nevertheless, in the framework of the DFT, such a local potential $v_{xc}[n](\boldsymbol{r})$ exists at least formally against our intuition. The functional $E_{xc}[n]$, as a mathematical entity, can describe the variation of the exchange-correlation energy given as a result of a complex nonlocal response to the variation of the density at $\boldsymbol{r}$. In fact, the problem in the KS-DFT arises from the approximation which underlies the functional development, and not from the KS-DFT formalism itself. The major series of the approximate E_{xc} functional is based on the local-density approximation (LDA) [5, 6, 8, 11] and it refers only to the local and semi-local properties of the system, namely the electron density $n(\boldsymbol{r})$ and its gradient $|\nabla n(\boldsymbol{r})|$ or second derivative $\nabla^2 n(\boldsymbol{r})$ at most [6, 7, 12–15]. Obviously, such an approach cannot realize the energy change in E_{xc} associated with the nonlocal response to the density fluctuation at the distal site. Beyond the framework of the approximation, is there any method to realize the nonlocal effect by the local potential? In fact, through an *implicit* approach termed optimized-effective potential (OEP) method [7, 16, 17], such a local potential can be made possible. The OEP potential is conceptually quite important because it can be regarded as a proof-of-principle for the KS-DFT. Unfortunately, however, it is not efficient in the practical applications because it requires the self-consistent field (SCF) calculation to build the mean field (effective potential) that is combined with the SCF for the wave functions. The detailed explanation for the OEP approach will be provided in a later subsection.

Static Correlation

Inclusion of the nonlocal effect of the electrons is essential for a quantum mechanical approach to describe the dissociation of a chemical bond even in the simplest molecule, e.g. a hydrogen molecule H_2. The energy due to the nonlocal effect associated with a bond dissociation is often referred to as the static correlation (SC). The SC plays a decisive role in describing the correct potential energy curves of chemical bonds [2, 4]. Actually, it makes a significant contribution to the total energy at the dissociation limit. It is well known that the SC can be described in a formally robust approach within the wave-function theory, where the total wave function is represented by a linear combination of a few electronic configurations with the major contributions. In other words, the long-range correlation between the two electrons on the stretched chemical bond can be properly described by means of the multi-configuration wave function. Unfortunately, however, it inevitably leads to a considerable increase in the computational costs. On the other hand, in the DFT, evaluation of the SC energy is known to be one of the most difficult problems, since the use of only a single electronic configuration is assumed in the KS-DFT. The failure of the KS-DFT is directly caused by the fact that the absolute of the exchange energy is erroneously decreased during the bond dissociation [18]. As noted above, the source of the problem in DFT lies in the fact that the approximate E_{xc} functional refers only to the local or semi-local property of electrons. It is quite unlikely that the potential formed by the exchange hole, that is split into two fragments at the atomic sites upon bond dissociation, can be described solely in terms of the density. Note, however, that the *perfect* E_{xc} functional is *in principle* capable of realizing the SC energy without introducing the multi-electronic configurations, although its explicit functional form is not yet known.

Density-Functional Theory Based on the Energy Electron Distribution

The above discussions on the SC motivates the writing of this book. Explicitly, the aim of this book is to provide the details of the recent developments on the new DFT formalism, which offers a simple and natural solution for incorporating the SC into a local potential. The basic idea in the new method is to project the electron density $n(\boldsymbol{r})$ onto another one-dimensional coordinate to obtain an electron distribution which serves as a new fundamental variable in the DFT [19]. The introduction of such a variable in the DFT for electrons had been motivated by a development of the theory of solutions by Matubayasi et al. [20] and the subsequent works [21, 22]. In the theory, the solute-solvent interaction potential is used as the function to define the 1-dimensional *coordinate* ϵ for a solvent molecule with respect to a solute of interest. Then, the solvent distribution function $\rho^e(\epsilon)$ serves to formulate the free energy change associated with the insertion of the solute into the solvent within the rigorous framework of the density-functional theory of solution. It was shown in the work by Takahashi in 2018 [19] that the DFT for electrons can also be founded in terms of the energy distribution $n^e(\epsilon)$ where ϵ is the energy coordinate defined as the interaction energy between an electron and the nuclei in the system of interest. The energy electron density $n^e(\epsilon)$ can be obtained by projecting the electron density $n(\boldsymbol{r})$ onto the coordinate ϵ. Of course, some of the information contents of the density $n(\boldsymbol{r})$ will be lost by the projection except for some special cases.

Then, what is the advantage of introducing such a new distribution function? Actually, the quantity $n^e(\epsilon)$ contains the information that is nonlocal in the real space, even though $n^e(\epsilon)$ itself is a local property at the energy coordinate ϵ. To see this more clearly, let us suppose the chemical bond of an H_2 molecule is stretched infinitely with the spin symmetry being maintained. The system of the two hydrogen atoms (H_A and H_B) in such a situation is known as a typical *entangled* system. Note that the population of the electron with spin σ $(= \alpha$ or $\beta)$ is exactly 0.5 on each atomic site, and the electronic energy of each atom is equal to that of an isolated single hydrogen atom. The KS-DFT calculation for atom H_A (or H_B) in the entangled system, however, cannot reproduce the energy of the single hydrogen atom because it only refers to the local property, i.e. the density on H_A (or H_B) within the framework of the LDA. In contrast, the energy electron density $n^e_\sigma(\epsilon)$ at coordinate ϵ contains the electron population on both atomic sites (H_A and H_B) simultaneously, and hence it gives exactly the same energy density as the isolated atom. Thus, using the novel DFT with the distribution $n^e_\sigma(\epsilon)$, the energy of the entangled two hydrogen atoms is, in principle, the twice the energy of the isolated atom. This property offers the major advantage of the new DFT based on the energy electron density.

In the following two subsections, the Hartree-Fock approximation and the Kohn-Sham approach are formulated in detail to show how the mean fields are created in the wave-function theory and in the DFT, respectively. The comparison between these two approaches elucidates the sources of the static correlation error (SCE) arising from the mean-field approximations. In Chap. 2, we then review the conventional approaches to conquer the SCE in these theories. It will be followed by a description of the DFT based on the new formalism utilizing the energy electron density $n^e(\epsilon)$ as a fundamental variable.

1.2 Hartree-Fock Approach

Hartree-Fock (HF) method is the most primitive approach in the field of quantum chemistry and its explanation can be found in any textbook [1, 2]. However, it constitutes the basis of the mean-field approximation in the wave-function theory, and the formulations to derive the HF equation can be used for later references. Thus, we dare to provide a concise review for the HF theory.

1.2.1 Independent Particle Model

The most important approximation in the HF method is that the electrons in the system are assumed to be mutually independent. With this in mind, let us construct the total wave function Ψ_{HF} for $2N$ electrons. By virtue of the assumption, it is justified to consider the wave function φ_i for each electron, termed as one-electron wave function. With the set of $\{\varphi_i\}(i = 1, 2, \cdots, N)$, Ψ_{HF} can be expressed as

$$\Psi_{\mathrm{HF}}(\boldsymbol{x}_1, \boldsymbol{x}_2, \cdots, \boldsymbol{x}_{2N}) = \mathcal{A}[\varphi_1(\boldsymbol{r}_1)\alpha(\omega_1)\varphi_1(\boldsymbol{r}_2)\beta(\omega_2)\cdots\varphi_N(\boldsymbol{r}_{2N})\beta(\omega_{2N})] \quad (1.1)$$

where the closed shell structure for the electrons is assumed and $\mathcal{A}$ represents the operator to make the wave-function anti-symmetric and normalized. In Eq. (1.1) $\boldsymbol{x}_i$ denotes the set of the position vector $\boldsymbol{r}_i$ and the spin coordinate ω_i for the ith electron, that is, $\boldsymbol{x}_i = (\boldsymbol{r}_i, \omega_i)$. On the other hand, we consider the *exact* Hamiltonian $\mathbf{H}$ where the $2N$ electrons are fully interacting with each other. Explicitly, $\mathbf{H}$ is given by

$$\mathbf{H} = \mathbf{T} + \mathbf{V}_{\mathrm{ee}} + \upsilon_{\mathrm{ext}} \quad (1.2)$$

Each operator on the right-hand side of the equation is defined by

$$\mathbf{T} = \sum_{i=1}^{2N} -\frac{1}{2}\nabla_i^2 \quad (1.3)$$

$$\mathbf{V}_{\mathrm{ee}} = \sum_{i>j}^{2N} \frac{1}{|\boldsymbol{r}_i - \boldsymbol{r}_j|} = \sum_{i>j}^{2N} \frac{1}{r_{ij}} \quad (1.4)$$

$$\upsilon_{\mathrm{ext}} = -\sum_{i=1}^{2N}\sum_{a=1}^{N_A} \frac{Z_a}{|\boldsymbol{r}_i - \boldsymbol{R}_a|} \quad (1.5)$$

In Eq. (1.5), it is supposed that the system of interest consists of N_A atoms and $\boldsymbol{R}_a$ and Z_a are the position vector and the charge of the ath nucleus, respectively. It is readily confirmed that Eq. (1.1) can be an eigenfunction of a Hamiltonian $\mathbf{H}_1$ which includes only one-electron operators, obtained by excluding the Coulomb repulsions $\mathbf{V}_{\mathrm{ee}}$ between electrons in Eq. (1.2). Explicitly, $\mathbf{H}_1$ is given by

$$\mathbf{H}_1 = \sum_{i=1}^{2N} \mathbf{h}_1(i) \quad (1.6)$$

Assuming that the system of interest consists of N_A atoms, $\mathbf{h}_1(i)$ for ith electron is expressed as

$$\mathbf{h}_1(i) = -\frac{1}{2}\nabla_i^2 - \sum_{a=1}^{N_A} \frac{Z_a}{|\boldsymbol{r}_i - \boldsymbol{R}_a|} \quad (1.7)$$

The full Hamiltonian $\mathbf{H}$ of Eq. (1.2) is given by sum of the one-electron and the two-electron Hamiltonians, thus,

$$\mathbf{H} = \mathbf{H}_1 + \mathbf{H}_2 \tag{1.8}$$

The two-electron Hamiltonian $\mathbf{H}_2$ is identical to $\mathbf{V}_{\text{ee}}$ in Eq. (1.4)

$$\mathbf{H}_2 = \mathbf{V}_{\text{ee}} \tag{1.9}$$

In the following, it is assumed that the orthonormalization conditions for the wave functions

$$\langle \varphi_i | \varphi_j \rangle = \delta_{ij} \quad (i, j = 1, 2, \cdots, N) \tag{1.10}$$

are fulfilled. The operator $\mathcal{A}$ can be represented with a Slater determinant, thus,

$$\Psi_{\text{HF}}(\boldsymbol{x}_1, \boldsymbol{x}_2, \cdots, \boldsymbol{x}_{2N}) = \frac{1}{\sqrt{2N!}} \begin{vmatrix} \varphi_1(1)\alpha(1) & \varphi_1(1)\beta(1) \cdots & \varphi_N(1)\beta(1) \\ \varphi_1(2)\alpha(2) & \varphi_2(2)\beta(2) & \\ \vdots & \ddots & \vdots \\ \varphi_1(2N)\alpha(2N) & \cdots & \varphi_N(2N)\beta(2N) \end{vmatrix} \tag{1.11}$$

where shorthand notations are used for the position vector and the spin coordinate. For the following formulation, however, it is convenient to rewrite the Slater determinant in an equivalent form

$$\Psi_{\text{HF}}(\boldsymbol{x}_1, \boldsymbol{x}_2, \cdots, \boldsymbol{x}_{2N}) = \frac{1}{\sqrt{2N!}} \sum_{\nu=1}^{2N!} (-1)^{p_\nu} \times \mathcal{P}_\nu[\varphi_1(1)\alpha(1)\varphi_1(2)\beta(2) \cdots \varphi_N(2N)\beta(2N)] \tag{1.12}$$

where $\mathcal{P}_\nu$ represents the permutation operator for the $2N$ electrons and p_ν is the number of exchanges to realize the permutation $\mathcal{P}_\nu$. It is important to note that the parity of p_ν is uniquely determined for a given $\mathcal{P}_\nu$. Thus, the sign of $(-1)^{p_\nu}$ can also be determined uniquely.

1.2.2 Hartree-Fock Energy

Our major interest is to obtain the mean field for the one-electron wave functions $\{\varphi_i\}$ that minimizes the expectation value for the *full* Hamiltonian $\mathbf{H}$. To this end, we formulate the Hartree-Fock energy $E_{\text{HF}} = \langle \Psi_{\text{HF}} | \mathbf{H} | \Psi_{\text{HF}} \rangle$ in terms of the wave function $\{\varphi_i\}$. For simplicity, we divide the energy E_{HF} into two contributions due to the one-electron and the two-electron Hamiltonians,

$$E_{\mathrm{HF}} = E_1^{\mathrm{HF}} + E_2^{\mathrm{HF}} = \langle \Psi_{\mathrm{HF}} | \mathbf{H}_1 | \Psi_{\mathrm{HF}} \rangle + \langle \Psi_{\mathrm{HF}} | \mathbf{H}_2 | \Psi_{\mathrm{HF}} \rangle \tag{1.13}$$

We first consider the energy E_1^{HF}. As shown in Eq. (1.6), $\mathbf{H}_1$ is given by sum of the contributions from $2N$ electrons. However, by virtue of the symmetry of the wave function Ψ_{HF}, it can be written in a reduced form, thus,

$$E_1^{\mathrm{HF}} = 2N \times \langle \Psi_{\mathrm{HF}} | \mathbf{h}_1(1) | \Psi_{\mathrm{HF}} \rangle \tag{1.14}$$

where the expectation value only for electron 1 is considered and it is multiplied by $2N$. Now we substitute the explicit form of Ψ_{HF} in Eq. (1.12) into Eq. (1.14), which gives

$$\begin{aligned} E_1^{\mathrm{HF}} = \frac{1}{(2N-1)!} \int d\tau \sum_{\mu=1}^{2N!} (-1)^{p_\mu} \mathcal{P}_\mu [\varphi_1(1)\alpha(1)\varphi_1(2)\beta(2) \cdots \varphi_N(2N)\beta(2N)]^* \\ \times \mathbf{h}_1(1) \sum_{\nu=1}^{2N!} (-1)^{p_\nu} \mathcal{P}_\nu [\varphi_1(1)\alpha(1)\varphi_1(2)\beta(2) \cdots \varphi_N(2N)\beta(2N)] \end{aligned} \tag{1.15}$$

where $d\tau$ represents $d\boldsymbol{x}_1 d\boldsymbol{x}_2 \cdots d\boldsymbol{x}_{2N}$. Note that $(2N!)^2$ terms are included in the expansion of Eq. (1.15). However, because of the orthonormalization condition of Eq. (1.10) and the fact that $\mathbf{h}_1(1)$ concerns only electron 1, the terms vanish unless $\mu = \nu$. Thus, only $2N!$ terms remain in Eq. (1.15). Suppose that electron 1 is being fixed at a wave function φ_i, then the number of permutations of the rest of the electrons is $(2N-1)!$. Note that each of them gives the same energy contribution to E_1^{HF}. Hence, E_1^{HF} can be reduced to

$$E_1^{\mathrm{HF}} = 2 \sum_{i=1}^{N} \langle \varphi_i | \mathbf{h}_1(1) | \varphi_i \rangle \tag{1.16}$$

where the factor 2 on the right-hand side is to consider the two spin states (α and β) for electron 1. For later references, we introduce a definition; $H_{ii} = \langle \varphi_i | \mathbf{h}_1 | \varphi_i \rangle$. Then, we have

$$E_1^{\mathrm{HF}} = 2 \sum_{i=1}^{N} H_{ii} \tag{1.17}$$

The expectation value E_2^{HF} in Eq. (1.13) for the two-electron operator Eq. (1.9) can also be obtained in a similar manner as the energy E_1^{HF} although it will be more complicated. E_2^{HF} can be written as

$$E_2^{\mathrm{HF}} = \frac{2N(2N-1)}{2} \times \left\langle \Psi_{\mathrm{HF}} \left| \frac{1}{r_{12}} \right| \Psi_{\mathrm{HF}} \right\rangle \tag{1.18}$$

In Eq. (1.18), only the repulsion between electrons 1 and 2 is considered and the value is multiplied by the number of electron pairs in the $2N$ electrons. By substituting Eq. (1.12) into the above expression, we get

$$E_2^{\mathrm{HF}} = \frac{1}{2(2N-2)!} \int d\tau \sum_{\mu=1}^{2N!} (-1)^{p_\mu} \mathcal{P}_\mu [\cdots \varphi_i(1)\alpha(1) \cdots \varphi_j(2)\alpha(2) \cdots]^* \\ \times \frac{1}{r_{12}} \sum_{\nu=1}^{2N!} (-1)^{p_\nu} \mathcal{P}_\nu [\cdots \varphi_i(1)\alpha(1) \cdots \varphi_j(2)\alpha(2) \cdots] \tag{1.19}$$

In the notation of Eq. (1.19), it is supposed that electrons 1 and 2 occupy the wave functions φ_i and φ_j, respectively. For the sake of clarity of the discussion, we first assume that the two electrons are placed on the different spatial orbitals ($i \neq j$). Then, the other case ($i = j$) will be discussed later. In the expansion of Eq. (1.19), the terms vanish unless $\mathcal{P}_\mu = \mathcal{P}_\nu$ or $\mathcal{P}_\mu = \mathcal{P}_{(1,2)}\mathcal{P}_\nu$, where $\mathcal{P}_{(1,2)}$ represents the exchange of the electrons 1 and 2. Note that the permutation $\mathcal{P}_\mu$ can differ from $\mathcal{P}_\nu$ by the exchange $\mathcal{P}_{(1,2)}$, since the operator $1/r_{12}$ includes the coordinates of the electrons 1 and 2. Below we first formulate the energy $E_2^{\mathrm{HF}}(\mathcal{P}_\mu = \mathcal{P}_\nu)$. Assuming that electrons 1 and 2 are being fixed on φ_i and φ_j, respectively, the number of permutations for the remaining $2N-2$ electrons is $(2N-2)!$. It should also be noted that there are four different combinations of spin orientations for the two electrons, that is,

$$\begin{cases} \alpha(1)^*\alpha(2)^*\alpha(1)\alpha(2) \\ \alpha(1)^*\beta(2)^*\alpha(1)\beta(2) \\ \beta(1)^*\alpha(2)^*\beta(1)\alpha(2) \\ \beta(1)^*\beta(2)^*\beta(1)\beta(2) \end{cases}$$

Note that integrating each term with respect to the spin coordinates ω_1 and ω_2 leads to 1 since the spin wave functions are assumed to be orthonormalized. With these considerations, we have

$$E_2^{\mathrm{HF}}(\mathcal{P}_\mu = \mathcal{P}_\nu) = 2\sum_{i \neq j}^{N} \int d\boldsymbol{r}_1 d\boldsymbol{r}_2\, \varphi_i^*(1)\varphi_j^*(2) \frac{1}{r_{12}} \varphi_i(1)\varphi_j(2) \tag{1.20}$$

For the case of $\mathcal{P}_\mu = \mathcal{P}_{(1,2)}\mathcal{P}_\nu$, we also have four different combinations of spin orientations as follows:

$$\begin{cases} \alpha(1)^*\alpha(2)^*\alpha(2)\alpha(1) \\ \alpha(1)^*\beta(2)^*\alpha(2)\beta(1) \\ \beta(1)^*\alpha(2)^*\beta(2)\alpha(1) \\ \beta(1)^*\beta(2)^*\beta(2)\beta(1) \end{cases}$$

Note, however, that the second and the third terms in these combinations vanish when one integrates over the spin coordinates since $\langle\alpha|\beta\rangle = \langle\beta|\alpha\rangle = 0$.

$$E_2^{\mathrm{HF}}(\mathcal{P}_\mu = \mathcal{P}_{(1,2)}\mathcal{P}_\nu) = -\sum_{i\neq j}^{N} \int d\boldsymbol{r}_1 d\boldsymbol{r}_2\, \varphi_i^*(1)\varphi_j^*(2)\frac{1}{r_{12}}\varphi_i(2)\varphi_j(1) \tag{1.21}$$

We now consider the remaining case that electrons 1 and 2 occupy the same spatial orbital, namely $i = j$. From Fermi's exclusion principle, it is not forgiven that the two electrons have the same spin orientation. Then, the following two different combinations of spin orientations are possible,

$$\begin{cases} \alpha(1)^*\beta(2)^*\alpha(1)\beta(2) \\ \beta(1)^*\alpha(2)^*\beta(1)\alpha(2) \end{cases}$$

Accordingly, the energy $E_2^{\mathrm{HF}}(i = j)$ is given by

$$E_2^{\mathrm{HF}}(i = j) = \sum_{i}^{N} \int d\boldsymbol{r}_1 d\boldsymbol{r}_2\, \varphi_i^*(1)\varphi_i^*(2)\frac{1}{r_{12}}\varphi_i(1)\varphi_i(2) \tag{1.22}$$

Here we introduce the integrals J_{ij} and K_{ij} for notational convenience, which are, respectively, defined as

$$J_{ij} = \int d\boldsymbol{r}_1 d\boldsymbol{r}_2\, \varphi_i^*(1)\varphi_i(1)\frac{1}{r_{12}}\varphi_j^*(2)\varphi_j(2) \tag{1.23}$$

and

$$K_{ij} = \int d\boldsymbol{r}_1 d\boldsymbol{r}_2\, \varphi_i^*(1)\varphi_j(1)\frac{1}{r_{12}}\varphi_j^*(2)\varphi_i(2) \tag{1.24}$$

Note that the orders of the electrons in these definitions are different from those in Eqs. (1.20) and (1.21) although they are equivalent. The integrals J and K are often referred to as Coulomb integral, and exchange integral, respectively. Anyway, with these definitions, the energy E_2^{HF}, given by the sum of the energies in Eqs. (1.20), (1.21), and (1.22), can be expressed as

$$E_2^{\mathrm{HF}} = 2\sum_{i\neq j}^{N} J_{ij} - \sum_{i\neq j}^{N} K_{ij} + \sum_{i}^{N} J_{ii} \tag{1.25}$$

From the definitions of Eqs. (1.23) and (1.24), it is readily recognized that $J_{ii} = K_{ii}$ and $J_{ii} = 2J_{ii} - K_{ii}$. Then, we get the simplest form for the energy

$$E_2^{\mathrm{HF}} = 2\sum_{i,j}^{N} J_{ij} - \sum_{i,j}^{N} K_{ij} \tag{1.26}$$

Finally, the total electronic energy E_{HF} is obtained by the sum of Eqs. (1.17) and (1.26), which reads

$$E_{\mathrm{HF}} = 2\sum_{i}^{N} H_{ii} + 2\sum_{i,j}^{N} J_{ij} - \sum_{i,j}^{N} K_{ij} \tag{1.27}$$

For later references, we define the kinetic energy E_{kin}, the Hartree energy E_{H}, the HF exchange energy E_x^{HF}, and the external energy E_{ext} that constitute the HF total energy E_{HF}, thus,

$$E_{\mathrm{HF}} = E_{\mathrm{kin}} + E_{\mathrm{H}} + E_x^{\mathrm{HF}} + E_{\mathrm{ext}} \tag{1.28}$$

Explicitly, each of the terms on the right is given by

$$E_{\mathrm{kin}} = 2\sum_{i}^{N} \langle \varphi_i | -\frac{1}{2}\nabla^2 | \varphi_i \rangle \tag{1.29}$$

$$E_{\mathrm{H}} = 2\sum_{i,j}^{N} J_{ij} = 2\sum_{i,j}^{N} \int d\boldsymbol{r}_1 d\boldsymbol{r}_2\, \varphi_i^*(1)\, \varphi_i(1) \frac{1}{r_{12}} \varphi_j^*(2)\, \varphi_j(2) \tag{1.30}$$

$$E_x^{\mathrm{HF}} = -\sum_{i,j}^{N} K_{ij} = -\sum_{i,j}^{N} \int d\boldsymbol{r}_1 d\boldsymbol{r}_2\, \varphi_i^*(1)\, \varphi_j(1) \frac{1}{r_{12}} \varphi_j^*(2)\, \varphi_i(2) \tag{1.31}$$

$$E_{\mathrm{ext}} = 2\sum_{i}^{N} \langle \varphi_i | \hat{\upsilon}_{\mathrm{ext}} | \varphi_i \rangle = \int d\boldsymbol{r}\, n(\boldsymbol{r})\, \upsilon_{\mathrm{ext}}(\boldsymbol{r}) \tag{1.32}$$

where $n(\boldsymbol{r})$ is the electron density and is given by

$$n(\boldsymbol{r}) = 2\sum_{i}^{N} \varphi_i(\boldsymbol{r})\, \varphi_i^*(\boldsymbol{r}) \tag{1.33}$$

1.2.3 Minimization of HF Energy

The HF energy expressed by Eq. (1.27) is the basis of the mean field $\hat{\upsilon}_{\mathrm{HF}}$ given by the HF approximation. The potential $\hat{\upsilon}_{\mathrm{HF}}$ will appear in the equations to be fulfilled by the one-electron wave functions $\{\varphi_i\}$ at the minimum of the energy E_{HF}. Therefore, we next consider the minimization of Eq. (1.27) under the constraint of Eq. (1.10). In the following discussions, the functional derivative plays an important role. For an infinitesimal variation of a wave function from φ_i to $\varphi_i + \delta\varphi_i$, the variation δE_{HF} is of our major concern. Actually, δE_{HF} within the first order of the variation is expressed in terms of the functional derivative, thus,

$$\delta E_{\rm HF} = \int d\boldsymbol{r}\ \frac{\delta E_{\rm HF}}{\delta \varphi_i(\boldsymbol{r})} \delta\varphi_i(\boldsymbol{r}) + c.c. \tag{1.34}$$

where $\frac{\delta E_{\rm HF}}{\delta \varphi_i(\boldsymbol{r})}$ is the functional derivative, and the quantity itself is a function of $\boldsymbol{r}$. We also note in Eq. (1.34) that the complex conjugate of $\delta\varphi_i(\boldsymbol{r})$ also makes a contribution to $\delta E_{\rm HF}$, whereby $\delta E_{\rm HF}$ becomes a real number. With these preparations, the functional derivative of each term in Eq. (1.27) can be readily obtained. For the energy $E_1^{\rm HF}$ given in Eqs. (1.16) and (1.17), the functional derivative with respect to $\varphi_i^*(\boldsymbol{r})$ becomes

$$\frac{\delta}{\delta \varphi_i^*(\boldsymbol{r})} \left(2 \sum_i^N H_{ii} \right) = 2 \langle \boldsymbol{r} | \mathbf{h}_1 | \varphi_i \rangle = 2\, \mathbf{h}_1(\boldsymbol{r}) \varphi_i(\boldsymbol{r}) \tag{1.35}$$

For the term with the Coulomb integrals J_{ij}, we obtain

$$\frac{\delta}{\delta \varphi_i^*(\boldsymbol{r})} \left(2 \sum_{i,j}^N J_{ij} \right) = 4 \left(\sum_j^N \int d\boldsymbol{r}_2 \frac{\varphi_j^*(\boldsymbol{r}_2)\varphi_j(\boldsymbol{r}_2)}{|\boldsymbol{r} - \boldsymbol{r}_2|} \right) \varphi_i(\boldsymbol{r}) \tag{1.36}$$

Note that the prefactor for J_{ij} becomes four after the differentiation in the right-hand side of Eq. (1.36). This is due to the fact that the summation is carried out both for the subscripts i and j. Similarly, the functional derivative for the term with the exchange integrals K_{ij} is given by

$$\frac{\delta}{\delta \varphi_i^*(\boldsymbol{r})} \left(-\sum_{i,j}^N K_{ij} \right) = -2 \left(\sum_j^N \int d\boldsymbol{r}_2 \frac{\varphi_j(\boldsymbol{r})\varphi_j^*(\boldsymbol{r}_2)}{|\boldsymbol{r} - \boldsymbol{r}_2|} \right) \varphi_i(\boldsymbol{r}_2) \tag{1.37}$$

Now it is possible to express $\delta E_{\rm HF}$ within the first order of the change $\delta\varphi_i$ in terms of φ_i. It reads

$$\delta E_{\rm HF} = 2 \int d\boldsymbol{r}\ \delta\varphi_i^*(\boldsymbol{r}) \times \left(\mathbf{h}_1(\boldsymbol{r})\varphi_i(\boldsymbol{r}) + \upsilon_{\rm H}(\boldsymbol{r})\varphi_i(\boldsymbol{r}) + \int d\boldsymbol{r}' \upsilon_x(\boldsymbol{r}, \boldsymbol{r}')\varphi_i(\boldsymbol{r}') \right) + c.c. \tag{1.38}$$

where the Hartree potential $\hat{\upsilon}_{\rm H}$ and the exchange potential $\hat{\upsilon}_x$ are defined as, respectively,

$$\upsilon_{\rm H}(\boldsymbol{r}) = 2 \int d\boldsymbol{r}' \sum_j^N \frac{\varphi_j^*(\boldsymbol{r}')\varphi_j(\boldsymbol{r}')}{|\boldsymbol{r} - \boldsymbol{r}'|} \tag{1.39}$$

and

$$v_x(\boldsymbol{r}, \boldsymbol{r}') = -\sum_j^N \frac{\varphi_j(\boldsymbol{r})\varphi_j^*(\boldsymbol{r}')}{|\boldsymbol{r} - \boldsymbol{r}'|} \tag{1.40}$$

It is shown in these equations that the Hartree potential $\hat{v}_{\mathrm{H}}$ is local in the coordinate representation, but the exchange potential $\hat{v}_x$ is found to be nonlocal. Later we will discuss the nonlocality of the exchange potential $\hat{v}_x$ in more detail.

As mentioned above, the minimization of the energy E_{HF} must be subject to the orthonormalization condition expressed by Eq. (1.10). Such a constraint minimization can be performed through the standard Lagrange's multiplier method (see Appendix B). To do this we introduce the Lagrangian $L[\{\varphi_i\}]$ defined as

$$L[\{\varphi_i\}] = E_{\mathrm{HF}}[\{\varphi_i\}] - \sum_{i,j}^N \epsilon_{ij} \left(\langle \varphi_i|\varphi_j\rangle - \delta_{ij}\right) \quad (i, j = 1, 2, \cdots, N) \tag{1.41}$$

where ϵ_{ij} are the multipliers with the dimension of energy. Then, the constraint minimization can be equivalently replaced by the problem to solve the following equations:

$$\frac{\delta L}{\delta \varphi_i^*(\boldsymbol{r})} = 0 \quad (i = 1, 2, \cdots, N) \tag{1.42}$$

which means that the functional L is stationary with respect to the small variation of each wave function φ_i. The functional derivative of the term with Lagrange's multiplier in the Lagrangian becomes

$$\frac{\delta}{\delta \varphi_i^*(\boldsymbol{r})} \sum_{i,j}^N \epsilon_{ij} \left(\langle \varphi_i \mid \varphi_j\rangle - \delta_{ij}\right) = \sum_j^N \epsilon_{ij}\varphi_j(\boldsymbol{r}) \tag{1.43}$$

From Eqs. (1.38) and (1.43), we obtain the explicit form of Eq. (1.42) for orbital φ_i, thus,

$$\mathbf{h}_1(\boldsymbol{r})\varphi_i(\boldsymbol{r}) + v_{\mathrm{H}}(\boldsymbol{r})\varphi_i(\boldsymbol{r}) + \int d\boldsymbol{r}' v_x(\boldsymbol{r}, \boldsymbol{r}')\varphi_i(\boldsymbol{r}') = \sum_j^N \epsilon_{ij}\varphi_j(\boldsymbol{r}) \tag{1.44}$$

Note that in Eq. (1.44) the multiplier divided by 2 is redefined as ϵ_{ij} for the sake of notational simplicity. Equation (1.44) is referred to as the Hartree-Fock equation and it specifies the necessary condition to be satisfied by the wave functions $\{\varphi_i\}$ that minimize the Hartree-Fock energy E_{HF} given in Eq. (1.27). The left-hand side of the HF equation is regarded as an application of the operator $\hat{f}$, which includes the nonlocal operator $\hat{v}_x$, to the wave function φ_i. Then, Eq. (1.44) can be rewritten as

$$\left\langle \boldsymbol{r} \middle| \hat{f} \middle| \varphi_i \right\rangle = \sum_j^N \epsilon_{ij}\varphi_j(\boldsymbol{r}) \tag{1.45}$$

Multiplying $\varphi_j^*(\boldsymbol{r})$ from the left and integrating with respect to $\boldsymbol{r}$ leads to

$$\left\langle \varphi_j \middle| \hat{f} \middle| \varphi_i \right\rangle = f_{ji} = \epsilon_{ij} \tag{1.46}$$

Since $\hat{f}$ is an Hermite operator, it is possible to find a unitary transformation U to make the matrix for $\hat{f}$ diagonal, thus,

$$\sum_{i,j} (U^\dagger)_{ai} f_{ij} U_{jb} = \epsilon_{ab}\delta_{ab} \tag{1.47}$$

where the matrix U transforms $\{\varphi_i\}$ to form another orthonormalized set $\{\chi_a\}$, thus,

$$\langle \boldsymbol{r} | \chi_a \rangle = \sum_i \langle \boldsymbol{r} | \varphi_i \rangle \langle \varphi_i | \chi_a \rangle \tag{1.48}$$

The quantity $\langle \varphi_i | \chi_a \rangle$ in Eq. (1.48) defines the matrix element U_{ia}. Then, we achieve the *canonical* form of the HF equation instead of Eq. (1.44), thus,

$$\left\langle \boldsymbol{r} \middle| \hat{f} \middle| \chi_a \right\rangle = \epsilon_a \chi_a(\boldsymbol{r}) \tag{1.49}$$

In deriving Eq. (1.49), it is important to remind that the operator $\hat{f}$ itself is dependent on the wave functions $\{\varphi_i\}$ because it includes $\hat{v}_\mathrm{H}$ and $\hat{v}_x$ defined by Eqs. (1.39) and (1.40), respectively. However, it can be readily recognized through elementary calculations that these operators are not changed by any unitary transformations.

In a numerical calculation, one will use some basis set to represent the wave functions or operators. Then, the number of the wave functions obtained as the solution for the HF equation is in principle equal to that of the basis functions. Thus the remaining problem is how to construct the total wave function Ψ_HF from the obtained set of orbitals $\{\varphi_i\}$. In fact, we have no necessary and sufficient conditions to solve the problem. There is only an *ansatz* to construct Ψ_HF that one adopts the wave functions φ_i from the bottom to the Nth in the order of increasing eigenvalues. Anyway, the HF equation must be solved through a self-consistent manner so that the potentials $\hat{v}_\mathrm{H}$ and $\hat{v}_x$ formed by $\{\varphi_i\}$ are consistent with the resultant eigenfunctions. After such a self-consistent field (SCF) calculation, the mean field for the electrons is obtained as a result.

The discussion so far revealed that the anti-symmetrized wave function based on the independent particle model leads to the formation of the mean field as a result of minimizing the expectation value of the full Hamiltonian. Thus, the mean-field approximation is equivalent to the independent particle model in the wave-function theory. Therefore, it is by no means possible to cast the explicit correlations between electrons onto the mean field within the framework of the wave-function theory.

1.2.4 Beyond Mean-Field Approximation

It should be noticed that the deficiency of the HF method lies in the fact that the wave function Ψ_{HF} is merely based on the independent particle model. That is, the variational search of the wave function is constrained within the set of the anti-symmetric functions with the form of Eq. (1.11) or equivalently, Eq. (1.12) for the case of the spin non-polarized $2N$ electrons for example. In fact, to obtain the wave function of the fully correlated electrons, the variational search must be carried out within the set of *all* anti-symmetric functions. Thus, the error of the mean-field approximation (MFA) in the wave-function theory arises from the restriction of the region to be searched in the energy minimization. In Fig. 1.1, it is illustrated that the region of the variational search in the HF method is smaller than the set of all normalized anti-symmetric wave functions for $2N$ electrons.

The extension of the region for the variational search can be readily achieved by introducing the tensor-product space in terms of the vector spaces for the individual electrons. The construction of the tensor-product space is nothing but a mathematical procedure. The explanation can be given for the simplest example. Suppose that we have a real function $f(x, y)$ with two independent variables x and y, and we also have the orthonormalized set of functions $\{q_i(x)\}$ which forms a vector space $\mathbb{V}$ under the operation of addition. Then, we introduce a function $\overline{f}_y(x)$ by fixing the variable y in $f(x, y)$ at a certain value. Provided that the function $f(x, y)$ is a continuous, $\overline{f}_y(x)$ can be expanded by the set $\{q_i(x)\}$, thus,

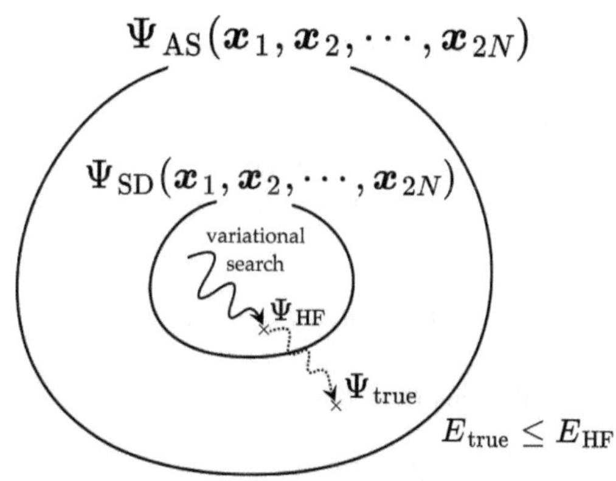

Fig. 1.1 The set of the normalized anti-symmetric wave functions Ψ_{AS} for $2N$ electrons. The variational search of the HF method is confined within the subset of wave functions Ψ_{SD} described with a single determinant. The true wave function Ψ_{true} must be constructed using the electronic configurations created, for example, through HF calculation. The electronic configurations serve as the basis functions to expand the many-body wave functions. In terms of mathematics, the HF wave functions compose the tensor-product space. The ground state energy E_{true} is always less than E_{HF} from the variational principle

$$\overline{f}_y(x) = \sum_i c_i(y) q_i(x) \tag{1.50}$$

where $c_i(y)$ is the coefficient for the function $q_i(x)$ and is dependent on y. The function $c_i(y)$ itself can also be expanded with the set $\{q_i(x)\}$,

$$c_i(y) = \sum_j c_{ij} q_j(y) \tag{1.51}$$

By substituting Eq. (1.51) into Eq. (1.50), we have

$$f(x, y) = \sum_{i,j} c_{ij} q_i(x) q_j(y) \tag{1.52}$$

where y is treated as a free variable. In this expression, $q_i(x)q_j(y)$ is an element of the tensor-product space denoted as $\mathbb{V} \otimes \mathbb{V}$. In general, the tensor-product space can be defined as follows. For the vectors $\mathbf{v}_{\mathrm{A}}^1, \mathbf{v}_{\mathrm{A}}^2 \in \mathbb{V}_{\mathrm{A}}$, and $\mathbf{v}_{\mathrm{B}} \in \mathbb{V}_{\mathrm{B}}$, when the distributive property;

$$\left(c_1\mathbf{v}_{\mathrm{A}}^1 + c_2\mathbf{v}_{\mathrm{A}}^2\right) \otimes \mathbf{v}_{\mathrm{B}} = c_1\mathbf{v}_{\mathrm{A}}^1 \otimes \mathbf{v}_{\mathrm{B}} + c_2\mathbf{v}_{\mathrm{A}}^2 \otimes \mathbf{v}_{\mathrm{B}} \in \mathbb{V}_{\mathrm{A}} \otimes \mathbb{V}_{\mathrm{B}} \tag{1.53}$$

holds (the same property should also be satisfied for the vectors in $\mathbb{V}_{\mathrm{B}}$), the set $\mathbb{V}_{\mathrm{A}} \otimes \mathbb{V}_{\mathrm{B}}$ constitutes the tensor-product space. Thus, it is apparent that $q_i(x)q_j(y) \in \mathbb{V} \otimes \mathbb{V}$.

The above discussion can be readily applied to the wave function for a many-electrons system. Provided that the HF equations of Eq. (1.49) for N orbitals are being solved using some basis functions, then it is possible to form a vector space $\mathbb{W}$ spanned by $\chi_a(\boldsymbol{r})\sigma(\omega)$ $(a = 1, 2, \cdots, M; \sigma = \alpha \text{ or } \beta)$ where M is the number of basis functions $(N < M)$. For the sake of simplicity, we use the notations $\chi_i(\boldsymbol{r}_1)\alpha(\omega_1) = \phi_i(1)$ and $\chi_i(\boldsymbol{r}_1)\beta(\omega_1) = \overline{\phi}_i(1)$. Then, the product, termed Hartree product, expressed by

$$\phi_i(1)\overline{\phi}_i(2)\phi_j(3)\cdots\overline{\phi}_k(2N) \quad (i, j, \cdots, k \leq M) \tag{1.54}$$

constitute an element of the tensor-product space Ω

$$\Omega = \mathbb{W}_1 \otimes \mathbb{W}_2 \otimes \cdots \otimes \mathbb{W}_{2N}, \tag{1.55}$$

where all the vector spaces $\mathbb{W}_i$ are identical, that is, $\mathbb{W} = \mathbb{W}_i \quad (1 \leq i \leq 2N)$. Note, however, that the product of Eq. (1.54) itself is not anti-symmetric and therefore cannot be used as a basis to represent a wave function for Fermions. Thus, we apply the operator $\mathcal{A}$ in Eq. (1.1) to make the product anti-symmetric,

$$\mathcal{A}[\phi_i(1)\overline{\phi}_i(2)\phi_j(3)\cdots\overline{\phi}_k(2N)] \quad (i, j, \cdots, k \leq M) \tag{1.56}$$

Explicitly, $\mathcal{A}$ can be expressed as

$$\mathcal{A} = \frac{1}{\sqrt{2N!}} \sum_{\nu=1}^{2N!} (-1)^{p_\nu} \mathcal{P}_\nu \tag{1.57}$$

Obviously, any element expressed by the form of Eq. (1.56) is included in the vector space Ω. Importantly, the operator $\mathcal{A}$ does not destroy the distributive property of Eq. (1.53). Actually, it can be readily confirmed that the following relation is satisfied:

$$\begin{aligned} \mathcal{A}\Big[\big(c_A \phi_i^A(1) + c_B \phi_i^B(1)\big)\,\overline{\phi}_i(2)\phi_j(3)\cdots\overline{\phi}_k(2N)\Big] = \\ \mathcal{A}\Big[c_A \phi_i^A(1)\overline{\phi}_i(2)\phi_j(3)\cdots\overline{\phi}_k(2N)\Big] \\ + \mathcal{A}\Big[c_B \phi_i^B(1)\overline{\phi}_i(2)\phi_j(3)\cdots\overline{\phi}_k(2N)\Big] \end{aligned} \tag{1.58}$$

where ϕ_i^A and ϕ_i^B are supposed to be the elements in the vector space $\mathbb{W}$. Thus, the elements in the form of Eq. (1.56) compose the subspace Ω_{AS} of the vector space Ω. In the similar manner, it is also possible to construct the subspace Ω_S which consists of the symmetrized products

$$\mathcal{S}[\phi_i(1)\overline{\phi}_i(2)\phi_j(3)\cdots\overline{\phi}_k(2N)] \quad (i, j, \cdots, k \le M) \tag{1.59}$$

although Ω_S is not concerned with the wave functions for Fermions.

Now we are ready to perform the calculations for obtaining the wave function that considers electron correlations. That is, the variational calculation is carried out within the tensor-product subspace Ω_{AS}. Explicitly, the ground state wave function $\Psi_0(\boldsymbol{x}_1, \boldsymbol{x}_2, \cdots, \boldsymbol{x}_{2N})$ with the correlation shall be expanded as

$$|\Psi_0\rangle = C_0\,|\Phi_0\rangle + C_1\,|\Phi_1\rangle + \cdots + C_p\,|\Phi_p\rangle \quad (|C_0|^2 \ge |C_1|^2 \ge \cdots \ge |C_p|^2) \tag{1.60}$$

where $|\Phi_i\rangle$ is the ket vector corresponding to a slater determinant (SD) in the subspace Ω_{AS} and $p+1$ gives the number of SDs included in the expansion. Note that the set of the SDs is orthonormalized, that is, $\langle \Phi_i \,|\Phi_j \rangle = \delta_{ij}$ $(i, j = 0, 1, \cdots, p)$. Specifically, the term $|\Phi_0\rangle$, which gives the largest contribution to the wave function $|\Psi_0\rangle$, will be mostly coincident with the Hartree-Fock wave function. When the ground state is supposed to be singlet in spin state, a lot of SDs that do not have the singlet spin symmetry can be omitted from the outset in the numerical calculations. In practice, to reduce the number p, it is often the case to construct proper linear combinations of the SDs so that the resultant wave functions, often referred to as configuration state functions (CSFs), become spin eigenfunctions. Precisely, each CSF shall be strictly an eigenfunction of the spin operator $\mathbf{S}^2$. How to create the CSF for a given spin state is beyond the scope of this book. We refer readers to e.g. Ref. [23] in more details. For the sake of notational simplicity in the formulation, all the terms are included in the expansion of Eq. (1.60). Of course, even when the CSFs are not employed, the wave function Ψ_0 with the proper spin symmetry is obtained after the full Hamiltonian is diagonalized. The expansion coefficients $\{C_i\}$ in Eq. (1.60) can be obtained through the minimization of the expectation value of the energy $\langle \Psi_0 |\mathbf{H}| \Psi_0 \rangle$ with respect to $\{C_i\}$ under the constraint that $\sum_i^p |C_i|^2 = 1$. To this end, we introduce a Lagrangian L defined as

$$L(C_0, C_1, \cdots, C_p) = \langle \Psi_0 |\mathbf{H}| \Psi_0 \rangle - \epsilon \left(\sum_i^p |C_i|^2 - 1 \right) \tag{1.61}$$

At the local minimum of L, the differentiation of L with respect to each C_i should be zero, thus,

$$\frac{\partial L}{\partial C_i^*} = 0 \quad (i = 1, 2, \cdots, p) \tag{1.62}$$

which leads to

$$\frac{\partial L}{\partial C_i^*} = \sum_j^p \langle \Phi_i |\mathbf{H}| \Phi_j \rangle C_j - \epsilon C_i = 0 \tag{1.63}$$

It can be rewritten in a simpler form

$$\sum_j^p H_{ij} C_j = \epsilon C_i \tag{1.64}$$

by the definition of $H_{ij} = \langle \Phi_i |\mathbf{H}| \Phi_j \rangle$ for the matrix element. The matrix with the element H_{ij} is referred to as CI (configuration interaction) matrix. Fortunately, the CI matrix becomes a sparse matrix by virtue of the orthonormalization condition of the one-electron wave functions $\{\phi_i\}$. Actually, the matrix elements between the electronic configurations Φ_i and Φ_j becomes zero in principle when these wave functions differ from each other by three or more constituent one-electron wave functions. This contributes to the substantial reduction of the computational cost associated with the CI calculations. Since the CI matrix is Hermite, it can be diagonalized by a unitrary transformation $\mathbf{U}$, thus,

$$\mathbf{U}^\dagger \mathbf{H} \mathbf{U} = \begin{pmatrix} \epsilon_0 & & & 0 \\ & \epsilon_1 & & \\ & & \ddots & \\ 0 & & & \epsilon_p \end{pmatrix} \tag{1.65}$$

where each row in the matrix $\mathbf{U}$ constitutes the eigenvector corresponding to each eigenvalue ϵ_i. We thus obtain the coefficients $\{C_i\}$ for the expansion of Eq. (1.60) in the first column of $\mathbf{U}$.

We compare in Fig. 1.2 the HF method based on the independent particle model with the CI method. For later discussion, we consider here the electron density $n(\boldsymbol{r})$ as a representative local property of the system. $n(\boldsymbol{r})$ for an N-electron system is explicitly given by

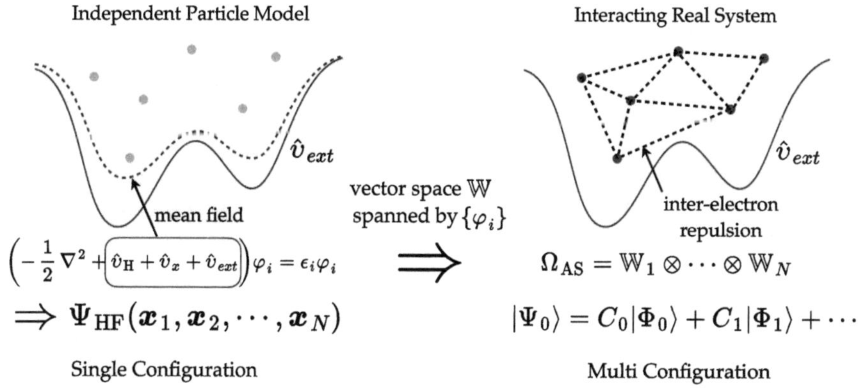

Fig. 1.2 Comparison between *left*: the mean-field approximation (MFA) and *right*: the configuration interaction (CI) method for the calculation of the system with N electrons. On the basis of the independent particle model the mean field for each electron is generated through the SCF calculation. Simultaneously, the set of one electron wave functions $\{\varphi_i\}$ is obtained, which spans a vector space $\mathbb{W}$. The anti-symmetrized products of the elements in $\mathbb{W}$ form a tensor-product space Ω_{AS}. In the CI calculation, the ground state wave function Ψ_0 is given by the sum of multi configurations in Ω_{AS}, while the HF wave function Ψ_{HF} is represented with a single element

$$n_0(\boldsymbol{r}) = N\int d\omega_1 d\boldsymbol{x}_2\cdots d\boldsymbol{x}_N \Psi_0(\boldsymbol{x}_1,\boldsymbol{x}_2,\cdots,\boldsymbol{x}_N)\Psi_0^*(\boldsymbol{x}_1,\boldsymbol{x}_2,\cdots,\boldsymbol{x}_N) \tag{1.66}$$

For the special case that the wave function is represented by a single slater determinant, Eq. (1.66) can be reduced to

$$n_0(\boldsymbol{r}) = \sum_i^N |\phi_i(\boldsymbol{r})|^2 \tag{1.67}$$

where the sum is taken over both spins. As was discussed in the previous paragraph, a wave function for the fully correlated electrons is to be represented with multi-electronic configurations in the space Ω_{AS} provided by the HF method. Therefore, the resultant electron density is, of course, different from the density $n_{\mathrm{HF}}(\boldsymbol{r})$ corresponding to a single configuration. Thus, it is clear that the wave function derived from the mean-field approximation is not capable of reproducing the electron density of the real interacting system. Now, let us change our way of thinking. So far, we have always consider the Hamiltonian of the real system as a matter of course. What happens if one changes the Hamiltonian itself? For a given density $n_0(\boldsymbol{r})$ corresponding to a ground state wave function Ψ_0, is it possible to build some *effective* Hamiltonian $\mathbf{H}_{\mathrm{eff}}$ for the independent particles which yield n_0 through Eq. (1.67)? Even further, is it possible to derive the ground state energy E_0 and other properties of the real system from the ground state wave function Ψ_{ref} for the reference system? These discussions are summarized in Fig. 1.3 to help readers' understandings. It is anticipated that $\mathbf{H}_{\mathrm{eff}}$ includes a sort of 'mean field' for electrons, which, of course, differs

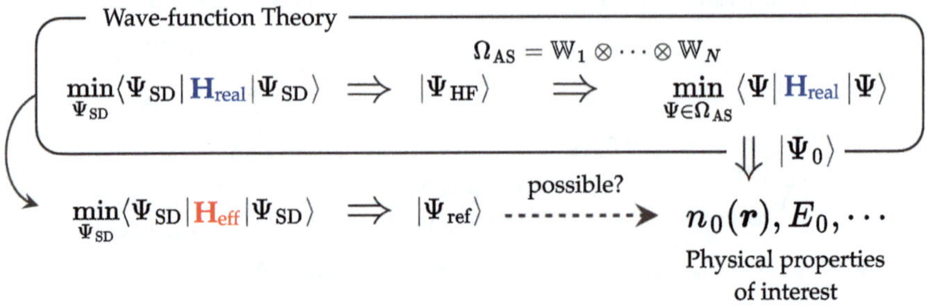

Fig. 1.3 In the wave-function theory, the energy minimization with respect to wave function is always performed for the Hamiltonian (Eq. (1.8)) of the real interacting system. In the first step, the variational search is carried out within the set of Slater determinants (SD) providing Ψ_{HF} at the energy minimum. In the second step, the search region for Ψ is extended to Ω_{AS} spanned by the HF solution, which delivers the exact wave function Ψ_0 and physical properties at the complete basis set limit. In the lower scheme, we consider another approach that can achieve the same result by introducing an effective Hamiltonian $\mathbf{H}_{eff}$. The resultant wave function Ψ_{ref} is different from Ψ_0

from that provided in the HF theory. One might think that changing the Hamiltonian $\mathbf{H}_{real}$ to $\mathbf{H}_{eff}$ makes no sense since it is a manipulation of the system itself even if such a Hamiltonian could actually be realized. In fact, however, the idea constitutes the basis of the Kohn-Sham DFT which will be introduced in the next subsection. The modified system is often referred to as the 'non-interacting reference system'. How the approach can be justified theoretically is one of the main subject in the discussion.

1.3 Kohn-Sham Approach

As revealed in the previous subsection, even a one-electron property, e.g. electron density $n(\boldsymbol{r})$, for the system with fully correlated N electrons cannot be described with a single electronic configuration obtained through a variational search for a full Hamiltonian that defines the system of interest. However, the method proposed by Kohn and Sham in 1965 [8] is to construct the ground state properties by optimizing the one-electron wave functions for the electrons on the local potential. How can we make it possible? It is of crucial importance to examine the theoretical basis underlying the approach. Actually, it needed a rather long history to justify the entire framework of the Kohn-Sham method. The first impact was made by Hohenberg and Kohn in 1964 [9], where they made a proof for a theorem that there exists the one-to-one correspondence between the set of the external potentials v_{ext} and the set of the *resultant* electron densities $n(\boldsymbol{r})$ for the ground states apart from degeneracies. As will be shown in the following, the proof is based on the 'proof by contradiction'. It is interesting to note that the way of the proof was also employed later to formulate the density-functional theory (DFT) for *solutions* [24]. Thus, the DFT for electrons has a great relevance with the DFT for solutions where the distribution function of the solvent molecules around a solute of interest plays a role in determining the solvation free energy of the solute in solution [25]. In fact, the new formalism of the DFT for electrons [19], to be provided in Chap. 3, also has its origin in the recent progress in the DFT for solutions [20]. Lastly, it should also be emphasized that the Hohenberg-Kohn theorem also provides the theoretical basis that the optimized-effective potential (OEP) method [16, 17, 26, 27] can also be considered a method of the Kohn-Sham DFT. The details of the OEP method will be provided in Sect. 1.4.

1.3.1 Hohenberg-Kohn Theorem

The Hohenberg-Kohn (HK) theorem [9] proved in 1964 plays a decisive role in the DFT. It is thought that HK theorem provides the justification for the DFT which had been at most considered as a model in the beginning of the history of the DFT. Actually, it was not until 1964 that an amount of efforts were devoted to the discussions to reinforce the DFT framework and to the developments of the approximate exchange-correlation functionals. Here we make a brief review of the proof of the HK theorem for the later references in Chap. 3 which provides a new formalism of DFT. To this end, suppose that we have a system with N electrons under the influence of an external potential $v_{\text{ext}}(\boldsymbol{r})$ with $\boldsymbol{r}$ the coordinate of the electrons. Then, the Hamiltonian $\hat{\mathbf{H}}$ of the system can be written as

$$\hat{\mathbf{H}} = \hat{\mathbf{T}} + \hat{\mathbf{V}}_{\text{ee}} + \hat{v}_{\text{ext}} \tag{1.68}$$

where $\hat{\mathbf{T}}$ and $\hat{\mathbf{V}}_{\text{ee}}$ are, respectively, the operators for the kinetic energy and the electron-electron repulsion energy. The contents of the operators $\hat{\mathbf{T}}$, $\hat{\mathbf{V}}_{\text{ee}}$ and $\hat{v}_{\text{ext}}$ are explicitly given in Eqs. (1.3)–(1.5), although the number of electrons is changed to N in the present discussion. Suppose that Ψ_0 is the non-degenerate ground state wave function for this Hamiltonian, then we get the following Schrödinger equation:

$$\hat{\mathbf{H}}\,|\Psi_0\rangle = E_0\,|\Psi_0\rangle \tag{1.69}$$

The electron density $n_0(\boldsymbol{r})$ for the wave function can be obtained using Eq. (1.66). Thus, the density $n_0(\boldsymbol{r})$ of the ground state can be uniquely determined from the given external potential $v_{\text{ext}}(\boldsymbol{r})$. How about the opposite direction? Provided that the ground state density $n_0(\boldsymbol{r})$ is given, is it possible to derive uniquely the corresponding external potential? The answer is actually 'yes'. To prove the unique correspondence from the density to the potential, we make an assumption that 'there exists two different external potentials v_{ext} and v'_{ext} that yield the same ground state electron density n_0 except for degeneracy'. We anticipate this will lead to a contradiction when the assumption is not correct. To this end, let us introduce another Hamiltonian $\hat{\mathbf{H}}'$ for the N electrons bound to the v'_{ext} potential,

$$\hat{\mathbf{H}}' = \hat{\mathbf{T}} + \hat{\mathbf{V}}_{\text{ee}} + \hat{v}'_{\text{ext}} \tag{1.70}$$

which yields the corresponding Schrödinger equation

$$\hat{\mathbf{H}}' \left|\Psi_0'\right\rangle = E_0' \left|\Psi_0'\right\rangle \tag{1.71}$$

From the variational principle, the following inequality can be derived:

$$\begin{aligned} E_0 = \langle\Psi_0|\,\hat{\mathbf{H}}\,|\Psi_0\rangle < \left\langle\Psi_0'\right|\hat{\mathbf{H}}\left|\Psi_0'\right\rangle &= E_0' + \left\langle\Psi_0'\right|\hat{\mathbf{H}} - \hat{\mathbf{H}}'\left|\Psi_0'\right\rangle \\ &= E_0' + \int d\boldsymbol{r} n_0(\boldsymbol{r}) \left(\upsilon_{\text{ext}}(\boldsymbol{r}) - \upsilon_{\text{ext}}'(\boldsymbol{r})\right) \end{aligned} \tag{1.72}$$

where we use the assumption to derive the last equality. For the energy E_0', it is also possible to construct a similar inequality

$$\begin{aligned} E_0' = \left\langle\Psi_0'\right|\hat{\mathbf{H}}'\left|\Psi_0'\right\rangle < \langle\Psi_0|\,\hat{\mathbf{H}}'\,|\Psi_0\rangle &= E_0 + \langle\Psi_0|\,\hat{\mathbf{H}}' - \hat{\mathbf{H}}\,|\Psi_0\rangle \\ &= E_0 + \int d\boldsymbol{r} n_0(\boldsymbol{r}) \left(\upsilon_{\text{ext}}'(\boldsymbol{r}) - \upsilon_{\text{ext}}(\boldsymbol{r})\right) \end{aligned} \tag{1.73}$$

Sum of Eqs. (1.72) and (1.73) leads to a contradiction

$$E_0 + E_0' < E_0' + E_0 \tag{1.74}$$

Thus, it is proved that a ground state electron density n_0 corresponds to a unique external potential υ_{ext}. This is the famous Hohenberg-Kohn (HK) theorem [5, 7, 9], which is considered as the basis of the electronic DFT.

It is quite important to note here that the Hohenberg-Kohn (HK) theorem only proves the one-to-one correspondence between the set of υ_{ext} and the set of υ-representable densities. The υ-rep. density is defined as the density that is derived from a ground state wave function for an external potential through Eq. (1.66). This fact was thought to be problematic in the subsequent theoretical developments since the condition for determining whether a density is υ-rep. or not is not yet known. Anyway, let us discuss what will be brought about from the HK theorem. Since the ground state wave function has the one-to-one correspondence to the external potential, there also exists one-to-one correspondence between the set of the wave functions and the set of υ-rep. electron densities as a direct consequence of the HK theorem. The relationships among the three quantities are summarized in Fig. 1.4. It is clear that the variational search in the set of the ground state wave functions Ψ_0 can be equivalently replaced by that in the set of υ-rep. densities. This fact serves as the basis of the DFT for electrons and justifies the subsequent developments.

1.3.2 Hohenberg-Kohn Functional

It is of particular importance to remind that the proof given in the previous subsection only applies to the 'υ-representable' electron densities, that is, a subset of the electron densities associated with the ground state wave functions. In fact, it was implicitly

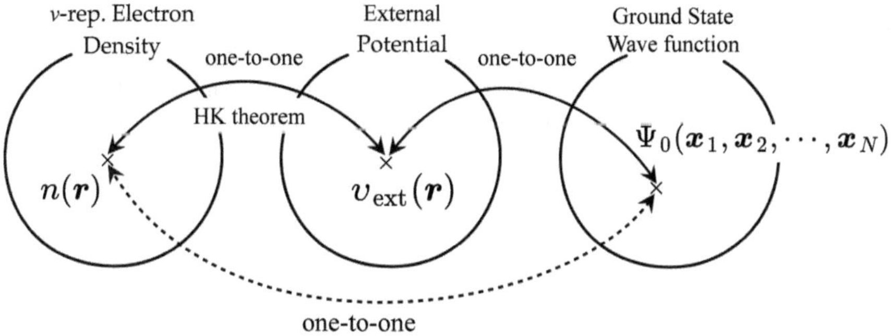

Fig. 1.4 From the Hohenberg-Kohn (HK) theorem, for a given v-representable electron density $n(\boldsymbol{r})$, it is possible to determine uniquely the external potential which yields $n(\boldsymbol{r})$ as the ground state density. Since the correspondence between the set of v_{ext} and the set of the ground state wave functions Ψ_0 is one-to-one, for the density $n(\boldsymbol{r})$, the corresponding wave function Ψ_0 can also be determined uniquely

assumed in the proof that the electron densities are being v-representable, although there exists a lot of electron densities that are *not* v-representable. With this fact in mind, we consider the energy functional $E_{\text{HK}}[n]$ based on the HK theorem. E_{HK} is expressed as

$$E_{\text{HK}}[n] = F_{\text{HK}}[n] + \int d\boldsymbol{r}\; n(\boldsymbol{r}) v_{\text{ext}}(\boldsymbol{r}) \tag{1.75}$$

where the functional F_{HK} is defined as

$$F_{\text{HK}}[n] = \left\langle \Psi_0[n] \left| \hat{\mathbf{T}} + \hat{\mathbf{V}}_{\text{ee}} \right| \Psi_0[n] \right\rangle . \tag{1.76}$$

where the operators $\hat{\mathbf{T}}$ and $\hat{\mathbf{V}}_{\text{ee}}$ are, respectively, the kinetic and electron-electron repulsion energies as shown in the previous subsection. Note that the domain of definition of the functional $F_{\text{HK}}[n]$ as well as $E_{\text{HK}}[n]$ is the set of the v-rep. densities. In Eq. (1.76), $\Psi_0[n]$ represents the ground state wave function which corresponds to a certain v-rep. electron density $n(\boldsymbol{r})$. Since n is supposed to be v-representable, $\Psi_0[n]$ is uniquely determined from the density n due to the one-to-one correspondence (Fig. 1.4). F_{HK} is a functional of the electron density and it serves the sum of the kinetic and the electron repulsion energies of the system with the electron density n. It is apparent in Eq. (1.76) that the domain of definition of F_{HK} is restricted to the set of v-rep. density. However, this constraint gives rise to a serious problem in the variational calculations expressed as

$$E_0 = \min_{n \to v_{\text{rep}}} \left\{ F_{\text{HK}}[n] + \int d\boldsymbol{r}\, n(\boldsymbol{r}) v_{\text{ext}}(\boldsymbol{r}) \right\} \tag{1.77}$$

because the bound of the subset of the v-rep. densities is not yet known. Fortunately, it was shown later that the problem can be formally circumvented through a mathematical technique referred to as Levy's constraint search [10]. The details will be discussed in Sect. 1.3.4. The minimization of Eq. (1.77) can be achieved when the density n becomes n_0 that uniquely corresponds to the external potential v_{ext}. Of course, it is quite straightforward to define the individual functionals for the energies $\hat{\mathbf{T}}$ and $\hat{\mathbf{V}}_{\text{ee}}$, thus,

$$\begin{aligned} T_{\text{HK}}[n] &= \left\langle \Psi_0[n] \left| \hat{\mathbf{T}} \right| \Psi_0[n] \right\rangle \\ V_{\text{ee}}^{\text{HK}}[n] &= \left\langle \Psi_0[n] \left| \hat{\mathbf{V}}_{\text{ee}} \right| \Psi_0[n] \right\rangle \end{aligned} \tag{1.78}$$

with n being a v-rep. density.

1.3.3 HK Theorem and Legendre Transformation

Here we consider the mathematical background of the Hohenberg-Kohn (HK) theorem and the HK functional. To this end, we show again the equation for the variational minimization using the HK functional F_{HK},

$$E_0[v_{\text{ext}}] = \inf_{n \to v\text{-rep.}} \left\{ F_{\text{HK}}[n] + \int d\boldsymbol{r}\ v_{\text{ext}}(\boldsymbol{r}) n(\boldsymbol{r}) \right\} \tag{1.79}$$

where v_{ext} is the external potential specifying the system of interest. Of course, the ground state energy E_0 can be regarded as a functional of v_{ext}. When the energy minimum E_0 is achieved, the density n coincides with $n[v_{\text{ext}}]$ that uniquely corresponds to v_{ext}. It is interesting to note that Eq. (1.79) has the form of the Legendre transformation from $F_{\text{HK}}[n]$ to $E_0[v_{\text{ext}}]$. We refer readers also to Ref. [28], which provides a comprehensive explanation for the HK theorem and the Legendre transformation. Since the functional $E_0[v_{\text{ext}}]$ is known to be concave with respect to v_{ext}, it is possible to construct its Legendre transform $F_L[n]$, thus,

$$-F_L[n] = \inf_{v_{\text{ext}}} \left\{ \int d\boldsymbol{r} n(\boldsymbol{r})\, v_{\text{ext}}(\boldsymbol{r}) - E_0[v_{\text{ext}}] \right\} \tag{1.80}$$

where the density n as an argument of F_L must be v-representable. It should also be noted that the potential v_{ext} is a local property and the density n conjugate to the potential is also local. Figure 1.5 shows a schematic illustration of the Legendre transformation of $E_0[v_{\text{ext}}]$ to generate the Hohenberg-Kohn functional F_{HK} as a functional of v-rep. density n. Note that F_L in Eq. (1.80) corresponds to F_{HK} in the figure. As shown in the figure, the concavity of the functional $E_0[v_{\text{ext}}]$ is essential for the Legendre transformation. The concavity is represented as

$$E_0\left[v_\lambda\right] \geq (1-\lambda)\, E_0\left[v_0\right] + \lambda E_0\left[v_1\right] \tag{1.81}$$

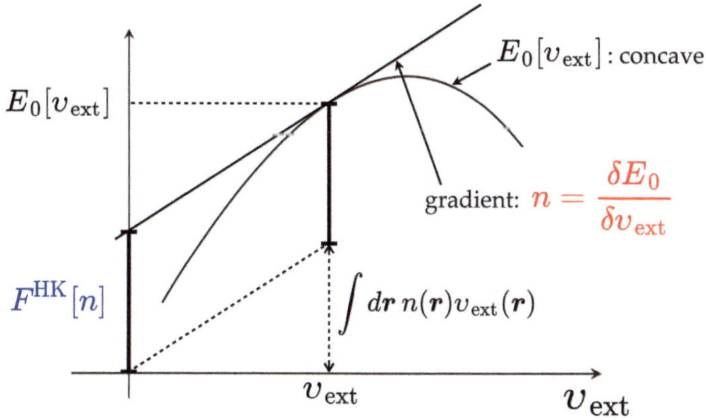

Fig. 1.5 Schematic illustration of the Legendre transformation of the energy functional $E_0[v_{\rm ext}]$ to the Hohenberg-Kohn functional $F_{\rm HK}[n]$ given as a functional of the v-representable density n. As shown in the main text, density n is given as the gradient of E_0 with respect to $v_{\rm ext}$. Then, the intercept of the tangent at $v_{\rm ext}$ gives the functional $F_{\rm HK}[n]$

where λ $(0 \leq \lambda \leq 1)$ is a coupling parameter and v_λ is defined as

$$v_\lambda = (1-\lambda)\, v_0 + \lambda v_1 \tag{1.82}$$

Suppose that the external potentials v_0 and v_1 correspond to the Hamiltonians H_0 and H_1, respectively, and each Hamiltonian H_i $(i = 0, 1)$ yields the ground state wave functions Ψ_i. Then the left-hand side of Eq. (1.81) can be expressed as

$$\begin{aligned} E_0[v_\lambda] &= E_0[\Psi_\lambda] \\ &= \langle \Psi_\lambda | (1-\lambda) H_0 + \lambda H_1 | \Psi_\lambda \rangle \\ &= (1-\lambda) \langle \Psi_\lambda | H_0 | \Psi_\lambda \rangle + \lambda \langle \Psi_\lambda | H_1 | \Psi_\lambda \rangle \end{aligned} \tag{1.83}$$

On the other hand, the right-hand side becomes

$$\begin{aligned} (1-\lambda) E_0[v_0] + \lambda E_0[v_1] &= (1-\lambda) E_0[\Psi_0] + \lambda E_0[\Psi_1] \\ &= (1-\lambda) \langle \Psi_0 | H_0 | \Psi_0 \rangle + \lambda \langle \Psi_1 | H_1 | \Psi_1 \rangle \end{aligned} \tag{1.84}$$

From the variational principle, we have relations

$$\begin{aligned} \langle \Psi_\lambda | H_0 | \Psi_\lambda \rangle &\geq \langle \Psi_0 | H_0 | \Psi_0 \rangle \\ \langle \Psi_\lambda | H_1 | \Psi_\lambda \rangle &\geq \langle \Psi_1 | H_1 | \Psi_1 \rangle \end{aligned} \tag{1.85}$$

Thus, the concavity of Eq. (1.81) is proved.

In the transformation, it is also important to recognize that the density $n(\boldsymbol{r})$ is given as the functional derivative of the energy E_0 with respect to $v_{\rm ext}$, thus,

$$n(\boldsymbol{r}) = \frac{\delta E_0[v_{\text{ext}}]}{\delta v_{\text{ext}}(\boldsymbol{r})} \tag{1.86}$$

This equation can be derived in a similar manner to that used in the proof of the Hellmann-Feynman theorem. Suppose that we have the wave function $\Psi_0 + \delta\Psi$ for the small variation of the external potential from $\hat{v}_{\text{ext}}$ to $\hat{v}_{\text{ext}} + \delta\hat{v}_{\text{ext}}$. The variation of the energy δE_0 up to the first order of $\delta\hat{v}_{\text{ext}}$ is then written as

$$\delta E_0 = \left\langle \Psi_0 \left| \delta\hat{v}_{\text{ext}} \right| \Psi_0 \right\rangle + \left(\left\langle \delta\Psi_0 \left| \hat{T} + \hat{V}_{ee} + \hat{v}_{\text{ext}} \right| \Psi_0 \right\rangle + c.c. \right) \tag{1.87}$$

Since the external potential is supposed to be local, the first term on the right is just given by

$$1\text{st} = \int d\boldsymbol{r} n(\boldsymbol{r}) \delta v_{\text{ext}}(\boldsymbol{r}) \tag{1.88}$$

The second term becomes

$$2\text{nd} = \left\langle \delta\Psi_0 \left| \hat{H}_0 \right| \Psi_0 \right\rangle = E_0 \left\langle \delta\Psi_0 \left| \Psi_0 \right. \right\rangle = 0 \tag{1.89}$$

The second equality comes from the normalization condition that $\left\langle \Psi_0 \left| \Psi_0 \right. \right\rangle = 1$. Then, we get

$$\delta E_0 = \int d\boldsymbol{r} n(\boldsymbol{r}) \delta v_{\text{ext}}(\boldsymbol{r}) \tag{1.90}$$

which directly leads to Eq. (1.86). In the above discussion, it is of essential importance that the potential $\hat{v}_{\text{ext}}$ is supposed to be local, otherwise the Legendre transformation of Eq. (1.80) cannot be carried out. The Legendre transform of $F_L[n]$ is given by

$$E_0[v_{\text{ext}}] = \inf_{n \to v\text{-rep.}} \left\{ F_L[n] + \int d\boldsymbol{r} n(\boldsymbol{r}) v_{\text{ext}}(\boldsymbol{r}) \right\} \tag{1.91}$$

This expression is equivalent to Eq. (1.79). It means that the Hohenberg-Kohn functional $F_{\text{HK}}[n]$ is given as the Legendre transform of the functional $E_0[v_{\text{ext}}]$. In other words, the HK functional $F_{\text{HK}}[n]$ is equivalent to $E_0[v_{\text{ext}}]$ through the Legendre transform.

It is also useful to formulate the functional derivative $\frac{\delta F_{\text{HK}}[n]}{\delta n(\boldsymbol{r})}$ as a counterpart of Eq. (1.86). Let $n(\boldsymbol{r})$ be a v-rep. density, then the functional $F_{\text{HK}}[n]$ can be expressed as

$$F_{\text{HK}}[n] = E_0[n] - \int d\boldsymbol{r} n(\boldsymbol{r}) v_{\text{ext}}[n](\boldsymbol{r}) \tag{1.92}$$

$E_0[n]$ is the ground state energy of the system with the density n. It should also be noticed that the potential v_{ext} corresponds to the density n through the one-to-one mapping. $F_{\text{HK}}[n]$ can be rewritten as

$$F_{\mathrm{HK}}[n] = \langle \Psi_0[n] | \hat{T} + \hat{V}_{ee} + \hat{v}_{\mathrm{ext}}[n] | \Psi_0[n] \rangle - \int d\boldsymbol{r} n(\boldsymbol{r}) v_{\mathrm{ext}}[n](\boldsymbol{r}) \quad (1.93)$$

For an infinitesimal variation δn of n, $\Psi[n]$ and $v_{\mathrm{ext}}[n]$ vary slightly by $\delta\Psi$ and δv, respectively. Then the variation δF_{HK} up to the first order of δn is evaluated as

$$\begin{aligned} \delta F_{\mathrm{HK}} &= E_0 \langle \delta\Psi_0 | \Psi_0 \rangle + \langle \Psi_0 | \delta\hat{v}_{\mathrm{ext}} | \Psi_0 \rangle \\ &\quad - \int d\boldsymbol{r} \delta n(\boldsymbol{r}) v_{\mathrm{ext}}(\boldsymbol{r}) - \int d\boldsymbol{r} n(\boldsymbol{r}) \delta v_{\mathrm{ext}}(\boldsymbol{r}) \end{aligned} \quad (1.94)$$

The first term in the right-hand side is zero as shown in Eq. (1.89), and the second and the fourth terms cancel each other. Thus, we obtain the functional derivative

$$\frac{\delta F_{\mathrm{HK}}[n]}{\delta n(\boldsymbol{r})} = -v_{\mathrm{ext}}(\boldsymbol{r}) \quad (1.95)$$

It can be recognized that Eq. (1.95) is related to the Legendre transformation of Eq. (1.91). The relationship between the functionals $E_0[v_{\mathrm{ext}}]$ and $F_{\mathrm{HK}}[n]$ is summarized in a scheme in Fig. 1.6.

It is worth noting that the functional $F_{\mathrm{HK}}[n]$ is a *universal* functional of n in principle, which means that the functional allows to provide the sum of the kinetic energy T and the electronic repulsion energy V_{ee} solely from the density n without any reference to the external potential of the system. This fact is quite encouraging because an approximate functional is anticipated to be useful for various systems provided that the functional is able to produce the exact value of $T + V_{ee}$ for a certain model system. Remind, however, that the domain of definition of the functional $F_{\mathrm{HK}}[n]$ is the v-representable density. The problem is that the conditions are not yet known for determining whether the density is v-rep. or not. Hence, even when the explicit form of $F_{\mathrm{HK}}[n]$ is given somehow, it might not be guaranteed that the variational search will be done within the set of v-rep. densities. Later, a solution was given to this problem in a quite mathematical approach. The solution is referred to as Levy's constraint search method [10], whereby the region of the variational

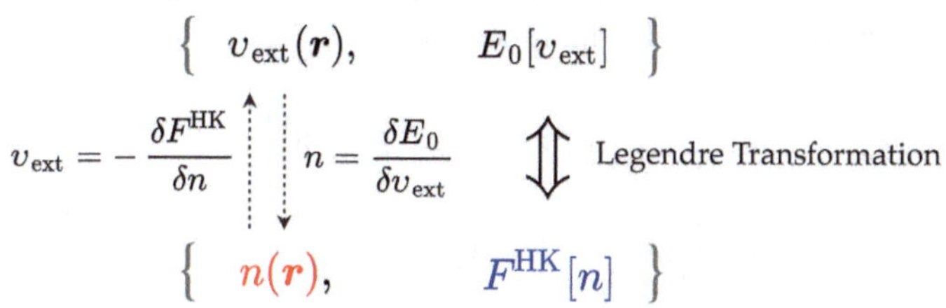

Fig. 1.6 Legendre transformations between the potential functional $E_0[v_{\mathrm{ext}}]$ and the Hohenberg-Kohn functional $F_{\mathrm{HK}}[n]$. The v-representable density n and the potential v_{ext} are given as the slopes (derivatives) of the functionals E_0 and F_{HK} with respect to v_{ext} and n, respectively. The electron density n in the scheme must be v-representable

search can be formally extended to the set of the N-representable densities. N-rep. density is defined as the density derived from an anti-symmetric wave function for N electrons. Fortunately, the conditions for the N-rep. density are well established. In the next subsection, the details of Levy's approach will be described. Then, in the subsequent subsection, we derive the mean field realized in the Kohn-Sham DFT.

1.3.4 Levy's Constraint Search

The Hohenberg-Kohn theorem noted in the previous subsection is a milestone in the development of the electronic density-functional theory (DFT). However, as mentioned, the theorem applies only to the subset of electron densities, namely, '$\upsilon-$representable' electron densities of which boundary was not yet known explicitly. This could be a critical issue in the foundation of the theoretical basis of the DFT. To overcome the problem the method of Levy's constraint search was developed in 1979 [10], which made it possible to formulate an energy functional in terms of the '$N-$representable' electron densities. '$N-$rep.' density is defined as the electron distribution derived from an anti-symmetric wave function for an N-electrons system. Importantly, the boundary of the '$N-$rep.' density is explicitly identified. It is also important to note that Levy's constraint search method can be extended to the novel theory of DFT based on the energy electron density $n^e(\epsilon)$ defined on the energy coordinate ϵ, for which details will be provided in Chap. 3. In this subsection, we first make a review of Levy's method within the conventional DFT framework.

The ground state energy E_0 of the system with external potential $\upsilon_{\text{ext}}(\boldsymbol{r})$ can be obtained through a minimization

$$E_0 = \min_{\Psi \to N} \left\langle \Psi \left| \hat{T} + \hat{V}_{ee} + \hat{\upsilon}_{\text{ext}} \right| \Psi \right\rangle \tag{1.96}$$

where $\Psi \to N$ represents the condition that the variational search of Ψ is to be performed within the set of anti-symmetric wave functions for the systems with N electrons. It can be readily understood that Eq. (1.96) is equivalent to the Schrödinger equation for obtaining E_0 and the corresponding eigenfunction Ψ_0. Equation (1.96) can be rewritten using two-fold minimization

$$\begin{aligned} E_0 &= \min_{n \to N_{\text{rep}}} \left\{ \min_{\Psi \to n} \left\langle \Psi \left| \hat{T} + \hat{V}_{ee} + \hat{\upsilon}_{\text{ext}} \right| \Psi \right\rangle \right\} \\ &= \min_{n \to N_{\text{rep}}} \left\{ \min_{\Psi \to n} \left\langle \Psi \left| \hat{T} + \hat{V}_{ee} \right| \Psi \right\rangle + \int d\boldsymbol{r}\ \upsilon_{\text{ext}}(\boldsymbol{r}) n(\boldsymbol{r}) \right\} \end{aligned} \tag{1.97}$$

The minimization of the outer parenthesis in Eq. (1.97) is performed in the subset of '$N-$rep.' density n, while the inner minimization is for the wave function of which density is being constrained to the density n. The first term in the parenthesis is actually an implicit functional of the '$N-$rep.' density n, thus,

$$F_{\text{Levy}}[n] = \min_{\Psi \to n} \left\langle \Psi \left| \hat{T} + \hat{V}_{ee} \right| \Psi \right\rangle \tag{1.98}$$

For a given density n, Eq. (1.98) identifies the wave function $\Psi_{\min}[n]$ which minimizes the expectation value of $\hat{T} + \hat{V}_{ee}$. Then it is possible to decompose the energy functional $F_{\text{Levy}}[n]$ into the contributions due to the kinetic and the electron repulsion energies, thus,

$$F_{\text{Levy}}[n] = T[n] + V_{ee}[n] \tag{1.99}$$

The individual functionals for $T[n]$ and $V_{ee}[n]$ are, respectively, defined as

$$T[n] = \left\langle \Psi_{\min}[n] | \hat{T} | \Psi_{\min}[n] \right\rangle \tag{1.100}$$

and

$$V_{ee}[n] = \left\langle \Psi_{\min}[n] | \hat{V}_{ee} | \Psi_{\min}[n] \right\rangle \tag{1.101}$$

$\Psi_{\min}[n]$ in these equations is the wave function that is obtained after the minimization of Eq. (1.98). It is now possible to reformulate the ground state energy E_0 in terms of the 'N−rep.' density n, thus,

$$E_0 = \min_{n \to N_{\text{rep}}} \left\{ F_{\text{Levy}}[n] + \int d\boldsymbol{r}\; v_{\text{ext}}(\boldsymbol{r}) n(\boldsymbol{r}) \right\} \tag{1.102}$$

This shows a clear contrast to Eq. (1.77) where the variational search is carried out within the subset of 'v−rep.' densities. The minimization in Eq. (1.102) requires that the electron density n is only N-representable. In the next subsection, we present a practical approach to perform the minimization of Eq. (1.102), where a set of one-electron wave functions will be introduced to represent a N-rep. density.

1.3.5 Kohn-Sham Method

The variational search of Eq. (1.102) shall be done under the constraint that the number of electrons is being fixed at N, that is,

$$\int d\boldsymbol{r}\; n(\boldsymbol{r}) = N \tag{1.103}$$

Then, we introduce a Lagrangian $L[n]$

$$L[n] = F_{\text{Levy}}[n] + \int d\boldsymbol{r} v_{\text{ext}}(\boldsymbol{r}) n(\boldsymbol{r}) - \mu \left(\int d\boldsymbol{r}\; n(\boldsymbol{r}) - N \right) \tag{1.104}$$

where the multiplier μ plays as a role of the chemical potential as will be shown below. Substituting Eqs. (1.100) and (1.101) to Eq. (1.104) and differentiating with respect to $n(\boldsymbol{r})$, we get the condition of the equilibrium

$$\frac{\delta L[n]}{\delta n(\boldsymbol{r})} = 0 \tag{1.105}$$

which gives

$$\frac{\delta T[n]}{\delta n(\boldsymbol{r})} + \frac{\delta V_{ee}[n]}{\delta n(\boldsymbol{r})} + v_{\text{ext}}(\boldsymbol{r}) = \mu \tag{1.106}$$

When the equality is fulfilled, the ground state energy E_0 is attained. The left-hand side of the equality is a function of $\boldsymbol{r}$, while the right-hand side is a constant. Thus, at the stationary state, the electrons feel flat potential with the value μ. When the system is supposed to be placed in the vacuum, the chemical potential is naturally taken as 0. On the other hand, when the system is being attached to an environmental system that plays a role of an electron reservoir, μ must be equal to the chemical potential ν of the environment, namely, $\mu = \nu$. Equation (1.106) describes the basic equation in DFT, although the explicit forms of $T[n]$ and $V_{ee}[n]$ are not given.

It is convenient to extract the classical Coulomb repulsion energy E_{H} termed Hartree energy from the electron repulsion energy V_{ee}, which defines the exchange-correlation energy E_{xc}, thus,

$$E_{xc}[n] = V_{ee}[n] - E_{\text{H}}[n] \tag{1.107}$$

where $E_{\text{H}}[n]$ is given by

$$E_{\text{H}}[n] = \frac{1}{2}\int d\boldsymbol{r}_1 d\boldsymbol{r}_2 \frac{n(\boldsymbol{r}_1)\, n(\boldsymbol{r}_2)}{|\boldsymbol{r}_1 - \boldsymbol{r}_2|} \tag{1.108}$$

It is easy to see that the sum of the Coulomb integrals $2\sum_{i,j} J_{ij}$ defined in Eq. (1.23) is identical to the Hartree energy $E_{\text{H}}[n]$. Anyway, Eq. (1.106) can be rewritten as

$$\frac{\delta T[n]}{\delta n(\boldsymbol{r})} + v_{\text{H}}[n](\boldsymbol{r}) + v_{xc}[n](\boldsymbol{r}) + v_{\text{ext}}(\boldsymbol{r}) = \mu \tag{1.109}$$

where the Hartree potential $v_{\text{H}}[n](\boldsymbol{r})$ is given by

$$v_{\text{H}}[n](\boldsymbol{r}) \equiv \frac{\delta E_{\text{H}}[n]}{\delta n(\boldsymbol{r})} = \int d\boldsymbol{r}' \frac{n(\boldsymbol{r}')}{|\boldsymbol{r} - \boldsymbol{r}'|} \tag{1.110}$$

and the exchange-correlation potential $v_{xc}[n](\boldsymbol{r})$ is defined as

$$v_{xc}[n](\boldsymbol{r}) \equiv \frac{\delta E_{xc}[n]}{\delta n(\boldsymbol{r})} \tag{1.111}$$

Note, however, that the solution of Eq. (1.106) or equivalently Eq. (1.109) is not useful because an adequate kinetic energy functional $T[n]$ is not currently available. As a notable feature of the KS-DFT [8], Kohn and Sham introduced a set of one-electron orbitals $\{\varphi_i\}$ to describe a part of the kinetic energy of the system. In other words, a major part of the kinetic energy of the real system is replaced by the sum of the kinetic energy $T_s[n]$ of the non-interacting system. In Sect. 1.2, it was assumed that there are N orbitals and $2N$ electrons with a closed shell structure for notational simplicity. However, it is supposed in the following that N electrons form a closed shell electronic structure (i.e. N is an even integer) in the system under consideration. In terms of the set of wave functions $\{\varphi_i\}(i = 1, 2, \cdots, N/2)$, $T_s[n]$ is given by

$$T_s[n] = \min_{\{\varphi_i\}\to n} \left\{ 2\sum_i^{N/2} \int d\boldsymbol{r}\varphi_i^*(\boldsymbol{r}) \left(-\frac{1}{2}\nabla^2\right) \varphi_i(\boldsymbol{r}) \right\} \tag{1.112}$$

Note that the minimization of the kinetic energy of the non-interacting system is performed under the constraint that

$$n(\boldsymbol{r}) = 2\sum_i^{N/2} \varphi_i(\boldsymbol{r})\varphi_i^*(\boldsymbol{r}) \tag{1.113}$$

for a given total electron density $n(\boldsymbol{r})$. Since the functional $T_s[n]$ is defined for the density described by a set of $N/2$ orbitals, it is of critical importance to examine whether such a density is capable of covering the entire domain of the N-rep. densities. Fortunately, by Harriman [29] it was proved in 1981 that any N-rep. density can be decomposed into the contributions from orthonormalized $N/2$ orbitals. Thus, the minimization with respect to the density given by N orbitals covers the entire N-rep. domain. The minimization is also subject to the orthonormalization conditions $\langle\varphi_i|\varphi_j\rangle = \delta_{ij}$ $(i, j = 1, 2, \cdots, N/2)$. Such a minimization can be treated by Lagrange's multiplier method. The formulation using the multiplier is worth describing for the later discussion. To this end we introduce the Lagrangian $L[\{\varphi_i\}]$,

$$\begin{aligned} L[\{\varphi_i\}] = 2\sum_i^{N/2} \int d\boldsymbol{r}\varphi_i^*(\boldsymbol{r}) \left(-\frac{1}{2}\nabla^2\right) \varphi_i(\boldsymbol{r}) + \int d\boldsymbol{r} q(\boldsymbol{r}) \left(2\sum_i^{N/2} \varphi_i(\boldsymbol{r})\varphi_i^*(\boldsymbol{r}) - n(\boldsymbol{r})\right) \\ - \sum_{i,j}^{N/2} \epsilon_{ij} \left(\langle\varphi_i|\varphi_j\rangle - \delta_{ij}\right) \end{aligned} \tag{1.114}$$

where $q(\boldsymbol{r})$ and ϵ_{ij} are Lagrange's multipliers. Since the constraint for the density is specified on each coordinate $\boldsymbol{r}$, the multiplier q is dependent on $\boldsymbol{r}$. The minimization with the constraints is achieved when $\frac{\delta L}{\delta\varphi_i(\boldsymbol{r})} = 0$ $(i = 1, 2, \cdots, N/2)$ is satisfied. This leads to the Schrödinger equations

$$\left(-\frac{1}{2}\nabla^2 + q(\boldsymbol{r})\right) \varphi_i(\boldsymbol{r}) = \epsilon_i\varphi_i(\boldsymbol{r}) \tag{1.115}$$

where it is assumed that a certain unitary transformation is applied to the set $\{\varphi_i\}$ so that the matrix ϵ_{ij} becomes diagonal. It is quite interesting to note that Lagrange's multiplier $q(\boldsymbol{r})$ plays a role as a local potential acting on $\{\varphi_i\}$. Since $q(\boldsymbol{r})$ yields the given density n after the solution of the Schrödinger equations, the potential $q(\boldsymbol{r})$ can be regarded as a functional of n, i.e. $q[n](\boldsymbol{r})$. Equation (1.115) serves a necessary condition to be fulfilled by the set of $\{\varphi_i\}$ at the minimum of the quantity in the parenthesis of Eq. (1.112). The electron density n is supposed to be constructed using a certain set of N orbitals out of those provided as solutions of Eq. (1.115). It is an issue of great interest to derive numerically the potential $q(\boldsymbol{r})$ for a given density $n(\boldsymbol{r})$. Several methods were proposed [30–33] to realize it. In fact, the method has a theoretical relevance to the 'optimized-effective potential' to be described in more detail in Sect. 1.4. The difference $\Delta T[n]$ of the kinetic energy between the interacting and the non-interacting systems is given by

$$\Delta T[n] = T[n] - T_s[n] \tag{1.116}$$

We then incorporate the energy $\Delta T[n]$ due to the electron correlations into $E_{xc}[n]$ to get $\overline{E}_{xc}[n]$

$$\overline{E}_{xc}[n] = \Delta T[n] + E_{xc}[n] \tag{1.117}$$

With these definitions, let us reformulate the ground state energy E_0, thus,

$$E_0 = \min_{n \to N-\text{rep.}} \left\{ T_s[n] + E_{\text{H}}[n] + \overline{E}_{xc}[n] + \int d\boldsymbol{r} v_{\text{ext}}(\boldsymbol{r}) n(\boldsymbol{r}) \right\} \tag{1.118}$$

Substituting Eq. (1.112)–(1.118), we have

$$\begin{aligned} E_0 = \min_{n \to N-\text{rep.}} \Bigg\{ & \min_{\{\varphi_i\} \to n} \left\{ 2 \sum_i^{N/2} \int d\boldsymbol{r} \varphi_i^*(\boldsymbol{r}) \left(-\frac{1}{2} \nabla_i^2 \right) \varphi_i(\boldsymbol{r}) \right\} \\ & + E_{\text{H}}[n] + \overline{E}_{xc}[n] + \int d\boldsymbol{r} v_{\text{ext}}(\boldsymbol{r}) n(\boldsymbol{r}) \Bigg\} \end{aligned} \tag{1.119}$$

The minimization with respect to the density can be equivalently replaced by the variational search in the set of orthonormalized $N/2$ wave functions $\{\varphi_i\}$,

$$\begin{aligned} E_0 = \min_{\{\varphi_i\}} \Bigg\{ & 2 \sum_i^{N/2} \int d\boldsymbol{r} \varphi_i^*(\boldsymbol{r}) \left(-\frac{1}{2} \nabla_i^2 \right) \varphi_i(\boldsymbol{r}) \\ & + E_{\text{H}}[n] + \overline{E}_{xc}[n] + \int d\boldsymbol{r} v_{\text{ext}}(\boldsymbol{r}) n(\boldsymbol{r}) \Bigg\} \end{aligned} \tag{1.120}$$

Note that the domain of the density given by $\{\varphi_i\}$ in the minimization of Eq. (1.120) is the same as that of Eq. (1.119) by virtue of the method of Harriman's construction. The minimization subject to the orthonormalization condition can be again formulated by Lagrange's multiplier method, which leads to the Kohn-Sham equations,

$$\left(-\frac{1}{2}\nabla^2 + v_{\text{eff}}[n](\boldsymbol{r})\right)\varphi_i(\boldsymbol{r}) = \epsilon_i \varphi_i(\boldsymbol{r}) \tag{1.121}$$

where the effective potential $v_{\text{eff}}[n](\boldsymbol{r})$ is defined as

$$v_{\text{eff}}[n](\boldsymbol{r}) \equiv v_{\text{H}}[n](\boldsymbol{r}) + \overline{v}_{xc}[n](\boldsymbol{r}) + v_{\text{ext}}(\boldsymbol{r}) \tag{1.122}$$

The potential $\overline{v}_{xc}[n](\boldsymbol{r})$ is the functional derivative of $\overline{E}_{xc}[n]$ with respect to $n(\boldsymbol{r})$. The Kohn-Sham equation of Eq. (1.121) must be solved self-consistently. That is, the SCF cycle should be repeated until the effective potential $v_{\text{eff}}[n](\boldsymbol{r})$ becomes consistent with the electron density given by the solution of the equation. It is interesting to note that Eq. (1.121) has the same form as Eq. (1.115). Actually, the potential $q[n_0](\boldsymbol{r})$ in Eq. (1.115) for the ground state density n_0 coincides with the effective potential $v_{\text{eff}}[n_0](\boldsymbol{r})$ in Eq. (1.121).

As mentioned in Sect. 1.2, the electron density $n(\boldsymbol{r})$ in the Hartree-Fock method is constructed employing the orbitals from the bottom to the $N/2$th in the order of increasing eigenvalues. This convention is also adopted in the KS approach. However, the method imposes a restriction on the domain of the variational search. That is, the search is to be restricted to a set of non-interacting ground state (v-rep.) densities. Therefore, if the density n_0 of the real interacting system lies outside of this set, there is no way to obtain the density n_0 through the standard KS-DFT approach. This issue is recognized as the non-interacting v-representability of the density although it is not considered as a serious problem so far in the context of the numerical calculations at least.

1.3.6 Approximate Functionals

One might wonder how the exchange-correlation functional $\overline{E}_{xc}[n]$ can be constructed without referring to the wave functions of the system of interest. The history of the functional development is longer than that of the density-functional theory (DFT) itself. Actually, the first exchange functional was developed by Slater [11] in 1951 before the proof was given for the Hohenberg-Kohn (HK) theorem in 1964 [9]. Slater made a simplification on the exchange or Fermi hole distribution to reduce the computational cost for the Hartree-Fock (HF) exchange energy. To illustrate the idea, we first define the exchange (Fermi) hole function [5]. The HF exchange energy E_x^{HF} defined in Eq. (1.31) can be expressed by the sum of the spin contributions $E_{x\sigma}^{\text{HF}}$ ($\sigma = \alpha$ or β). For the closed shell system with N electrons, $E_{x\sigma}^{\text{HF}}$ for spin σ can be simply written in terms of the first-order density matrix $n_{1\sigma}(\boldsymbol{r}_1, \boldsymbol{r}_2)$ for spin σ, thus,

$$E_{x\sigma}^{\text{HF}} = -\int d\boldsymbol{r}_1 d\boldsymbol{r}_2 \frac{1}{r_{12}} |n_{1\sigma}(\boldsymbol{r}_1, \boldsymbol{r}_2)|^2 \tag{1.123}$$

where $n_{1\sigma}(\boldsymbol{r}_1, \boldsymbol{r}_2)$ is defined as

$$n_{1\sigma}(\boldsymbol{r}_1, \boldsymbol{r}_2) = \sum_i^{N/2} \varphi_{i\sigma}(\boldsymbol{r}_1)\varphi_{i\sigma}^*(\boldsymbol{r}_2) \tag{1.124}$$

Note that the diagonal element of the matrix $n_{1\sigma}(\boldsymbol{r}, \boldsymbol{r})$ represents the electron density $n_\sigma(\boldsymbol{r})$ for the spin σ. Now we define the exchange (or Fermi) hole function $n_{x\sigma}(\boldsymbol{r}_1, \boldsymbol{r}_2)$ for spin σ as

$$n_{x\sigma}(\boldsymbol{r}_1, \boldsymbol{r}_2) = -\frac{|n_{1\sigma}(\boldsymbol{r}_1, \boldsymbol{r}_2)|^2}{n_{1\sigma}(\boldsymbol{r}_1, \boldsymbol{r}_1)} = -\frac{|n_{1\sigma}(\boldsymbol{r}_1, \boldsymbol{r}_2)|^2}{n_\sigma(\boldsymbol{r}_1)} \tag{1.125}$$

Then, $E_{x\sigma}^{\mathrm{HF}}$ can be expressed as

$$E_{x\sigma}^{\mathrm{HF}} = \int d\boldsymbol{r}_1 d\boldsymbol{r}_2 \frac{1}{r_{12}} n_\sigma(\boldsymbol{r}_1)\, n_{x\sigma}(\boldsymbol{r}_1, \boldsymbol{r}_2) = \frac{1}{2}\int d\boldsymbol{r}_1 d\boldsymbol{r}_2 \frac{1}{r_{12}} n(\boldsymbol{r}_1)\, n_{x\sigma}(\boldsymbol{r}_1, \boldsymbol{r}_2) \tag{1.126}$$

A physical meaning of this equation is that the energy $E_{x\sigma}^{\mathrm{HF}}$ can be viewed as an integration of the classical Coulomb interaction between the electron density n at $\boldsymbol{r}_1$ and the exchange hole $n_{x\sigma}$ associated with the reference electron placed at $\boldsymbol{r}_1$. A graphical representation of the exchange hole function $n_{x\sigma}$ of a real system can be found e.g. in Refs. [34, 35]. For later references, it is helpful to change the form of Eq. (1.126) as,

$$\begin{aligned} E_{x\sigma}^{\mathrm{HF}} &= \int d\boldsymbol{r}_1 n_\sigma(\boldsymbol{r}_1) \int d\boldsymbol{r}_2 \frac{1}{r_{12}} n_{x\sigma}(\boldsymbol{r}_1, \boldsymbol{r}_2) \\ &= \int d\boldsymbol{r}_1 n_\sigma(\boldsymbol{r}_1)\, \epsilon_{x\sigma}(\boldsymbol{r}_1) \end{aligned} \tag{1.127}$$

The quantity $\epsilon_{x\sigma}(\boldsymbol{r}_1)$ with the dimension of energy is often referred to as exchange energy density at $\boldsymbol{r}_1$. Note that the exchange-correlation energy E_{xc} can also be written in the form of the integration of the classical Coulomb interaction between the density at $\boldsymbol{r}_1$ and the exchange-correlation hole n_{xc} associated with $\boldsymbol{r}_1$ [5]. The development of an approximate exchange functional is to design $\epsilon_{x\sigma}$ on the basis of the known physical properties fulfilled by the exchange hole function $n_{x\sigma}$ of the real system. As an important property of the exchange hole function, a sum rule is satisfied for any position of $\boldsymbol{r}_1$, thus,

$$\begin{aligned} \int d\boldsymbol{r}_2 n_{x\sigma}(\boldsymbol{r}_1, \boldsymbol{r}_2) &= -\frac{1}{n_\sigma(\boldsymbol{r}_1)} \int d\boldsymbol{r}_2\, |n_{1\sigma}(\boldsymbol{r}_1, \boldsymbol{r}_2)|^2 \\ &= -\frac{1}{n_\sigma(\boldsymbol{r}_1)} \int d\boldsymbol{r}_2 \left(\sum_i^{N/2} \varphi_i(\boldsymbol{r}_1)\varphi_i^*(\boldsymbol{r}_2)\right)\left(\sum_j^{N/2} \varphi_j^*(\boldsymbol{r}_1)\varphi_j(\boldsymbol{r}_2)\right) \\ &= -\frac{1}{n_\sigma(\boldsymbol{r}_1)} \sum_{i,j}^{N/2} \varphi_i(\boldsymbol{r}_1)\varphi_j^*(\boldsymbol{r}_1) \int d\boldsymbol{r}_2 \varphi_i^*(\boldsymbol{r}_2)\varphi_j(\boldsymbol{r}_2) \end{aligned}$$

$$= -\frac{1}{n_\sigma(\boldsymbol{r}_1)} \sum_{i,j}^{N/2} \varphi_i(\boldsymbol{r}_1)\varphi_j^*(\boldsymbol{r}_1)\delta_{ij}$$
$$= -1 \tag{1.128}$$

where the subscripts σ for the wave functions are omitted for the sake of simplicity. The sum rule of the exchange hole function $n_{x\sigma}$ is essential in the development of the approximate exchange functional E_x since it serves as a guide that the functional should obey. It is also important to recognize in Eq. (1.125) that the hole depth at the reference point $\boldsymbol{r}_1$ is equal to the spin density $n_\sigma(\boldsymbol{r}_1)$ with the negative sign, that is,

$$n_{x\sigma}(\boldsymbol{r}_1) = n_{x\sigma}(\boldsymbol{r}_1, \boldsymbol{r}_1) = -n_\sigma(\boldsymbol{r}_1) \tag{1.129}$$

This relation is also important for the construction of an approximate exchange functional.

1.3.6.1 Local-density Approximation

Utilizing these facts (Eqs. (1.128) and (1.129)) as guides, it is possible to make a rough estimation of the exchange energy E_x in terms of the density as discussed in Ref. [11]. Actually, the equations are regarded as the most important physical properties that an approximate exchange hole function satisfies (Fig. 1.7).

Assuming that the electron density $n_\sigma(\boldsymbol{r}_1)$ at a rcference point $\boldsymbol{r}_1$ is given as $n_{1\sigma}$, it is possible to estimate the radius r_0 of the sphere Ω that encloses the exchange hole for the reference point. For simplicity, it is assumed that the sphere Ω has its center at $\boldsymbol{r}_1$ and the hole density inside Ω is $-n_1$ everywhere. Then, from the sum rule of Eq. (1.128), the radius r_0 for this model can be obtained as

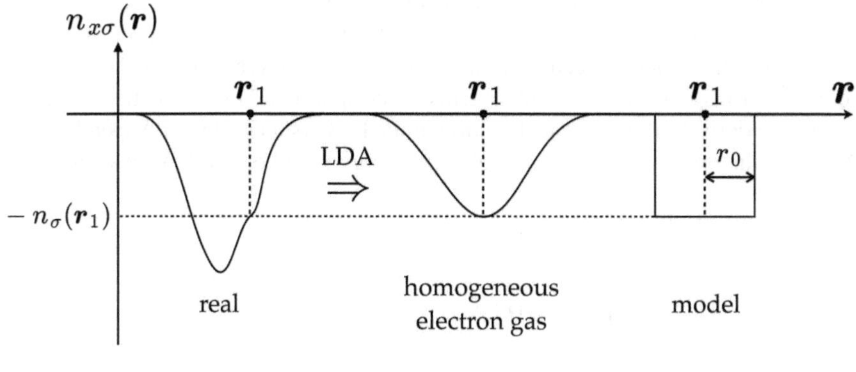

Fig. 1.7 Schematic illustration of the exchange holes $n_{x\sigma}(\boldsymbol{r})$ for the real system, homogeneous electron gas (HEG), and the model system associated with the reference electron placed at $\boldsymbol{r}_1$. r_0 is the radius of the model hole with the constant value $-n_\sigma(\boldsymbol{r}_1)$ within the hole

$$r_0 = \left(\frac{3}{4\pi}\right)^{\frac{1}{3}} n_{1\sigma}^{-\frac{1}{3}} \quad (1.130)$$

The Coulomb potential $\epsilon_{x\sigma}(\boldsymbol{r}_1)$ created at $\boldsymbol{r}_1$ by the simplified hole is then given by

$$\epsilon_{x\sigma}^{\text{model}}(\boldsymbol{r}_1) = -\int_0^{r_0} n_{1\sigma} \times 4\pi r^2 \frac{1}{r} dr = -2\left(\frac{3}{4}\right)^{\frac{2}{3}} \pi^{\frac{1}{3}} n_{1\sigma}^{\frac{1}{3}} = -C_x^{\text{model}} n_{1\sigma}^{\frac{1}{3}} \quad (1.131)$$

Remind that the exchange energy is given by the integration of the Coulomb interaction between the density and the associated exchange hole as shown in Eq. (1.126). And also remind that $n_{1\sigma}$ is equal to $n_\sigma(\boldsymbol{r}_1)$. Then, the exchange energy $E_{x\sigma}^{\text{model}}$ is expressed as

$$\begin{aligned} E_{x\sigma}^{\text{model}} &= -\frac{1}{2}\int d\boldsymbol{r}\, n(\boldsymbol{r})\, \epsilon_{x\sigma}^{\text{model}}(\boldsymbol{r}) \\ &= -C_x^{\text{model}} \int d\boldsymbol{r}\, n_\sigma(\boldsymbol{r})^{\frac{4}{3}} \\ &= -2.4179\cdots \int d\boldsymbol{r}\, n_\sigma(\boldsymbol{r})^{\frac{4}{3}} \end{aligned} \quad (1.132)$$

The quantity $\epsilon_{x\sigma}(\boldsymbol{r})$ is often referred to as the exchange energy density of electron with spin σ and it represents the exchange energy per an electron placed at $\boldsymbol{r}$. In fact, the simplified exchange hole model described above illustrates the basic framework of the local-density approximation (LDA). A characteristic of the model is that the hole distribution is spherically symmetric around a reference electron. With this property and the sum rule of the hole, the exchange energy at $\boldsymbol{r}$ becomes proportional to $n_\sigma(\boldsymbol{r})^{4/3}$. It is, of course, unlikely that the exchange hole population is constant inside the sphere Ω even when the hole function is spherically symmetric around the reference point. However, it is worth noting that Eq. (1.132) captures the essence of the LDA exchange hole, except for the coefficient C_x^{model}. In the full version of the LDA approach, the exchange hole function of the homogeneous electron gas (HEG) is applied to the density at each position in the system of interest. In fact, the exchange hole $n_{x\sigma}^{\text{HEG}}(\boldsymbol{r})$ is exactly described in terms of the spherical Bessel function j_1 of first order. The derivation of the hole function in the HEG is presented in Appendix D. In Ref. [11], the explicit form of the exchange energy based on the LDA was given by

$$\begin{aligned} E_{x\sigma}^{\text{LDA}} &= -\frac{1}{2}\int d\boldsymbol{r}\, n(\boldsymbol{r})\, \epsilon_{x\sigma}^{\text{LDA}}(\boldsymbol{r}) \\ &= -2^{\frac{1}{3}} C_x^{\text{LDA}} \int d\boldsymbol{r}\, n_\sigma(\boldsymbol{r})^{\frac{4}{3}} \\ &= -2^{\frac{1}{3}} \frac{3}{4}\left(\frac{3}{\pi}\right)^{\frac{1}{3}} \int d\boldsymbol{r}\, n_\sigma(\boldsymbol{r})^{\frac{4}{3}} = -0.93052\cdots \int d\boldsymbol{r}\, n_\sigma(\boldsymbol{r})^{\frac{4}{3}} \end{aligned} \quad (1.133)$$

where the factor $2^{\frac{1}{3}}$ is the spin scaling factor of which derivation is provided in Appendix A. It can be recognized that the exchange energy of Eq. (1.132) based on the simplified exchange hole model shows a good agreement with Eq. (1.133) given by the exchange hole of the HEG except for the coefficient. The performance of Eq. (1.133) was investigated e.g. in Ref. [36] by computing the atomization energies of relatively small 32 molecules by means of the LDA-based KS-DFT method although rather small LCAO basis set (6-31G*) was used. The computed atomization energies range from 5.6 kcal/mol (Li_2) to 604.7 kcal/mol (C_2H_6). The results were compared with those obtained by experiments. Surprisingly enough, the mean deviation of the LDA values was found to be only -19.1 kcal/mol and the mean absolute deviation was 25.0 kcal/mol. These results were found to be much better than those of the HF calculations. Actually, the mean deviation of the HF method was evaluated as -85.8 kcal/mol for the same molecular set. Of course, this result does not mean that the LDA is superior to the HF method since Eq. (1.133) is nothing but an approximation to the HF method. However, the results are quite encouraging because they shows that LDA, despite of its great simplicity, can describe the chemical bonds in a variety of systems at least qualitatively. It is expected that the exchange hole model provided by the HEG will be appropriate in the inner region of a molecule. Note, however, that the model does not work properly in the outer region because the reference electron placed in the region will be significantly separated from the center of the exchange hole. This problem has a relevance with the fact that the exchange-correlation functional based on LDA tends to overestimate the atomization energies of molecules. The generalized gradient approximation (GGA) provides an effective solution for such a problem.

1.3.6.2 Generalized Gradient Approximation

To improve the LDA-based functional, the inhomogeneity of the electron density of the real system must be taken into consideration. The first discussion was already given by Hohenberg and Kohn (HK) in Ref. [37]. In the following, a concise review of the gradient expansion of a universal functional by HK will be provided. They considered an energy portion $G[n]$ of the electronic energy $E[n]$ of an electron density $n(\boldsymbol{r})$. Explicitly, $G[n]$ is defined by the sum of the kinetic energy T and the exchange-correlation energy E_{xc}. Then we consider a local energy $g[n](\boldsymbol{r})$ defined by

$$G[n] = \int d\boldsymbol{r} g[n](\boldsymbol{r}) \tag{1.134}$$

This form was already provided for the HF exchange energy E_x^{HF} as shown in Eq. (1.127). It is assumed (*not* proved) that $g[n](\boldsymbol{r})$, a *universal* functional of n, can be expressed by a gradient expansion at $\boldsymbol{r}$,

$$
\begin{aligned}
g[n](\boldsymbol{r}) = g_0(n(\boldsymbol{r})) &+ \sum_{i=x,y,z} g_i(n(\boldsymbol{r}))\nabla_i n(\boldsymbol{r}) \\
&+ \sum_{i,j=x,y,z} \left\{ g_{ij}^{(1,1)}(n(\boldsymbol{r}))\nabla_i n(\boldsymbol{r})\nabla_j n(\boldsymbol{r}) \right. \\
&\left. + g_{ij}^{(2)}(n(\boldsymbol{r}))\nabla_i\nabla_j n(\boldsymbol{r}) \right\} + \cdots
\end{aligned}
\tag{1.135}
$$

The term $g_0(n(\boldsymbol{r}))$ is a function of the scalar quantity $n(\boldsymbol{r})$ and it represents the energy density at $\boldsymbol{r}$ when the density is completely homogeneous. Similarly, other coefficients, $g_i, g_{ij}^{(1,1)}, g_{ij}^{(2)}$ are also the functions of $n(\boldsymbol{r})$. Since $n(\boldsymbol{r})$ is invariant with respect to the rotation about $\boldsymbol{r}$, these coefficients are also invariant. Then the substantial simplification can be made to Eq. (1.135), thus,

$$
g[n](\boldsymbol{r}) = g_0(n(\boldsymbol{r})) + \left\{ g_2^{(1,1)}\nabla n \cdot \nabla n + g_2^{(2)}\nabla^2 n \right\} + \text{terms of order } \nabla^4 \tag{1.136}
$$

where the subscript of g represents the number of operations of the gradient operator. To see in more detail about the simplification, we consider the portion of the expansion with the coefficients $g_{ij}^{(1,1)}$, thus,

$$
\begin{aligned}
&\sum_{i,j=x,y,z} \left\{ g_{ij}^{(1,1)}(n(\boldsymbol{r}))\nabla_i n(\boldsymbol{r})\nabla_j n(\boldsymbol{r}) + \cdots \right. \\
&= \cdots + g_{xy}^{(1,1)}\frac{\partial n}{\partial x}\frac{\partial n}{\partial y} + g_{xz}^{(1,1)}\frac{\partial n}{\partial x}\frac{\partial n}{\partial z} + g_{yz}^{(1,1)}\frac{\partial n}{\partial y}\frac{\partial n}{\partial z} + \cdots
\end{aligned}
\tag{1.137}
$$

When the density $n(\boldsymbol{r})$ is rotated π e.g. around the z axis, the signs of the values $\frac{\partial n}{\partial x}$ and $\frac{\partial n}{\partial y}$ are reversed. Therefore, the coefficients $g_{xz}^{(1,1)}$ and $g_{yz}^{(1,1)}$ must be zero. It is also possible to show that the other nondiagonal terms vanish in the same way. Here it is quite important to notice that the terms of first-order ∇ does not contribute to the expansion. It is found that only the terms of the second order and higher orders make contributions to improve the value given by LDA. This underlies the fact that the LDA works rather nicely in spite of its rude approximation.

Becke provided the gradient expansion of the exchange hole function described in Eq. (1.125) [38]. As shown in Eq. (1.127) the exchange energy density $\epsilon_{x\sigma}(\boldsymbol{r}_1)$ is given by the Coulomb interaction between an electron with spin σ placed at $\boldsymbol{r}_1$ and the exchange hole function $n_{x\sigma}(\boldsymbol{r}_1, \boldsymbol{r}_2)$ associated with $\boldsymbol{r}_1$. A schematic illustration of $n_{x\sigma}$ is provided in Fig. 1.8 to help readers' understandings. As shown in Eq. (1.127), the integration of $n_{x\sigma}(\boldsymbol{r}_1, \boldsymbol{r}_2)$ over the sphere Ω_s is essential in the computation of the exchange energy density at $\boldsymbol{r}_1$ since the Coulomb potential is a function of only the distance $s = |\boldsymbol{r}_2 - \boldsymbol{r}_1|$. Hence, a Taylor expansion is considered for the spherical average of the hole function $n_{x\sigma}(\boldsymbol{r}_1, \boldsymbol{r}_2)$ with respect to small s. Actually, the Taylor expansion of the hole function along the vector $\boldsymbol{s} = \boldsymbol{r}_2 - \boldsymbol{r}_1$ can be expressed using the operator $\exp(\boldsymbol{s} \cdot \nabla_2)$ as shown below,

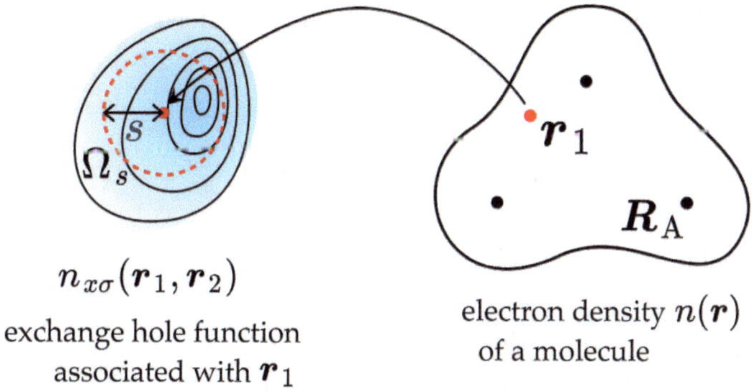

Fig. 1.8 Right: Illustration of a molecule with electron density $n(\boldsymbol{r})$. $\boldsymbol{R}_\mathrm{A}$ is the coordinate of a nucleus of A in the molecule. $\boldsymbol{r}_1$ is the coordinate of the reference point placed in the molecule. Left: Illustration of the contour map for the exchange hole function $n_{x\sigma}(\boldsymbol{r}_1, \boldsymbol{r}_2)$ as a function of $\boldsymbol{r}_2$ associated with the point $\boldsymbol{r}_1$. s is the radius of a sphere Ω_s that has the center at $\boldsymbol{r}_1$

$$\begin{aligned}
&\exp\left(\boldsymbol{s}\cdot\nabla_2\right) n_{x\sigma}\left(\boldsymbol{r}_1, \boldsymbol{r}_2\right)\big|_{\boldsymbol{r}_2=\boldsymbol{r}_1} \\
&\quad = \exp\left(x_s\frac{\partial}{\partial x_2}\right)\exp\left(y_s\frac{\partial}{\partial y_2}\right)\exp\left(z_s\frac{\partial}{\partial z_2}\right) n_{x\sigma}\left(\boldsymbol{r}_1, \boldsymbol{r}_2\right)\big|_{\boldsymbol{r}_2=\boldsymbol{r}_1} \\
&\quad = \left(1 + x_s\frac{\partial}{\partial x_2} + \frac{1}{2}x_s^2\frac{\partial^2}{\partial x_2^2} + \cdots\right) \\
&\qquad \times \left(1 + y_s\frac{\partial}{\partial y_2} + \frac{1}{2}y_s^2\frac{\partial^2}{\partial y_2^2} + \cdots\right) \\
&\qquad \times \left(1 + z_s\frac{\partial}{\partial z_2} + \frac{1}{2}z_s^2\frac{\partial^2}{\partial z_2^2} + \cdots\right) n_{x\sigma}\left(\boldsymbol{r}_1, \boldsymbol{r}_2\right)\big|_{\boldsymbol{r}_2=\boldsymbol{r}_1}
\end{aligned} \tag{1.138}$$

where it is supposed that $\boldsymbol{s} = (x_s, y_s, z_s)$ and $\boldsymbol{r}_2 = (x_2, y_2, z_2)$ and ∇_2 applies only to $\boldsymbol{r}_2$. Since the gradient vector $\nabla_2 n_{x\sigma}\left(\boldsymbol{r}_1, \boldsymbol{r}_2\right)\big|_{\boldsymbol{r}_2=\boldsymbol{r}_1}$ is constant irrespective of the vector $\boldsymbol{s}$, we denote $\boldsymbol{g} = \nabla$ for simplicity. Then the spherical average $\langle\exp(\boldsymbol{s}\cdot\boldsymbol{g})\rangle_{\Omega_s}$ of the exchange hole over the sphere Ω_s is computed using the spherical coordinate (s, θ, ϕ), thus,

$$\begin{aligned}
\langle\exp\left(\boldsymbol{s}\cdot\boldsymbol{g}\right)\rangle_{\Omega_s} &= \frac{1}{4\pi s^2}s^2\int_0^{2\pi} d\phi\int_0^{\pi}\exp\left(\boldsymbol{s}\cdot\boldsymbol{g}\right)\sin\theta d\theta \\
&= \frac{1}{4\pi}2\pi\int_0^{\pi}\exp\left(sg\cos\theta\right)\sin\theta d\theta \\
&= \frac{1}{2}\int_{-sg}^{sg} dt\frac{\exp t}{sg} = \frac{1}{2}\frac{\exp\left(sg\right) - \exp\left(-sg\right)}{sg} \\
&= \frac{\sinh\left(sg\right)}{sg}
\end{aligned} \tag{1.139}$$

where the scalar quantity g is the magnitude of the vector $\boldsymbol{g}$ and θ is taken as the angle between $\boldsymbol{s}$ and $\boldsymbol{g}$. The Taylor expansion of $\langle \exp(\boldsymbol{s} \cdot \boldsymbol{g}) \rangle_{\Omega_s} = \sinh(sg)/sg$ is readily given by

$$\langle \exp(\boldsymbol{s} \cdot \boldsymbol{g}) \rangle_{\Omega_s} = 1 + \frac{1}{3!} s^2 g^2 + \frac{1}{5!} s^4 g^4 + \cdots \tag{1.140}$$

Substituting this expression to Eq. (1.138), we obtain

$$\begin{aligned} &\exp(\boldsymbol{s} \cdot \nabla_2)\, n_{x\sigma}(\boldsymbol{r}_1, \boldsymbol{r}_2)\big|_{\boldsymbol{r}_2=\boldsymbol{r}_1} \\ &= \left(1 + \frac{1}{3!} s^2 \nabla_2^2 + \frac{1}{5!} s^4 \nabla_2^4 + \cdots\right) n_{x\sigma}(\boldsymbol{r}_1, \boldsymbol{r}_2)\Big|_{\boldsymbol{r}_2=\boldsymbol{r}_1} \end{aligned} \tag{1.141}$$

It is important in the equation to note that the first-order term with respect to s does not appear in the expansion, which corresponds to the fact that the first-order gradient term does not contribute to the gradient expansion in Eq. (1.135). As a reasonable approximation, we consider the terms up to the second order of s in Eq. (1.141). The second term of the right-hand side becomes,

$$\begin{aligned} \nabla_2^2 n_{x\sigma}(\boldsymbol{r}_1, \boldsymbol{r}_2)\Big|_{\boldsymbol{r}_2=\boldsymbol{r}_1} &= \frac{1}{n_\sigma(\boldsymbol{r}_1)} \times \\ &\quad \nabla_2 \left(\nabla_2 \left(\sum_i^{N/2} \varphi_i(\boldsymbol{r}_1) \varphi_i^*(\boldsymbol{r}_2) \times \sum_j^{N/2} \varphi_j^*(\boldsymbol{r}_1) \varphi_j(\boldsymbol{r}_2) \right) \right) \Bigg|_{\boldsymbol{r}_2=\boldsymbol{r}_1} \\ &= \frac{1}{n_\sigma(\boldsymbol{r}_1)} \nabla_2 \left(\sum_i^{N/2} \varphi_i(\boldsymbol{r}_1) \nabla_2 \varphi_i^*(\boldsymbol{r}_2) \left(\sum_j^{N/2} \varphi_j^*(\boldsymbol{r}_1) \varphi_j(\boldsymbol{r}_2) \right) \right. \\ &\quad \left. + \left(\sum_i^{N/2} \varphi_i(\boldsymbol{r}_1) \varphi_i^*(\boldsymbol{r}_2) \right) \sum_j^{N/2} \varphi_j^*(\boldsymbol{r}_1) \nabla_2 \varphi_j(\boldsymbol{r}_2) \right) \Bigg|_{\boldsymbol{r}_2=\boldsymbol{r}_1} \\ &= \sum_i^{N/2} \varphi_i(\boldsymbol{r}_1) \nabla_1^2 \varphi_i^*(\boldsymbol{r}_1) + c.c. \\ &\quad + \frac{2}{n_\sigma(\boldsymbol{r}_1)} \left| \sum_i^{N/2} \varphi_i(\boldsymbol{r}_1) \nabla_1 \varphi_i^*(\boldsymbol{r}_1) \right|^2 \end{aligned} \tag{1.142}$$

To get a simpler form of this equation, we consider the first and second derivatives of the spin density $n_\sigma(\boldsymbol{r})$. The first derivative is obtained as

$$\nabla n_\sigma(\boldsymbol{r}) = \sum_i^{N/2} \varphi_i(\boldsymbol{r}) \nabla \varphi_i^*(\boldsymbol{r}) + c.c. \tag{1.143}$$

Then the last term of Eq. (1.142) can be rewritten as

$$\frac{2}{n_\sigma(\boldsymbol{r})}\left|\sum_i^{N/2}\varphi_i(\boldsymbol{r})\nabla\varphi_i^*(\boldsymbol{r})\right|^2=\frac{(\nabla n_\sigma(\boldsymbol{r}))^2}{2n_\sigma(\boldsymbol{r})} \tag{1.144}$$

On the other hand, the second derivative of the density is computed as

$$\nabla^2 n_\sigma(\boldsymbol{r})=\sum_i^{N/2}\varphi_i(\boldsymbol{r}_1)\nabla_1^2\varphi_i^*(\boldsymbol{r}_1)+c.c.$$
$$+2\sum_i^{N/2}|\nabla\varphi_i(\boldsymbol{r})|^2 \tag{1.145}$$

Now it is possible to rewrite the Taylor expansion of Eq. (1.141), that is, the spherically averaged exchange hole $\overline{n}_{x\sigma}(\boldsymbol{r}_1, s)$ around $\boldsymbol{r}_1$, thus,

$$\begin{aligned}&\overline{n}_{x\sigma}(\boldsymbol{r}_1,s)\\&\quad=\exp(\boldsymbol{s}\cdot\nabla_2)\,n_{x\sigma}(\boldsymbol{r}_1,\boldsymbol{r}_2)|_{\boldsymbol{r}_2=\boldsymbol{r}_1}\\&\quad=n_\sigma(\boldsymbol{r}_1)+\frac{1}{6}\left(\nabla^2 n_\sigma(\boldsymbol{r}_1)-2\tau_\sigma(\boldsymbol{r}_1)+\frac{(\nabla n_\sigma(\boldsymbol{r}_1))^2}{2n_\sigma(\boldsymbol{r}_1)}\right)s^2+\cdots\end{aligned} \tag{1.146}$$

where we introduce the kinetic energy density τ_σ defined by

$$\tau_\sigma(\boldsymbol{r})=\sum_i^{N/2}|\nabla\varphi_i(\boldsymbol{r})|^2 \tag{1.147}$$

It might be helpful to consider for a moment why this term is referred to as kinetic energy density. To this end, we start from the kinetic energy T_z along the z axis. Since the total kinetic energy is merely given by the sum of the kinetic energies along the three axes, consideration of only a single axis is sufficient without loss of generality. T_z can be formulated as

$$\begin{aligned}&\sum_i^{N/2}\int_{-\infty}^{\infty}dz\varphi_i^*(z)\left(-\frac{1}{2}\frac{\partial^2}{\partial z^2}\right)\varphi_i(z)\\&\quad=-\frac{1}{2}\sum_i^{N/2}\int_{-\infty}^{\infty}dz\varphi_i^*(z)\frac{\partial}{\partial z}\varphi_i'(z)\\&\quad=-\frac{1}{2}\sum_i^{N/2}\left\{\left[\varphi_i^*(z)\varphi_i'(z)\right]_{-\infty}^{\infty}-\int_{-\infty}^{\infty}dz\frac{\partial}{\partial z}\varphi_i^*(z)\frac{\partial}{\partial z}\varphi_i(z)\right\}\\&\quad=\frac{1}{2}\sum_i^{N/2}\int_{-\infty}^{\infty}dz\frac{\partial}{\partial z}\varphi_i^*(z)\frac{\partial}{\partial z}\varphi_i(z)\end{aligned} \tag{1.148}$$

Thus $\sum_i \frac{\partial}{\partial z}\varphi_i^*(z)\frac{\partial}{\partial z}\varphi_i(z)$ can be regarded as the kinetic energy density at z.

As shown in Eq. (1.136), the leading term in the gradient expansion of the energy density $g[n](\boldsymbol{r})$ includes the square of magnitude of the gradient ∇n and the second derivative $\nabla^2 n$. Although $g[n](\boldsymbol{r})$ includes the kinetic and the correlation energy in addition to the exchange energy, Eq. (1.136) shows a reasonable correspondence to the expansion in Eq. (1.146) for the exchange hole with respect to the radius s of the sphere Ω_s. From both equations, it can be seen that the variation e.g. of the exchange energy density with respect to the small variation of the density in the vicinity of the reference point $\boldsymbol{r}_1$ can be approximately described with the magnitude of the gradient of the density at $\boldsymbol{r}_1$. These analyses constitute the basis of the generalized gradient approximation (GGA). Equation (1.146) also provides a useful insight in the development of approximate exchange functional. It shows that the occurrence of a density gradient ∇n at $\boldsymbol{r}_1$ deepens the exchange hole depth. This is equivalent to state that increase in the magnitude of ∇n at $\boldsymbol{r}_1$ shrinks the width of the hole. Hence the exchange energy is lowered due to the gradient because the sum rule of Eq. (1.128) is fulfilled. Thus the exchange energy tends to create an imbalance in the electron density. In general, the GGA functional for the exchange energy $E_{x\sigma}^{\mathrm{GGA}}[n_\sigma]$ can be written in the form

$$E_{x\sigma}^{\mathrm{GGA}}[n_\sigma] = E_{x\sigma}^{\mathrm{LDA}}[n_\sigma] - \int d\boldsymbol{r}\; n_\sigma(\boldsymbol{r})^{\frac{4}{3}} F(q_\sigma) \tag{1.149}$$

where the LDA functional $E_{x\sigma}^{\mathrm{LDA}}$ is defined in Eq. (1.133), and the quantity q_σ in the function $F(q_\sigma)$ is a dimensionless variable defined by

$$q_\sigma = \frac{|\nabla n_\sigma(\boldsymbol{r})|}{n_\sigma(\boldsymbol{r})^{\frac{4}{3}}} \tag{1.150}$$

It can be readily confirmed that q_σ is dimensionless and it measures the inhomogeneity of the electrons at $\boldsymbol{r}$. It should be noticed that the smaller the electron density, the greater the inhomogeneity for gradients of the same magnitude. Therefore, q_σ tends to be larger in the outer regions of atoms or molecules in general. According to the gradient expansion of Eq. (1.136), the simplest choice of the function $F(q_\sigma)$ is obviously,

$$F(q_\sigma) = \alpha q_\sigma^2 = \alpha \frac{|\nabla n_\sigma(\boldsymbol{r})|^2}{n_\sigma(\boldsymbol{r})^{\frac{8}{3}}} \tag{1.151}$$

with α an empirical parameter. However, this functional is divergent for the large value of q_σ. To suppress the divergence, the functional form shall be changed to

$$F(q_\sigma) = \frac{\beta q_\sigma^2}{1 + \gamma q_\sigma^2} \tag{1.152}$$

which converges to β/γ asymptotically. The function is characterized by the property that it increases monotonically with increasing q_σ. Using Eq. (1.152) as a template,

Becke developed in 1988 a functional by imposing another physical property to be fulfilled by the exact functional [13]. Hereafter, we refer to this functional as Becke88 functional for simplicity. The real exchange energy density $\epsilon_x(\boldsymbol{r})$ asymptotically behaves as $-1/r$ for a large distance r between the reference point $\boldsymbol{r}$ and a molecule of interest. In atoms or molecules, it is well known that their electron densities converge to $\exp(-\alpha r)$ for large r, where α is a parameter related to the ionization potential of the system. Thus the LDA-based exchange energy density $\epsilon_{x\sigma}^{\text{LDA}}$ decays exponentially in the outer regions of molecules. This is considered to be a major deficiency of $E_{x\sigma}^{\text{LDA}}$. The explicit form of Becke's functional is expressed as

$$F^{\text{B88}}\left(q_\sigma\right) = \frac{\beta q_\sigma^2}{1 + 6\beta q_\sigma \sinh^{-1} q_\sigma} \tag{1.153}$$

At a glance, it seems quite curious that F^{B88} includes the function $\sinh^{-1}(x) = \log(x + \sqrt{x^2+1})$. As described above, the electron density obeys the function $\exp(-\alpha r)$ for large r. Hence, in the outer region of the system of interest, q_σ can be approximately expressed by

$$q_\sigma^{\text{outer}}(r) = \alpha \exp\left(\frac{1}{3}\alpha r\right) \tag{1.154}$$

It is important to note that $q_\sigma^{\text{outer}}(r)$ itself diverges for large r. Inserting this expression to Eq. (1.153) and applying some reasonable approximations, it is recognized that $F^{\text{B88}} \to -1/(2r)$ for each spin σ independent of the choice of β. As a whole, the exchange energy density ϵ_x recovers the $-1/r$ asymptotic behavior in the Becke88 functional. It is also worth noting that $F^{\text{B88}}(q_\sigma)$ includes only the terms with even powers of the density gradient, which can be readily confirmed from the Taylor expansion of $\sinh^{-1}(x)$. As favored, this property is consistent with the gradient expansion of Eq. (1.136). The undetermined parameter β in Eq. (1.153) was empirically determined by the least-square fittings to the exact HF exchange energies of the six noble gas atoms (from He to Rn), from which β was obtained as 0.0042 a.u. Unfortunately, it might not be possible to give any physical meaning to the parameter. One may be surprised by the fact that the GGA functional was developed only through a mathematical constructions that are to impose the conditions satisfied by the exact exchange functional. Actually, the functional was not based on any physical model. In 1989, Becke and Roussel developed a completely different type of the exchange-correlation functional on the basis of an explicit hole model in inhomogeneous system [39]. The detailed description of this model will be provided in Chap. 2.

A detailed review of the performance of the GGA functional in comparison with the LDA functional is not a purpose of this book. However, it may be useful to discuss the improvement provided by the GGA correction to the LDA functional. Computation of atomization energies E_{atm} of a set of molecules is often performed as benchmark test of an exchange and correlation functional E_{xc}. E_{atm} is defined as the

Table 1.1 Mean deviations (MDs) and mean absolute deviations (MADs) for the 32 small neutral molecules from TABLE V in Ref. [36]. Units are kcal/mol. Slater's exchange [42] indicated by 'S' is the LDA-based exchange energy of Eq. (1.133) multiplied by 2/3 as an empirical parameter. The notation 'B' indicates the Becke88 functional [13]. 6-31G* basis set was used for the calculations

	SVWN	SLYP	BVWN	BLYP	HF	QCISD
MD	35.6	37.7	0.1	1.0	−85.8	−28.8
MAD	35.7	38.3	4.4	5.6	85.9	28.8

energy needed to decompose a molecule into the constituent isolated atoms. Johnson *et al.* performed a series of calculations [36] for 32 neutral molecules employing the Becke88 functional, and the results were compared with those obtained by the LDA functional and wave-function theories (HF, MP2, and QCISD). Some of the data is summarized in Table 1.1. First of all, it should be noted that the correlation energy E_c besides E_x also affects the energy E_{atm}, although E_c is smaller than E_x by an order of magnitude in general. Actually, the exclusion of E_c tends to underestimate E_{atm}. Hence, the energy E_c computed at the same level of theory will be incorporated into E_{atm} in the following discussions to extract solely the effect of the GGA correction to the LDA energy. Explicitly, E_c was computed by the LDA-based functional developed by Vosko, Wilk, and Nusair(VWN) [40] for the exact homogeneous electron gas. E_c was also evaluated using the GGA-corrected functional developed by Lee et al. [41]. The value of the mean deviation (MD) was calculated as the average of the deviations $E_{\text{atm}}(\text{comp.}) - E_{\text{atm}}(\text{expt.})$, and the mean absolute deviation(MAD) was that of the absolute deviations. It is shown in the table that Slater's exchange functional [42] combined with VWN or LYP correlation functional (SVWN or SLYP) seriously overestimates E_{atm}. Both MD and MAD for these functional are larger than 35 kcal/mol. Among the 32 molecules in the test set, ethane molecule (C_2H_6) shows the largest value of the atomization energy $E_{\text{atm}} = 666.3$ kcal/mol(experimental value taken from Ref. [43]). The SVWN and SLYP functional overestimate the energy by 85.8 and 84.9 kcal/mol, respectively. On the other hand, it was found that the GGA functional (Becke88) drastically improves the results. Actually, E_{atm} values were obtained as 664.9 and 660.9 kcal/mol by BVWN and BLYP functionals, respectively. The MD and MAD values for these GGA-corrected functionals decrease to less than 1.0 and 5.6 kcal/mol, respectively. Thus the improvement given by the GGA correction to the LDA functional is quite striking. It is somewhat surprising that the evaluation by QCISD underestimates the E_{atm} values, although it is much better than the HF results. It should also be noted that the computational cost in QCISD is much larger than that in BVWN or BLYP calculation. The inadequacy of the correlation energy by the QCISD calculation might be attributed to the small number of electronic configurations included in the CI calculation and also to the use of the poor basis set. Anyway, it was demonstrated in Table 1.1, the GGA-corrected DFT functional is superior to the method based on the wave-function theory at least in the calculation of E_{atm}.

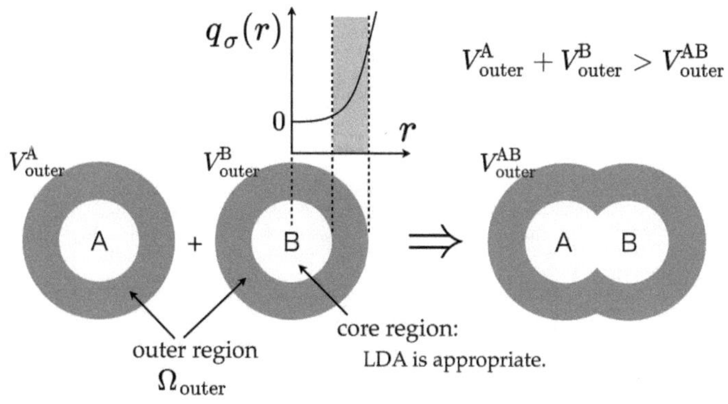

Fig. 1.9 Illustration of the outer regions before(left) and after(right) the covalent bond formation between atoms A and B. The shaded regions represent the outer region, while the white circles show the core regions where the LDA works nicely. The inhomogeneity parameter q_σ in Eq. (1.150) stays almost zero within the core regions

In Table 1.1, it is recognized that the LDA exchange functionals (SVWN and SLYP) almost always overestimate the atomization energies of the 32 molecules. This can be understood from the fact that MD is positive and its value is almost equal to the MAD value. It is worth discussing the mechanism underlying the overestimations in LDA and how the GGA functional can extinguish the errors. A notable feature of atoms or molecules compared to bulk systems is that they have outer regions around themselves. Let us take a closer look on what happens when two atoms make a chemical reaction to form a molecule. Figure 1.9 shows a schematic illustration of the outer regions before and after the formation of the chemical bond between the two atoms A and B. The parameter q_σ in Eq. (1.150), which measures the inhomogeneity of the electron density, becomes larger in such regions. Consequently, the *real* exchange energy density ϵ_x is lowered in the outer region than that evaluated by the LDA functional because of the shrink of the exchange hole distribution as discussed above. As illustrated in Fig. 1.9, the volume $V_{\rm outer}$ of the outer region $\Omega_{\rm outer}$ is decreased ($V^{\rm A}_{\rm outer} + V^{\rm B}_{\rm outer} > V^{\rm AB}_{\rm outer}$) since the outer regions cover each other after the bond formation. Thus, the error in the absolute exchange energy $E_x^{\rm LDA}$ before reaction is larger than that after the reaction, which leads to the relative destabilization in the exchange energy of the reactant compared to the product. It is shown in the table that the GGA functional form in Eq. (1.153) with the optimized set of parameters delivers almost perfect correction to LDA. After the release of the Becke88 functional, a lot of GGA-based exchange and correlation functional E_{xc} were developed, and the KS-DFT is established as a reliable and efficient computational tool in the fields of chemistry and physics. Note, however, that a further improvement can be made for the E_{xc} functional as will be shown in the next subsection.

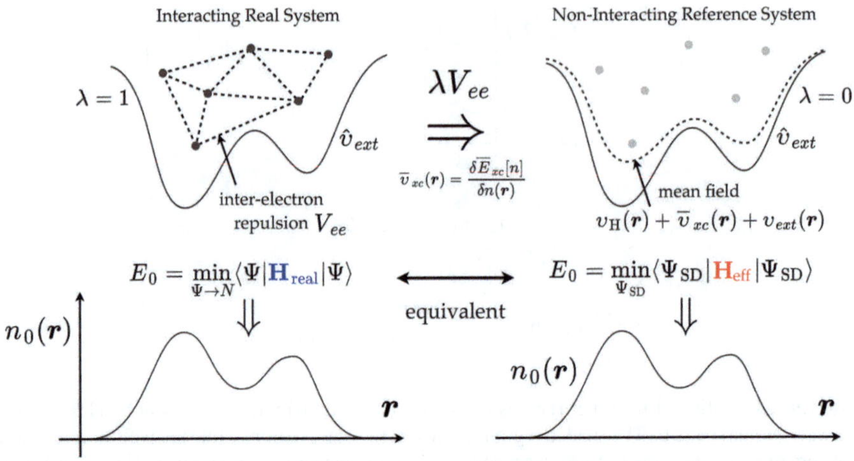

Fig. 1.10 Comparison of the wave-function theory (left) with the Kohn-Sham(KS) DFT approach (right). In the KS-DFT method, the real Hamiltonian $\mathbf{H}_{\mathrm{real}}$ is distorted to an effective Hamiltonian $\mathbf{H}_{\mathrm{eff}}$, thereby obtaining the density n_0 and the energy E_0 of the interacting real system from the wave function Ψ_{SD} (Slater determinant) of the non-interacting reference system. The quantity λ is a parameter introduced to represent the gradual change of the system from the interacting ($\lambda = 1$) to non-interacting ($\lambda = 0$) system

1.3.6.3 Hybrid Functional

As discussed in Sect. 1.3.5, the KS-DFT approach is able to construct the exact ground state electron density $n_0(\boldsymbol{r})$ and the energy E_0 provided that the exact universal functional $\overline{E}_{xc}$ is given and n_0 is non-interacting v-representable. We provide in Fig. 1.10 the illustration to make comparison of the KS-DFT with the wave-function theory. The picture shown on the left represents a real system where the electrons in the system are fully interacting, while the interactions are turned off in the non-interacting reference system shown on the right. The non-interacting system realizes the ground state density n_0 of the real system under the influence of the effective Hamiltonian $\mathbf{H}_{\mathrm{eff}}$. As an essential ingredient, $\mathbf{H}_{\mathrm{eff}}$ includes the exchange-correlation potential $\overline{v}_{xc}[n_0](\boldsymbol{r})$ in the mean field. Here, we introduce a coupling parameter λ to express the gradual change of the system from the non-interacting ($\lambda = 0$) to fully interacting ($\lambda = 1$) system with the density being fixed at n_0. Along the variation of λ from 0 to 1, the potential $\overline{v}_{xc}^{\lambda}[n_0](\boldsymbol{r})$ also varies accordingly.

To formulate $\overline{v}_{xc}^{\lambda}[n_0](\boldsymbol{r})$, let us consider the functional $F_\lambda[n]$ for N-rep. density n,

$$F_\lambda[n] = \min_{\Psi_\lambda \to n} \left\langle \Psi_\lambda \middle| \hat{T} + \lambda \hat{V}_{ee} \middle| \Psi_\lambda \right\rangle \tag{1.155}$$

where the minimization of the energy $\hat{T} + \lambda \hat{V}_{ee}$ with respect to Ψ_λ is performed with the constraint that

$$N \int d\omega_1 d\boldsymbol{x}_2 \cdots d\boldsymbol{x}_N \Psi_\lambda(\boldsymbol{x}_1, \cdots, \boldsymbol{x}_N) \Psi_\lambda^*(\boldsymbol{x}_1, \cdots, \boldsymbol{x}_N) - n(\boldsymbol{r}) = 0 \quad (1.156)$$

N is the number of electrons in the system of interest and $\boldsymbol{x}$ represents the set of the spatial coordinate $\boldsymbol{r}$ and the spin coordinate ω. The minimization is also subject to the normalization condition that

$$\langle \Psi_\lambda | \Psi_\lambda \rangle = 1 \quad (1.157)$$

Then, the problem is treated with Lagrange's multiplier method. To do this, we introduce the Lagrangian $L[\Psi_\lambda]$

$$\begin{aligned} L[\Psi_\lambda] = & \left\langle \Psi_\lambda \middle| \hat{T} + \lambda \hat{V}_{ee} \middle| \Psi_\lambda \right\rangle \\ & + \int d\boldsymbol{r} q_\lambda(\boldsymbol{r}) \left(\left\langle \Psi_\lambda \middle| \hat{n}(\boldsymbol{r}) \middle| \Psi_\lambda \right\rangle - n(\boldsymbol{r}) \right) - E_\lambda \left(\langle \Psi_\lambda | \Psi_\lambda \rangle - 1 \right) \end{aligned} \quad (1.158)$$

where $\hat{n}(\boldsymbol{r})$ is the density operator to produce the electron density at $\boldsymbol{r}$. At the stationary state of L, we have a Schrödinger equation

$$\left(\hat{T} + \lambda \hat{V}_{ee} + \hat{q}_\lambda[n] \right) \left| \Psi_\lambda^n \right\rangle = E_\lambda \left| \Psi_\lambda^n \right\rangle \quad (1.159)$$

Obviously, $q_\lambda[n](\boldsymbol{r})$ plays as a role of the effective potential (or mean field) to make the density derived from the wave function Ψ_λ becomes n that is given as an argument of Eq. (1.155).

Now we suppose that the density n is the ground state density n_0 for the system with an external potential $v_{\text{ext}}(\boldsymbol{r})$. By subtracting $\hat{v}_{\text{H}}[n_0]$ and $\hat{v}_{\text{ext}}$ from $\hat{q}_\lambda[n_0]$, we get the exchange-correlation potential $\overline{v}_{xc}^\lambda[n_0](\boldsymbol{r})$, thus,

$$\overline{v}_{xc}^\lambda[n_0](\boldsymbol{r}) = q_\lambda[n_0](\boldsymbol{r}) - v_{\text{H}}[n_0](\boldsymbol{r}) - v_{\text{ext}}(\boldsymbol{r}) \quad (1.160)$$

For the ground state density n_0, the potential $\overline{v}_{xc}^\lambda[n_0](\boldsymbol{r})$ can be regarded as the portion of the mean field added to the external potential at the coupling parameter λ. It is thus possible to formulate the *local* potential $\overline{v}_{xc}^\lambda[n_0](\boldsymbol{r})$ for the intermediately interacting systems. Provided that the wave function $\left| \Psi_\lambda^{n_0} \right\rangle$ is obtained as the solution of Eq. (1.159), the functional $F_\lambda[n_0]$ of Eq. (1.155) can be rewritten as

$$F_\lambda[n_0] = \left\langle \Psi_\lambda^{n_0} \middle| \hat{T} + \lambda \hat{V}_{ee} \middle| \Psi_\lambda^{n_0} \right\rangle \quad (1.161)$$

It is helpful to perform the coupling-parameter integration for $F_\lambda[n_0]$ [5]. Actually, it leads to

$$\int_0^1 d\lambda \frac{dF_\lambda[n_0]}{d\lambda} = F_1[n_0] - F_0[n_0] \quad (1.162)$$

By the definition of Eq. (1.161), $F_1[n_0]$ is given by

$$F_1[n_0] = \left\langle \Psi_1^{n_0} \middle| \hat{T} + \hat{V}_{ee} \middle| \Psi_1^{n_0} \right\rangle \tag{1.163}$$

The quantity $F_1[n_0]$ is exactly the same as $F^{\mathrm{HK}}[n_0]$ in Eq. (1.76). On the other hand, $F_0[n_0]$ is written as

$$F_0[n_0] = \left\langle \Psi_0^{n_0} \middle| \hat{T} \middle| \Psi_0^{n_0} \right\rangle \tag{1.164}$$

Since $\Psi_0^{n_0}$ is the wave function for the ground state of the non-interacting system, $F_0[n_0]$ is the same as $T_s[n_0]$ in Eq. (1.112). Therefore, we obtain,

$$\begin{aligned}
\int_0^1 d\lambda \frac{dF_\lambda[n_0]}{d\lambda} &= F_1[n_0] - F_0[n_0] \\
&= T\,[n_0] + V_{ee}\,[n_0] - T_s\,[n_0] \\
&= \Delta T\,[n_0] + E_{xc}\,[n_0] + V_{\mathrm{H}}\,[n_0] \\
&= \overline{E}_{xc}\,[n_0] + V_{\mathrm{H}}\,[n_0]
\end{aligned} \tag{1.165}$$

where the definitions of Eqs. (1.116) and (1.117) are used. It is quite important to note that the kinetic energy difference $\Delta T\,[n_0]$ between the interacting and non-interacting systems can be naturally incorporated into $\overline{E}_{xc}\,[n_0]$ through the coupling-parameter integration of $F_\lambda[n_0]$. As demonstrated in the works [44, 45] by A. D. Becke in 1993, the coupling-parameter integration can be formulated as

$$\begin{aligned}
\int_0^1 d\lambda \frac{dF_\lambda\,[n_0]}{d\lambda} &= \int_0^1 d\lambda \frac{d}{d\lambda} \left\langle \Psi_\lambda^{n_0} \middle| \hat{T} + \lambda \hat{V}_{ee} \middle| \Psi_\lambda^{n_0} \right\rangle \\
&= \int_0^1 d\lambda \left\langle \Psi_\lambda^{n_0} \middle| \hat{V}_{ee} \middle| \Psi_\lambda^{n_0} \right\rangle
\end{aligned} \tag{1.166}$$

where the Hellmann-Feynman theorem is utilized to derive the second equality. An elementary but promising approach to approximate the integration is to assume the linear dependence of the integrand in Eq. (1.166) on the parameter λ, which leads to

$$\int_0^1 d\lambda \frac{dF_\lambda\,[n_0]}{d\lambda} \approx \frac{1}{2} \left(\left\langle \Psi_0^{n_0} \middle| \hat{V}_{ee} \middle| \Psi_0^{n_0} \right\rangle + \left\langle \Psi_1^{n_0} \middle| \hat{V}_{ee} \middle| \Psi_1^{n_0} \right\rangle \right) \tag{1.167}$$

Since $\Psi_0^{n_0}$ is the wave function for the non-interacting system and is given by a Slater determinant, the first term on the right can be rewritten as

$$\left\langle \Psi_0^{n_0} \middle| \hat{V}_{ee} \middle| \Psi_0^{n_0} \right\rangle = E_x^{\mathrm{HF}}[\Psi_{\mathrm{SD}}^{n_0}] + V_{\mathrm{H}}[n_0] \tag{1.168}$$

where E_x^{HF} is the Hartree-Fock exchange energy (see Eq. (1.24)) and is given by

$$E_x^{\mathrm{HF}}\left[\Psi_{\mathrm{SD}}^{n_0}\right] = -\sum_{i,j}^{N/2} K_{ij} = -\sum_{i,j}^{N/2} \int d\boldsymbol{r}_1 d\boldsymbol{r}_2 \varphi_i^*(1)\varphi_j(1)\frac{1}{r_{12}}\varphi_j^*(2)\varphi_i(2) \quad (1.169)$$

In Eq. (1.169), the set $\{\varphi_i\}$ constitutes the Slater determinant $\Psi_{\mathrm{SD}}^{n_0}$ of the system of N electrons with closed shell structure. $V_{\mathrm{H}}[n_0]$ in Eq. (1.168) is the classical Coulomb energy of the density n_0 and explicitly given in the form of Eq. (1.108). Provided that the non-classical part of the second term on the right-hand side of Eq. (1.167) is represented by an exchange-correlation functional $E_{xc}[n_0]$,

$$\left\langle \Psi_1^{n_0} \middle| \hat{V}_{ee} \middle| \Psi_1^{n_0} \right\rangle = E_{xc}[n_0] + V_{\mathrm{H}}[n_0] \quad (1.170)$$

Then, we obtain a reasonable approximation to the coupling-parameter integration, thus,

$$\begin{aligned}\int_0^1 d\lambda \frac{dF_\lambda[n_0]}{d\lambda} &= \overline{E}_{xc}[n_0] + V_{\mathrm{H}}[n_0] \\ &\approx \frac{1}{2}\left(E_x^{\mathrm{HF}}[\Psi_{\mathrm{SD}}^{n_0}] + E_{xc}[n_0]\right) + V_{\mathrm{H}}[n_0]\end{aligned} \quad (1.171)$$

which directly leads to

$$\overline{E}_{xc}[n_0] \approx \frac{1}{2}\left(E_x^{\mathrm{HF}}[\Psi_{\mathrm{SD}}^{n_0}] + E_{xc}[n_0]\right) \quad (1.172)$$

Remind that $\overline{E}_{xc}[n_0]$ contains the kinetic energy difference $\Delta T = T - T_s$ as defined in Eq. (1.117). It is quite interesting in Eq. (1.172) that the half and half mixings [44] of the exact Hartree-Fock exchange energy E_x^{HF} and the density-functional exchange-correlation energy E_{xc} includes the quantity $\Delta T[n_0]$. A sophistication of the functional for $\overline{E}_{xc}$ can be made by improving the approximation for the coupling-parameter integration. Actually, A. D. Becke proposed to change the mixing coefficients of the ingredients in the GGA (generalized gradient approximation) functional and the HF exchange energy E_x^{HF} [45]. His original approach included three parameters, which were determined through the least-squares fit to the 56 atomization energies, 42 ionization potentials, 8 proton affinities, and 10 atomic energies of the first and second rows. Becke's three parameter model combined with the Lee-Yang-Parr (LYP) correlation functional [41] is termed B3LYP functional [45] and has been extensively utilized in a huge number of applications. In fact, the B3LYP functional is now regarded as a de facto standard in the application of the KS-DFT. The success of the B3LYP functional motivated theoretical physicists and chemists to develop a lot of functionals. A review of each development may be useful to the readers, but is beyond the scope of this book.

1.3.7 Summary

We close this subsection by summarizing the Sects. 1.2 and 1.3 In the wave-function theory, we introduced a Slater determinant $\Psi_{\rm SD}$ on the basis of the independent particle model, where the wave functions of the electrons are described with a set of N orbitals $\{\varphi_i\}$. Then, the Hartree-Fock energy $E_{\rm HF} = \langle\Psi_{\rm SD}|\,\mathbf{H}_{\rm real}\,|\Psi_{\rm SD}\rangle$ was formulated as an expectation value of the full Hamiltonian $\mathbf{H}_{\rm real}$. The minimization of $E_{\rm HF}$ with respect to $\{\varphi_i\}$ under the orthonormalization conditions leads to the formation of the nonlocal mean field felt by the independent particles. Since the Slater determinant is intrinsically anti-symmetric, the resultant mean field incorporates the effect of Pauli's exclusion principle. However, the explicit correlations among the electrons cannot be realized through the mean field because the variational search of the anti-symmetric wave function is confined within the set of *single* Slater determinants. To incorporates the electron correlation to the wave function, the domain of the variational search should be extended to the tensor-product space constructed using the HF wave functions as building blocks. Although the approach, termed configuration interaction (CI) method, is mathematically clear and straightforward, the number of Slater determinants included in the CI expansion drastically increases with the increase in the number of electrons in the system. As a consequence, the computational costs associated with the multi-configuration method becomes prohibitively large for large systems in general.

In the context of the density-functional theory (DFT), the mean field is constructed in a way completely different from the approach in the wave-function theory. To review how the mean field is generated in the DFT, let us first assume that the Hohenberg-Kohn density functional $F^{\rm HK}[n]$ is available. $F^{\rm HK}[n]$ delivers the sum of the kinetic energy and the repulsion energy between the electrons for the υ-representable electron density n as the argument. It should be noticed that the existence of the HK functional is guaranteed only *mathematically* by the Hohenberg-Kohn theorem [9] or more directly by the Legendre transformation of the functional $E_0[\upsilon_{\rm ext}]$ for the ground state energy as shown in Eq. (1.80). The minimization of the energy functional $E_0[n] = F^{\rm HK}[n] + \int d\boldsymbol{r} n(\boldsymbol{r})\upsilon_{\rm ext}(\boldsymbol{r})$ with respect to the density under the constraint that the number N of electrons is being fixed leads to the potential $\frac{\delta F^{\rm HK}[n]}{\delta n(\boldsymbol{r})} + \upsilon_{\rm ext}(\boldsymbol{r})$ for the density $n(\boldsymbol{r})$, which gives the chemical potential μ of the system at the stationary state. Thus, the SCF calculation for the DFT is to optimize the density so that the potential becomes constant regardless of the position $\boldsymbol{r}$ of the electrons. Kohn and Sham introduced a set of one-electron wave functions to describe the kinetic energy in the functional since any accurate kinetic energy functional $T^{\rm HK}[n]$ is not currently available in practice. Thus, the minimization of the energy with respect to the density $n(\boldsymbol{r})$ is equivalently replaced by the minimization using a set of wave functions $\{\varphi_i\}$ subject to the orthonormalization conditions. Through the SCF calculation, one obtains the effective potential, i.e. the mean field for each electron as defined in Eq. (1.122). Since the KS wave functions $\{\varphi_i\}$ must be regarded as the intermediate variables solely to span the electron density, the mean field has nothing to do with the potential obtained for the wave function of the independent particles.

The potential rather defines the non-interacting reference system which delivers the real density of the interacting system. Note that the potential originates solely from the functional $F^{\rm HK}[n]$ of the density and not from the functional of any reduced wave function. Thus, the mean field obtained by the KS-DFT has the potential in principle to provide the electronic properties of the ground state with comparable accuracy to the time-consuming wave-function theories with much lower computational costs. Actually, provided that the exact exchange-correlation functional $\overline{E}_{xc}[n]$ defined in Eq. (1.117) is available and the density n is non-interacting v-representable, one obtains the mean field which allows to generate the real density and the electronic energy of the system of interest. Thus, the success of the DFT depends solely on the developments of the accurate functional $\overline{E}_{xc}[n]$. This fact motivates a lot of theoretical physicists and chemists to make an efforts to develop and improve the functionals.

Lastly, we also make a remark on the locality of the exchange-correlation potential $\overline{v}_{xc}[n](\boldsymbol{r})$. It is somewhat surprising that the potential $\overline{v}_{xc}[n](\boldsymbol{r})$ is merely a local potential although it represents all the effects due to complex quantum correlations among the electrons. In fact, as was mentioned in Sect. 1.1, it is possible to incorporate the nonlocal effects of the electrons into the local potential. Remind that the potential $\overline{v}_{xc}[n](\boldsymbol{r})$ is defined as the ratio of the variation of the $\overline{E}_{xc}[n]$ energy to the small variation δn of the density at $\boldsymbol{r}$. The functional $\overline{E}_{xc}[n]$ is *in principle* capable of producing such a variation yielded as a consequence of the nonlocal electron correlations. It is sometimes misunderstood that the Kohn-Sham approach cannot adequately describe the electronic properties of systems where the nonlocal electron correlation plays a role. Underlying this misunderstanding is the confusion of the principle with the approximation applied to the construction of actual functionals. The most of the $\overline{E}_{xc}$ functionals are majorly based on the local-density approximation (LDA) which refers only to the density $n(\boldsymbol{r})$ to construct the local potential $\overline{v}_{xc}[n](\boldsymbol{r})$. It is quite unlikely that such an approximation framework makes it possible to describe the change in the $\overline{E}_{xc}$ energy given as a consequence of the nonlocal electron correlations. Thus, the construction of the potential $\overline{v}_{xc}[n](\boldsymbol{r})$ that realizes the nonlocal effect of the electrons is a critical issue in the development of the approximate functionals of DFT.

1.4 Optimized Effective Potential

As mentioned in the last subsection in the previous section, a major problem in the development of the DFT functional is how to incorporate the nonlocal correlation of the electrons into an approximate local potential, namely, the exchange-correlation potential $\overline{v}_{xc}[n](\boldsymbol{r})$. This problem constitutes a major subject of this book. Here we provide a review for an implicit approach termed optimized-effective potential (OEP) method that can be built into the framework of Kohn-Sham (KS) DFT. The concept of OEP was first introduced by Sharp and Horton in 1953 [16], who were motivated by Slater's work [11] in 1951. It should be noticed that the OEP approach appeared before the proof for the Hohenberg-Kohn theorem [9] was given. In the following, for a while, we consider a closed shell system with $2N$ electrons for simplicity. Sharp and Horton devised a method to optimize a *local* potential so that the set of eigenfunctions $\{\varphi_i\}$ $(i = 1, 2, \cdots, N)$ on the potential minimizes the Hartree-Fock energy $E_{\rm HF}$ expressed in the form of Eq. (1.27). The basic equation of OEP for the HF energy is expressed as

$$\frac{\delta E_{\rm HF}\left[\{\varphi_i\}\right]}{\delta v_{\rm eff}(\boldsymbol{r})} = 0 \tag{1.173}$$

where $v_{\rm eff}(\boldsymbol{r})$ is the local potential to be optimized. Equation (1.173) indicates that the energy $E_{\rm HF}$ as a functional of the wave functions $\{\varphi_i\}$ is stationary with respect to the small variation of the potential $v_{\rm eff}(\boldsymbol{r})$. Simultaneously it is also required that the wave functions $\{\varphi_i\}$ are the eigenfunctions for the potential $v_{\rm eff}$

$$\left(-\frac{1}{2}\nabla^2 + v_{\rm eff}(\boldsymbol{r})\right)\varphi_i(\boldsymbol{r}) = \epsilon_i\varphi_i(\boldsymbol{r}) \tag{1.174}$$

Thus, the OEP method involves dual minimizations with respect to the effective potential and the wave functions. The energy $E_0^{\rm HFOEP}$ obtained by the HF-OEP method can be concisely expressed by the following minimization:

$$E_0^{\rm HFOEP} = \min_{v_{\rm eff}} \langle \Psi_{\rm SD}\left[v_{\rm eff}\right] | \hat{T} + \hat{V}_{ee} + \hat{v}_{\rm ext} | \Psi_{\rm SD}\left[v_{\rm eff}\right] \rangle \tag{1.175}$$

where $\Psi_{\rm SD}[v_{\rm eff}]$ is the Slater determinant composed of the one-electron wave functions φ_i uniquely obtained as solutions of Eq. (1.174). Before we proceed to the

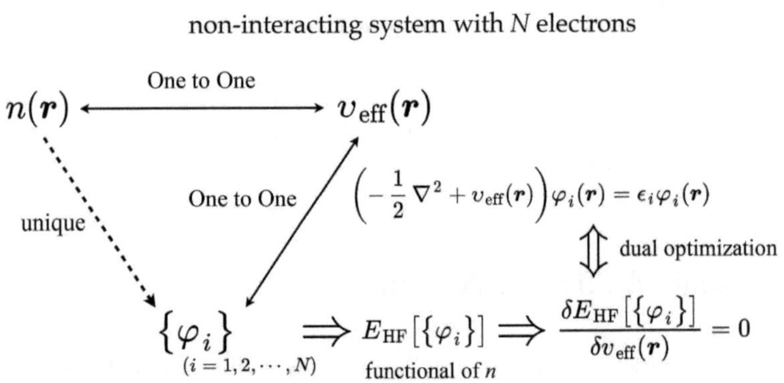

Fig. 1.11 Schematic illustration of the relations among the optimized-effective potential $v_{\rm eff}$, the non-interacting v-representable density n and the ground state wave functions $\{\varphi_i\}$ for the potential $v_{\rm eff}$. Due to the one-to-one correspondences, $\{\varphi_i\}$ can be uniquely determined from n. Thus, the Hartree-Fock energy $E_{\rm HF}[\{\varphi_i\}]$ can be viewed as a functional of the density n. To obtain the set of wave functions $\{\varphi_i\}$ that minimizes the energy $E_{\rm HF}$ the coupled dual optimizations have to be performed for the potential $v_{\rm eff}$ and the wave functions $\{\varphi_i\}$

details of the OEP method, let us discuss the relationship between OEP and KS-DFT. A scheme is presented in Fig. 1.11 to illustrate the principle of the OEP method for the HF energy. As shown in Eq. (1.174) the wave functions for OEP is constructed on the effective local potential $v_{\text{eff}}(\boldsymbol{r})$. There exists one-to-one correspondence between the set of the effective potential v_{eff} and the set of the ground state wave functions for the non-interacting system with N electrons. Note that the one-to-one correspondence is also established between the set of v_{eff} and the set of the non-interacting v-rep. densities n. In other words, a non-interacting v-rep. density uniquely defines the corresponding wave functions $\{\varphi_i\}$. Therefore, any functional in terms of $\{\varphi_i\}$ can be regarded as an implicit functional of density n. Thus, the Hartree-Fock energy $E_{\text{HF}}[\{\varphi_i\}]$ is just a functional of density and $\{\varphi_i\}$ are viewed as Kohn-Sham wave functions. But, how can we obtain the local potential v_{eff} for the HF energy functional? Remind that the differentiation of $E_{\text{HF}}[\{\varphi_i\}]$ with respect to $\varphi_i(\boldsymbol{r})$ only leads to the potential including the nonlocal HF exchange potential $v_x(\boldsymbol{r}, \boldsymbol{r}')$ defined in Eq. (1.40). Assuming that $\{\varphi_i\}$ are the ground state wave functions for a certain potential $v_{\text{eff}}(\boldsymbol{r})$ as shown in Eq. (1.174), the energy $E_{\text{HF}}[\{\varphi_i\}]$ is stationary with respect to the infinitesimal variation δv_{eff} of the potential v_{eff}, of which mathematical expression is given in Eq. (1.173). Of course, the coupled equations, namely Eqs. (1.173) and (1.174), must be solved self-consistently. Note that, in the ordinary KS-DFT approach, the potential is merely given by the functional differentiation of the exchange-correlation functional $\overline{E}_{xc}[n]$ with respect to $n(\boldsymbol{r})$ since $\overline{E}_{xc}[n]$ is given as an explicit functional of n.

A remaining problem in obtaining the OEP potential is the evaluation of the derivative of Eq. (1.173). Actually, it can be obtained by using the chain rule of the functional derivative, thus,

$$\frac{\delta E_{\text{HF}}[\{\varphi_i\}]}{\delta v_{\text{eff}}(\boldsymbol{r})} = 2\sum_i^{occ} \int d\boldsymbol{r}' \frac{\delta E_{\text{HF}}[\{\varphi_i\}]}{\delta \varphi_i^*(\boldsymbol{r}')} \frac{\delta \varphi_i^*(\boldsymbol{r}')}{\delta v_{\text{eff}}(\boldsymbol{r})} + c.c. \tag{1.176}$$

where the summation on the right-hand side is taken over all the occupied orbitals and the factor 2 is for the contributions from α and β spins. The derivative of E_{HF} with respect to $\varphi_i^*(\boldsymbol{r})$ is given by the sum of functional derivatives of Eqs. (1.35), (1.36), and (1.37). Since it is supposed that $\{\varphi_i\}$ are the eigenfunctions of Eq. (1.174), the variation of the wave function φ_i^* at $\boldsymbol{r}'$ due to the potential change δv_{eff} at $\boldsymbol{r}$ is expressed by the second-order perturbation theory

$$\frac{\delta \varphi_i^*(\boldsymbol{r}')}{\delta v_{\text{eff}}(\boldsymbol{r})} = -\sum_a^{vir} \frac{1}{\epsilon_a - \epsilon_i} \varphi_a^*(\boldsymbol{r}')\, \varphi_a(\boldsymbol{r})\, \varphi_i^*(\boldsymbol{r}) \tag{1.177}$$

where the indices i and a are for the occupied and the virtual orbitals, respectively. In the following, two paragraphs are devoted to describe the numerical approaches to implement the OEP method. An approximate approach suggested by Krieger et al. [46] is very famous and often utilized in the context of OEP. Although the efficiency of the KLI method is well established, its review does not meet the purpose of this book. We refer the readers to the reference for further study.

1.4.1 OEP Matrix Equation

The numerical implementation of the OEP will be one of the central subjects in the study of the mean-field approach. To the best of our knowledge, Talman and Shadwick [17] first constructed the OEPs for the HF exchange energy along the radial coordinate of atoms. The solution of OEP can be equivalently cast into the form of a matrix equation [47, 48]. Below we make a brief review of the formalism. To this end, we start the discussion with Eq. (1.176). In the following, it is assumed that the electronic structure of the system of interest is in a closed shell for simplicity. As shown in Eq. (1.28), the HF energy E_{HF} can be decomposed into the contributions from the kinetic energy E_{kin}, Hartree energy E_{H}, HF exchange energy E_x^{HF}, and the interaction energy E_{ext} between electrons and the external potential,

$$E_{\mathrm{HF}} = E_{\mathrm{kin}} + E_{\mathrm{H}} + E_x^{\mathrm{HF}} + E_{\mathrm{ext}} \tag{1.178}$$

Accordingly, the right-hand side of Eq. (1.176) can be rewritten as

$$\begin{aligned}\frac{\delta E_{\mathrm{HF}}\left[\{\varphi_i\}\right]}{\delta v_{\mathrm{eff}}(\boldsymbol{r})} = \sum_i^{occ} \int d\boldsymbol{r}' \Bigg(& \frac{\delta E_{\mathrm{kin}}}{\delta \varphi_i^*(\boldsymbol{r}')} + \frac{\delta E_{\mathrm{H}}}{\delta \varphi_i^*(\boldsymbol{r}')} \\ & + \frac{\delta E_x^{\mathrm{HF}}}{\delta \varphi_i^*(\boldsymbol{r}')} + \frac{\delta E_{\mathrm{ext}}}{\delta \varphi_i^*(\boldsymbol{r}')} \Bigg) \frac{\delta \varphi_i^*(\boldsymbol{r}')}{\delta v_{\mathrm{eff}}(\boldsymbol{r})} + c.c.\end{aligned} \tag{1.179}$$

Since it is supposed that Eq. (1.174) is fulfilled for the set of wave functions $\{\varphi_i\}$, the variation δE_{kin} with respect to $\delta\varphi_i$ is given by

$$\delta E_{\mathrm{kin}} = \left\langle \delta\varphi_i \left| \hat{T} \right| \varphi_i \right\rangle + c.c. = \left\langle \delta\varphi_i \left| -\hat{v}_{\mathrm{eff}} \right| \varphi_i \right\rangle + \epsilon_i \left\langle \delta\varphi_i \middle| \varphi_i \right\rangle + c.c. \tag{1.180}$$

where $\hat{T}$ is the kinetic energy operator for an electron. Due to the normalization condition imposed on φ_i, the variation $\langle \delta\varphi_i | \varphi_i \rangle + c.c.$ is equal to zero. Therefore, we get the relation for the terms in Eq. (1.179),

$$\begin{aligned}\frac{\delta E_{\mathrm{kin}}}{\delta \varphi_i^*(\boldsymbol{r})} + \frac{\delta E_{\mathrm{H}}}{\delta \varphi_i^*(\boldsymbol{r})} + \frac{\delta E_{\mathrm{ext}}}{\delta \varphi_i^*(\boldsymbol{r})} &= \left(-v_{\mathrm{eff}}(\boldsymbol{r}) + v_{\mathrm{H}}(\boldsymbol{r}) + v_{\mathrm{ext}}(\boldsymbol{r})\right)\varphi_i(\boldsymbol{r}) \\ &= -v_x^{\mathrm{HF}}(\boldsymbol{r})\,\varphi_i(\boldsymbol{r})\end{aligned} \tag{1.181}$$

Note that $v_x^{\mathrm{HF}}(\boldsymbol{r})$ is the local HF exchange potential that we want to get. At the minimum of E_{HF} with respect to the variation of v_{eff}, Eq. (1.173) is established. Then, the following relation is obtained:

$$\int d\boldsymbol{r}'\chi(\boldsymbol{r},\boldsymbol{r}')v_x^{\mathrm{HF}}\left(\boldsymbol{r}'\right)=\sum_i^{occ}\int d\boldsymbol{r}'\frac{\delta E_x^{\mathrm{HF}}}{\delta\varphi_i^*\left(\boldsymbol{r}'\right)}\frac{\delta\varphi_i^*\left(\boldsymbol{r}'\right)}{\delta v_{\mathrm{eff}}\left(\boldsymbol{r}\right)}+c.c. \tag{1.182}$$

where $\chi(\boldsymbol{r},\boldsymbol{r}')$ is a response function and is defined by

$$\chi(\boldsymbol{r},\boldsymbol{r}')=\frac{\delta n\left(\boldsymbol{r}\right)}{\delta v_{\mathrm{eff}}\left(\boldsymbol{r}'\right)} \tag{1.183}$$

In deriving Eq. (1.182), we use the relation

$$\frac{\delta n\left(\boldsymbol{r}'\right)}{\delta v_{\mathrm{eff}}\left(\boldsymbol{r}\right)}=\frac{\delta\left(2\sum_i^{occ}\varphi_i\left(\boldsymbol{r}'\right)\varphi_i^*\left(\boldsymbol{r}'\right)\right)}{\delta v_{\mathrm{eff}}\left(\boldsymbol{r}\right)}=\frac{2\sum_i^{occ}\varphi_i\left(\boldsymbol{r}'\right)\delta\varphi_i^*\left(\boldsymbol{r}'\right)}{\delta v_{\mathrm{eff}}\left(\boldsymbol{r}\right)}+c.c. \tag{1.184}$$

where the symmetry relation $\chi(\boldsymbol{r}',\boldsymbol{r})=\chi(\boldsymbol{r},\boldsymbol{r}')$ is also used. $\chi(\boldsymbol{r},\boldsymbol{r}')$ can be explicitly expressed in terms of the one-electron wave functions

$$\begin{aligned}\chi(\boldsymbol{r},\boldsymbol{r}')&=\frac{2\sum_i^{occ}\varphi_i\left(\boldsymbol{r}'\right)\delta\varphi_i^*\left(\boldsymbol{r}'\right)}{\delta v_{\mathrm{eff}}\left(\boldsymbol{r}\right)}+c.c.\\&=-2\sum_i^{occ}\sum_a^{vir}\frac{1}{\epsilon_a-\epsilon_i}\varphi_i\left(\boldsymbol{r}'\right)\varphi_a^*\left(\boldsymbol{r}'\right)\varphi_a\left(\boldsymbol{r}\right)\varphi_i^*\left(\boldsymbol{r}\right)+c.c.\end{aligned} \tag{1.185}$$

by using Eq. (1.177) derived from the perturbation theory. The differentiation of E_x^{HF} with respect to $\varphi_i(\boldsymbol{r})$ in the right-hand side of Eq. (1.182) is explicitly given by the derivative of Eq. (1.37). We also note that the explicit form of the derivative $\frac{\delta\varphi_i^*(\boldsymbol{r}')}{\delta v_{\mathrm{eff}}(\boldsymbol{r})}$ is provided in Eq. (1.177). Of course, the response function $\chi(\boldsymbol{r},\boldsymbol{r}')$ can also be obtained by a modification of Eq. (1.177). Anyway, Eq. (1.182) can be regarded as a matrix equation for the local potential v_x^{HF}. Provided that the matrix χ as well as potential v_x^{HF} are represented with some basis functions, v_x^{HF} can be obtained by applying the inverse of χ from the left to both sides of Eq. (1.182). However, in the construction of the inverse χ^{-1}, a numerical instability will arise from the fact described below [22]. Since the number N of the electrons in the system is supposed to be maintained, the following relations hold

$$\int d\boldsymbol{r}\chi(\boldsymbol{r},\boldsymbol{r}')=\int d\boldsymbol{r}'\chi(\boldsymbol{r},\boldsymbol{r}')=0 \tag{1.186}$$

for any $\boldsymbol{r}$ or $\boldsymbol{r}'$. These equation can be readily confirmed from the fact that φ_i is orthogonal to φ_a in Eq. (1.185). The relations show that the matrix has an eigenvector identified by $f(\boldsymbol{r})=c$ with c being an arbitrary constant. Actually, it is readily recognized that

$$\int d\boldsymbol{r}'\chi(\boldsymbol{r},\boldsymbol{r}')f\left(\boldsymbol{r}'\right)=c\int d\boldsymbol{r}'\chi(\boldsymbol{r},\boldsymbol{r}')=0 \tag{1.187}$$

Thus, it is proved that any invariant function can be an eigenfunction belonging to the null eigenvalue of the matrix $\chi(\boldsymbol{r}, \boldsymbol{r}')$. Suppose that χ is represented in the form of the spectral decomposition,

$$\chi(\boldsymbol{r}, \boldsymbol{r}') = \sum_k \gamma_k(\boldsymbol{r})\, g_k \gamma_k(\boldsymbol{r}') \tag{1.188}$$

where g_k and γ_k are the eigenvalue and the corresponding eigenvector of the matrix, respectively. The matrix χ is positive semi-definite, that is, $g_k \geq 0$ for all k. The positive semi-definiteness can be readily proved as shown in Appendix A of Ref. [33]. For the proof, we introduce a quantity Q defined as

$$Q = \int d\boldsymbol{r} d\boldsymbol{r}' q(\boldsymbol{r})\, \chi(\boldsymbol{r}, \boldsymbol{r}')\, q^*(\boldsymbol{r}') \tag{1.189}$$

where $q(\boldsymbol{r})$ is an arbitrary complex function. The necessary and sufficient condition for the positive semi-definiteness of χ is that $Q \geq 0$ always holds for all q. Using Eq. (1.185), Q can be expressed as

$$\begin{aligned} Q &= \int d\boldsymbol{r} d\boldsymbol{r}' \sum_i^{occ} \sum_a^{vir} \frac{1}{\epsilon_a - \epsilon_i} q(\boldsymbol{r})\, \varphi_i^*(\boldsymbol{r})\, \varphi_a(\boldsymbol{r})\, \varphi_a^*(\boldsymbol{r}')\varphi_i(\boldsymbol{r}') q^*(\boldsymbol{r}') \\ &= \int d\boldsymbol{r} d\boldsymbol{r}' \sum_i^{occ} \sum_a^{vir} \frac{1}{\epsilon_a - \epsilon_i} \Phi_{ia}^q(\boldsymbol{r})\, \Phi_{ia}^{q*}(\boldsymbol{r}') \\ &= \sum_i^{occ} \sum_a^{vir} \frac{1}{\epsilon_a - \epsilon_i} \left| \int d\boldsymbol{r}\; \Phi_{ia}^q(\boldsymbol{r}) \right|^2 \geq 0 \end{aligned} \tag{1.190}$$

where $\Phi_{ia}^q(\boldsymbol{r})$ is defined by $\Phi_{ia}^q(\boldsymbol{r}) = q(\boldsymbol{r})\, \varphi_i^*(\boldsymbol{r})\, \varphi_a(\boldsymbol{r})$. The inverse of χ is formally given by

$$\chi^{-1}(\boldsymbol{r}, \boldsymbol{r}') = \sum_k \gamma_k(\boldsymbol{r})\, g_k^{-1} \gamma_k(\boldsymbol{r}') \tag{1.191}$$

However, from the above discussion, the lowest eigenvalue is $g_0 = 0$ and hence, χ is not invertible. The Moore-Penrose inversion method for example can be used to perform the inversion, where the right-hand side of Eq. (1.191) is constructed excluding the eigenvector with the null eigenvector. The numerical instability will still arise from the eigenvectors with small eigenvalues g_k because they make significant contributions to the formation of χ^{-1}. However, even when the potential $v_x^{\mathrm{HF}}(\boldsymbol{r})$ is obtained, it includes an indefinite additive constant that cannot be determined from Eq. (1.182) by itself. This can be easily understood through the following discussion. Suppose that $v_x^{\mathrm{HF}}(\boldsymbol{r})$ is a solution of Eq. (1.182), then the shifted potential $v_x^{\mathrm{HF}}(\boldsymbol{r}) + c$ is also a solution because the relation

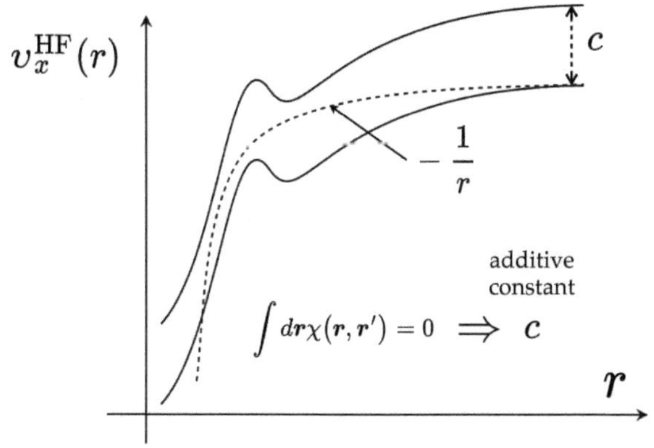

Fig. 1.12 Illustration of the optimized-effective potential $v_x^{\rm HF}(\boldsymbol{r})$ for exchange energy. Since the response function $\chi\left(\boldsymbol{r},\boldsymbol{r}'\right)$ in Eq. (1.182) has a null eigenvalue, the solution of the OEP matrix includes the indefinite additive constant c. The two real curves show the exchange potentials corresponding to the same eigenvalues of the response function $\chi\left(\boldsymbol{r},\boldsymbol{r}'\right)$. They differ from each other by a constant shift c. The amount of c must be determined so that the potential $v_x^{\rm HF}(\boldsymbol{r})$ reproduces the correct asymptotic behavior of $-1/r$

$$\int d\boldsymbol{r}'\chi(\boldsymbol{r},\boldsymbol{r}')\left(v_x^{\rm HF}\left(\boldsymbol{r}'\right)+c\right) = \int d\boldsymbol{r}'\chi(\boldsymbol{r},\boldsymbol{r}')v_x^{\rm HF}\left(\boldsymbol{r}'\right) \tag{1.192}$$

is satisfied by virtue of Eq. (1.186). In fact, the potential $v_x^{\rm HF}$ must be shifted so that it reproduces some correct physical property. Here we remind that the exchange potential can be viewed as the integration of the classical Coulomb interaction between the electron density $n(\boldsymbol{r}_1)$ and the corresponding exchange hole $n_{x\sigma}(\boldsymbol{r}_1,\boldsymbol{r}_2)$ as was shown in Eq. (1.126). We also remind the sum rule of the exchange hole function as shown in Eq. (1.128). Defining the quantity r roughly as the distance between the reference electron and the system of interest, the exchange potential should asymptotically approach $-1/r$ for large r. Thus, the additive constant must be determined so that the exchange potential $v_x^{\rm HF}(\boldsymbol{r})$ behaves as $-1/r$ asymptotically for the distance r. The issue of the indefinite additive constant associated with the OEP equation is summarized in Fig. 1.12 for readers' understandings.

We also make a remark about another approach of OEP. In another type of method [26] to obtain the OEP for the HF exchange energy, the local exchange potential $v_x^{\rm HF}$ is directly obtained by $v_x^{\rm HF}(\boldsymbol{r}) = \frac{\delta E_x^{\rm HF}[n]}{\delta n(\boldsymbol{r})}$ for each spin, which leads to

$$\begin{aligned} v_x^{\rm HF}(\boldsymbol{r}) &= \int d\boldsymbol{r}'\frac{\delta v_{\rm eff}\left(\boldsymbol{r}'\right)}{\delta n\left(\boldsymbol{r}\right)}\frac{\delta E_x^{\rm HF}[n]}{\delta v_{\rm eff}\left(\boldsymbol{r}'\right)} \\ &= \int d\boldsymbol{r}'\frac{\delta v_{\rm eff}\left(\boldsymbol{r}'\right)}{\delta n\left(\boldsymbol{r}\right)}\int d\boldsymbol{r}''\sum_i^{occ}\frac{\delta E_x^{\rm HF}[n]}{\delta\varphi_i\left(\boldsymbol{r}''\right)}\frac{\delta\varphi_i\left(\boldsymbol{r}''\right)}{\delta v_{\rm eff}\left(\boldsymbol{r}'\right)} + c.c. \end{aligned} \tag{1.193}$$

In Eq. (1.193), note that $E_x[n]$ is in fact a functional of orbitals $\{\varphi_i\}$, but regarded as an implicit functional of density. The second integrand on the right-hand side of the second equality can be described with the perturbation theory and the exchange potential given by Eq. (1.40). And the term $\frac{\delta v_{\text{eff}}(\boldsymbol{r}')}{\delta n(\boldsymbol{r})}$ is actually the inverse of the response function $\chi(\boldsymbol{r}, \boldsymbol{r}')$. Thus, this approach also requires the inversion of χ, which leads to the same problem noted above. In the next subsection, we introduce a method that is free from the matrix inversion problem.

1.4.2 Yang and Wu Approach

Here we review the method developed by Yang and Wu [27] in 2002 for the applications to general molecular systems. Their approach will be superior to the methods noted above in several points. First, it does not require the inversion of the response function and hence it is free from the numerical instability associated with the inversion. In addition, the form of the effective potential proposed in their approach is equipped with the long-range property by its construction. The central idea in the Yang and Wu method is to extract the potential of interest from the whole effective potential $v_{\text{eff}}(\boldsymbol{r})$ by introducing a reference potential. The potential $v_{\text{eff}}(\boldsymbol{r})$ is expressed by

$$v_{\text{eff}}(\boldsymbol{r}) = v_{\text{ext}}(\boldsymbol{r}) + v_{\text{ref}}(\boldsymbol{r}) + \sum_p c_p g_p(\boldsymbol{r}) \tag{1.194}$$

where v_{ref} is the Fermi-Amaldi (FA) potential [49] defined in terms of the Hartree potential v_{H} of the fixed density n_{ref} of a reference system,

$$v_{\text{ref}}(\boldsymbol{r}) = \frac{N-1}{N} v_{\text{H}}[n_{\text{ref}}](\boldsymbol{r}) = \frac{N-1}{N} \int d\boldsymbol{r}' \frac{n_{\text{ref}}(\boldsymbol{r}')}{|\boldsymbol{r} - \boldsymbol{r}'|} \tag{1.195}$$

The third term in the right-hand side of Eq. (1.194) is the potential to be optimized, and $\{g_p(\boldsymbol{r})\}$ is a set of the basis functions to represent the potential. When each $g_p(\boldsymbol{r})$ is supposed to be fixed during the optimization, the coefficients c_p solely serve as variables for the optimization. It is supposed that n_{ref} is to be constructed for the system with N electrons using e.g. the LDA functional. Thus, it is recognized that the potential is scaled to that of the system with $N-1$ electrons. As will be shown below, the introduction of the FA potential is intended to realize the long-range nature of the exchange potential using the short-range basis functions such as Gaussians. When the effective potential is given in the form of Eq. (1.194), the OEP equation of Eq. (1.173) can be expressed as

$$\frac{\partial E_{\mathrm{HF}}[\{\varphi_i\}]}{\partial c_p} = \sum_i^{occ} \int d\boldsymbol{r} d\boldsymbol{r}' \frac{\delta E_{\mathrm{HF}}[\{\varphi_i\}]}{\delta \varphi_i^*(\boldsymbol{r}')} \frac{\delta \varphi_i^*(\boldsymbol{r}')}{\delta v_{\mathrm{eff}}(\boldsymbol{r})} \frac{\partial v_{\mathrm{eff}}(\boldsymbol{r})}{\partial c_p} + c.c.$$
$$= \sum_i^{occ} \int d\boldsymbol{r} d\boldsymbol{r}' \frac{\delta E_{\mathrm{HF}}[\{\varphi_i\}]}{\delta \varphi_i^*(\boldsymbol{r}')} \frac{\delta \varphi_i^*(\boldsymbol{r}')}{\delta v_{\mathrm{eff}}(\boldsymbol{r})} g_p(\boldsymbol{r}) + c.c. \quad (1.196)$$

Substituting Eq. (1.177) into Eq. (1.196), we get a simpler expression

$$\frac{\partial E_{\mathrm{HF}}[\{\varphi_i\}]}{\partial c_p} = -2 \sum_i^{occ} \sum_a^{vir} \int d\boldsymbol{r}' \frac{\delta E_{\mathrm{HF}}[\{\varphi_i\}]}{\delta \varphi_i^*(\boldsymbol{r}')} \varphi_a^*(\boldsymbol{r}') \frac{\langle \varphi_i | g_p | \varphi_a \rangle}{\epsilon_a - \epsilon_i} + c.c. \quad (1.197)$$

Solution of this equation coupled with Eq. (1.174) yields the effective potential defined by

$$v_{\mathrm{eff}}(\boldsymbol{r}) = v_{\mathrm{ext}}(\boldsymbol{r}) + v_{\mathrm{H}}[n](\boldsymbol{r}) + v_x^{\mathrm{HF}}(\boldsymbol{r}) \quad (1.198)$$

By subtracting this equation from Eq. (1.194) we have the following expression:

$$v_x^{\mathrm{HF}}(\boldsymbol{r}) = \sum_p c_p g_p(\boldsymbol{r}) - \left(v_{\mathrm{H}}[n](\boldsymbol{r}) - v_{\mathrm{H}}[n_{\mathrm{ref}}](\boldsymbol{r})\right) - \frac{1}{N} v_{\mathrm{H}}[n_{\mathrm{ref}}](\boldsymbol{r}) \quad (1.199)$$

The second term in the right-hand side of Eq. (1.199) shows a short-range behavior when the reference density n_{ref} is reasonably close to the density n obtained. On the other hand, the third term asymptotically goes to $-1/r$ when r increases. Thus, by virtue of the fact that the FA potential of Eq. (1.195) is included beforehand in the effective potential, it is possible to ensure that the optimized exchange potential shows the correct asymptotic behavior. In the OEP approach by Yang and Wu, only the short-range part of the exchange potential is extracted and to be optimized using the short-range basis functions. In contrast to the method using matrix inversion, this approach no longer necessitates to shift the potential so that it realizes the correct asymptotic behavior. The coefficients c_p in Eq. (1.197) can be determined iteratively e.g. through a quasi-Newton method as was utilized in the implementation in Ref. [27].

1.4.3 Localized Hartree-Fock Method

Here we also introduce another type of optimized-effective potential method developed by Sala and Görling in 2001 [50] for the Hartree-Fock (HF) energy. Their approach is based on the assumption that the Slater determinant Ψ_{HF} given as an eigenfunction for the HF Hamiltonian can be obtained as the Slater determinant Ψ_{KS} optimized on a *local* exchange potential. The notation Ψ_{KS} indicates that Ψ_{KS} is the ground state wave function of a non-interacting reference system. At a glance, it seems that this approach, termed localized Hartree-Fock (LHF) method, is different from the OEP approach described above. However, LHF is, in principle, equivalent to the OEP method. This matter is worth discussing. Suppose that the minimization of $E_{\mathrm{HF}}[\{\varphi_i^{\mathrm{OEP}}\}]$ through OEP can achieve exactly the HF ground state energy $E_{\mathrm{HF}}[\{\varphi_i^{\mathrm{HF}}\}]$. Then the Slater determinant Ψ_{OEP} built from $\{\varphi_i^{\mathrm{OEP}}\}$ is exactly the same as Ψ_{HF} due to the variational principle. On the other hand, provided that the LHF method is successfully performed, the wave function $\Psi_{\mathrm{LHF}}(=\Psi_{\mathrm{KS}})$ can faithfully realize Ψ_{HF}. It is, thus, obvious that the OEP method is equivalent to the LHF approach. For the LHF or equivalently the OEP method to be successful, the HF density $n_0(\boldsymbol{r})$ must be non-interacting υ-representable as a necessary condition. The corresponding wave function Ψ_{KS} realizes the density $n_0(\boldsymbol{r})$ and simultaneously minimizes $E_{\mathrm{HF}}[\Psi_{\mathrm{KS}}]$ when and only when $\Psi_{\mathrm{KS}}=\Psi_{\mathrm{HF}}$ is fulfilled.

If the LHF approach is found to be successful, it directly serves as the proof-of-principle of the Kohn-Sham approach at least for the Hartree-Fock wave function as in the OEP approach. Before we discuss the LHF method in more details, let us first consider a primitive approach proposed by Slater [11] as an approximate LHF. In the following, we drop the suffix for spins for simplicity. To introduce Slater's approach, we first define the local exchange potential $\upsilon_{xi}(\boldsymbol{r})$ specific to the orbital φ_i, thus,

$$\upsilon_{xi}(\boldsymbol{r})=\frac{1}{\varphi_i(\boldsymbol{r})}\int d\boldsymbol{r}'\upsilon_x(\boldsymbol{r},\boldsymbol{r}')\varphi_i(\boldsymbol{r}') \tag{1.200}$$

where the nonlocal HF exchange potential υ_x is that defined in Eq. (1.40). It is readily recognized that multiplication of $\varphi_i(\boldsymbol{r})$ by $\upsilon_{xi}(\boldsymbol{r})$ is identical to the nonlocal operation of υ_x to φ_i. To eliminate the orbital dependence of the potential, we take the average sum of the potential

$$\upsilon_x^{\mathrm{Slater}}(\boldsymbol{r})=\sum_i^{occ}w_i(\boldsymbol{r})\upsilon_{xi}(\boldsymbol{r}) \tag{1.201}$$

where the weight function $w_i(\boldsymbol{r})$ is chosen as

$$w_i(\boldsymbol{r})=\frac{\varphi_i(\boldsymbol{r})\varphi_i^*(\boldsymbol{r})}{n(\boldsymbol{r})} \tag{1.202}$$

It is important to confirm that the HF exchange energy E_x^{HF} defined by Eq. (1.31) can be reproduced by the following calculation, that is,

$$E_x^{\mathrm{HF}}=\sum_a^{occ}\int d\boldsymbol{r}\varphi_a^*(\boldsymbol{r})\upsilon_x^{\mathrm{Slater}}(\boldsymbol{r})\varphi_a(\boldsymbol{r}) \tag{1.203}$$

We will refer the potential in Eq. (1.201) as Slater's localized HF exchange in the following discussion. The localization by Slater is thus quite reasonable and easy to understand. However, it is unlikely that the ground state wave function optimized on the potential $\upsilon_x^{\mathrm{Slater}}$ becomes identical to the HF wave function obtained on the nonlocal potential $\upsilon_x(\boldsymbol{r},\boldsymbol{r}')$.

Sala and Görling proposed an assumption [50] that there exists a local potential that yields the same ground state wave function as the one optimized on the nonlocal potential υ_x. To describe explicitly, let Ψ be a Slater determinant as an eigenfunction for the HF Hamiltonian. Then, Ψ satisfies the following Schrödinger equation:

$$\left(\hat{T} + \hat{V}_{\mathrm{H}} + \hat{V}_{\mathrm{ext}} + \hat{V}_x^{\mathrm{nloc}}\right)|\Psi\rangle = E_s^{\mathrm{HF}}|\Psi\rangle \tag{1.204}$$

where $\hat{T}$, $\hat{V}_{\mathrm{H}}$, and $\hat{V}_{\mathrm{ext}}$ are the many-electron operators for the kinetic energy, Hartree potential, and the external potential, respectively. Explicitly, $\hat{T}$ and $\hat{V}_{\mathrm{H}}$ in the real space is represented by

$$\hat{T} := \sum_i^N -\frac{1}{2}\nabla_i^2 \tag{1.205}$$

and

$$\hat{V}_{\mathrm{H}} := \sum_i^N \upsilon_{\mathrm{H}}(\boldsymbol{r}_i) \tag{1.206}$$

where υ_{H} is defined in Eq. (1.39). The nonlocal operator $\hat{V}_x^{\mathrm{nloc}}$ is given by the sum of the HF exchange operators, thus,

$$V_x^{\mathrm{nloc}}(\boldsymbol{r}_1, \boldsymbol{r}_2, \cdots, \boldsymbol{r}_N; \boldsymbol{r}'_1, \boldsymbol{r}'_2, \cdots, \boldsymbol{r}'_N) = \sum_i^N \upsilon_x^{\mathrm{nloc}}(\boldsymbol{r}_i, \boldsymbol{r}'_i) \tag{1.207}$$

$\upsilon_x^{\mathrm{nloc}}(\boldsymbol{r}, \boldsymbol{r}')$ is actually identical to the HF exchange $\upsilon_x(\boldsymbol{r}, \boldsymbol{r}')$ in Eq. (1.40). However, to avoid confusion in the following discussion, we attached the superscript 'nloc' to the notation υ_x. Note that the eigenvalue E_s^{HF} in Eq. (1.204) is just given by the sum of the orbital energies, i.e. $E_s^{\mathrm{HF}} = \sum_i^N \epsilon_i^{\mathrm{HF}}$ which are obtained as solutions of Eq. (1.49), and thus different from the expectation value E_{HF} for the real Hamiltonian given in the form of Eq. (1.27). According to the assumption by Sala and Görling [50], it is assumed that the same Slater determinant Ψ is also the eigenfunction of the Hamiltonian with some local exchange potential $\upsilon_x^{\mathrm{loc}}(\boldsymbol{r})$

$$\left(\hat{T} + \hat{V}_{\mathrm{H}} + \hat{V}_{\mathrm{ext}} + \hat{V}_x^{\mathrm{loc}}\right)|\Psi\rangle = E_s^{\mathrm{KS}}|\Psi\rangle \tag{1.208}$$

where $\hat{V}_x^{\mathrm{loc}}$ is given by

$$V_x^{\mathrm{loc}}(\boldsymbol{r}_1, \boldsymbol{r}_2, \cdots, \boldsymbol{r}_N) = \sum_i^N \upsilon_x^{\mathrm{loc}}(\boldsymbol{r}_i) \tag{1.209}$$

Again we note that the eigenvalue E_s^{KS} in Eq. (1.208) is also given by the sum of the eigenvalues of Kohn-Sham equation (Eq. (1.121)). To derive the equation identifying the target local potential $v_x^{\rm loc}(\boldsymbol{r})$, we multiply Eqs. (1.204) and (1.208) by Ψ^* from the left and integrate with respect to all spatial coordinates and spin variables except for $\boldsymbol{r}_1$. The right-hand sides of the resultant equations include the common quantity

$$\int d\tau' \Psi^* (\boldsymbol{x}_1, \boldsymbol{x}_2, \cdots, \boldsymbol{x}_N) \Psi (\boldsymbol{x}_1, \boldsymbol{x}_2, \cdots, \boldsymbol{x}_N) = \frac{1}{N} n (\boldsymbol{r}_1) \tag{1.210}$$

where the infinitesimal fraction $d\tau'$ associated with the integral stands for the variables $d\omega_1 d\boldsymbol{x}_2 \cdots d\boldsymbol{x}_N$. Subtracting the result of Eq. (1.204) from that of Eq. (1.208) we obtain

$$\begin{aligned} &\int d\tau' \Psi^* (\boldsymbol{x}_1, \boldsymbol{x}_2, \cdots, \boldsymbol{x}_N) V_x^{\rm loc} \Psi (\boldsymbol{x}_1, \boldsymbol{x}_2, \cdots, \boldsymbol{x}_N) \\ &+ \int d\tau' \langle \Psi | \boldsymbol{x}_1, \boldsymbol{x}_2, \cdots, \boldsymbol{x}_N \rangle \left\langle \boldsymbol{x}_1, \boldsymbol{x}_2, \cdots, \boldsymbol{x}_N \left| \hat{V}_x^{\rm nloc} \right| \Psi \right\rangle \\ &= \frac{1}{N} \left(E_s^{\rm KS} - E_s^{\rm HF} \right) n (\boldsymbol{r}_1) \end{aligned} \tag{1.211}$$

where $V_x^{\rm loc}$ is a shorthand notation of $V_x^{\rm loc}(\boldsymbol{r}_1, \boldsymbol{r}_2, \cdots, \boldsymbol{r}_N)$ and the same is true for $V_x^{\rm nloc}$. In formulating Eq. (1.211), we assume for simplicity that the Slater determinant Ψ is for the system with a closed shell electronic structure consisting of $N/2$ orbitals $\{\varphi_a\}$ $(a = 1, 2, \cdots, N/2)$. Remind that the operators $\hat{V}_x^{\rm nloc}$ and $\hat{V}_x^{\rm loc}$ are the one-electron operators as shown in Eqs. (1.207) and (1.209), respectively. Then, for the operators $v_x^{\rm loc}(\boldsymbol{r}_1)$ and $v_x^{\rm nloc}(\boldsymbol{r}_1, \boldsymbol{r}_1')$, the left-hand side of Eq. (1.211) becomes

$$l_1 = \frac{2}{N} \sum_a^{N/2} \left(\varphi_a^* (\boldsymbol{r}_1) v_x^{\rm loc} (\boldsymbol{r}_1) \varphi_a (\boldsymbol{r}_1) - \varphi_a^* (\boldsymbol{r}_1) \left\langle \boldsymbol{r}_1 \left| \hat{v}_x^{\rm nloc} \right| \varphi_a \right\rangle \right) \tag{1.212}$$

Next, we consider the operators $v_x^{\rm loc}(\boldsymbol{r}_i)$ and $v_x^{\rm nloc}(\boldsymbol{r}_i, \boldsymbol{r}_i')$ for $i \neq 1$ in the light-hand side of Eq. (1.211). It leads to the following two terms l_2 and l_3,

$$l_2 = \frac{4}{N} \sum_{a \neq b}^{N/2} \varphi_a^* (\boldsymbol{r}_1) \varphi_a (\boldsymbol{r}_1) \left\langle \varphi_b \left| \hat{v}_x^{\rm loc} - \hat{v}_x^{\rm nloc} \right| \varphi_b \right\rangle \tag{1.213}$$

$$l_3 = -\frac{2}{N} \sum_{a \neq b}^{N/2} \varphi_a^* (\boldsymbol{r}_1) \varphi_b (\boldsymbol{r}_1) \left\langle \varphi_b \left| \hat{v}_x^{\rm loc} - \hat{v}_x^{\rm nloc} \right| \varphi_a \right\rangle \tag{1.214}$$

In the construction of l_2, the order of the electrons in the left-hand side of the operator is exactly the same as that in the right-hand side, while the electrons 1 and 2 are exchanged in the construction of l_3. In the summations in l_2 and l_3, note that the cases of $a = b$ are excluded. For those cases, we have

$$l_4 = \frac{2}{N} \sum_{a}^{N/2} \varphi_a^* (\boldsymbol{r}_1)\, \varphi_a (\boldsymbol{r}_1) \left\langle \varphi_a \left| \hat{v}_x^{\mathrm{loc}} - \hat{v}_x^{\mathrm{nloc}} \right| \varphi_a \right\rangle \tag{1.215}$$

It can be readily understood that l_4 can be realized in the sum of l_2 and l_3 including the cases for $a = b$. As a result, the left-hand side of Eq. (1.211) becomes

$$\begin{aligned} \mathrm{lhs} = &\frac{1}{N} n (\boldsymbol{r}_1)\, v_x^{\mathrm{loc}} (\boldsymbol{r}_1) - \frac{2}{N} \sum_{a}^{N/2} \varphi_a^* (\boldsymbol{r}_1) \left\langle \boldsymbol{r}_1 \left| \hat{v}_x^{\mathrm{nloc}} \right| \varphi_a \right\rangle \\ &+ \frac{2}{N} n (\boldsymbol{r}_1) \sum_{b}^{N/2} \left\langle \varphi_b \left| \hat{v}_x^{\mathrm{loc}} - \hat{v}_x^{\mathrm{nloc}} \right| \varphi_b \right\rangle \\ &- \frac{2}{N} \sum_{a,b}^{N/2} \varphi_a^* (\boldsymbol{r}_1)\, \varphi_b (\boldsymbol{r}_1) \left\langle \varphi_b \left| \hat{v}_x^{\mathrm{loc}} - \hat{v}_x^{\mathrm{nloc}} \right| \varphi_a \right\rangle \end{aligned} \tag{1.216}$$

Now we consider the right-hand side of Eq. (1.211). It is easily recognized that

$$E_s^{\mathrm{KS}} - E_s^{\mathrm{HF}} = 2 \sum_{a}^{N/2} \left\langle \varphi_a \left| \hat{v}_x^{\mathrm{loc}} - \hat{v}_x^{\mathrm{nloc}} \right| \varphi_a \right\rangle \tag{1.217}$$

Using the relation in Eq. (1.210), the right-hand side of Eq. (1.211) reads

$$\mathrm{rhs} = \frac{2}{N} n (\boldsymbol{r}_1) \sum_{a}^{N/2} \left\langle \varphi_a \left| \hat{v}_x^{\mathrm{loc}} - \hat{v}_x^{\mathrm{nloc}} \right| \varphi_a \right\rangle \tag{1.218}$$

By equating Eqs. (1.216) and (1.218), we get the relation that identifies the target potential $v_x^{\mathrm{loc}} (\boldsymbol{r})$, that is,

$$\begin{aligned} v_x^{\mathrm{loc}} (\boldsymbol{r}) = &\frac{2}{n (\boldsymbol{r})} \sum_{a}^{N/2} \varphi_a^* (\boldsymbol{r}) \left\langle \boldsymbol{r} \left| \hat{v}_x^{\mathrm{nloc}} \right| \varphi_a \right\rangle \\ &+ \frac{2}{n (\boldsymbol{r})} \sum_{a,b}^{N/2} \varphi_a^* (\boldsymbol{r})\, \varphi_b (\boldsymbol{r}) \left\langle \varphi_b \left| \hat{v}_x^{\mathrm{loc}} \right| \varphi_a \right\rangle \\ &- \frac{2}{n (\boldsymbol{r})} \sum_{a,b}^{N/2} \varphi_a^* (\boldsymbol{r})\, \varphi_b (\boldsymbol{r}) \left\langle \varphi_b \left| \hat{v}_x^{\mathrm{nloc}} \right| \varphi_a \right\rangle \end{aligned} \tag{1.219}$$

Inspection of Eq. (1.219) is quite important to get deeper understanding of the LHF method. First, we remind that v_x^{nloc} is the same as $v_x \left(\boldsymbol{r}, \boldsymbol{r}' \right)$ in Eq. (1.200). Then, the first term on the right-hand side of Eq. (1.219) is found to be identical to Slater's local exchange potential defined by Eq. (1.201). Thus, the remaining two terms in Eq. (1.219) serve as corrections to the local exchange potential by Slater. Note that the right-hand side of Eq. (1.219) also includes the potential $\upsilon_x^{\mathrm{loc}}$. Then, it can be shown that $\upsilon_x^{\mathrm{loc}}(\boldsymbol{r})$ also includes the indefinite additive constant c. Suppose that $\upsilon_x^{\mathrm{loc}}$ is a solution of Eq. (1.219), then it is possible to prove that $\upsilon_x^{\mathrm{loc}} + c$ is also a solution. To show this we put the second term in Eq. (1.219) as $\upsilon_{x,\mathrm{2nd}}^{\mathrm{loc}}$ for the sake of simplicity. Then, the second term in Eq. (1.219) for the shifted potential $\upsilon_x^{\mathrm{loc}} + c$ becomes

$$
\begin{aligned}
\frac{2}{n(\boldsymbol{r})} & \sum_{a,b}^{N/2} \varphi_a^*(\boldsymbol{r}) \varphi_b(\boldsymbol{r}) \left\langle \varphi_b \left| \hat{\upsilon}_x^{\mathrm{loc}} + c \right| \varphi_a \right\rangle \\
&= \upsilon_{x,\mathrm{2nd}}^{\mathrm{loc}}(\boldsymbol{r}) + \frac{2}{n(\boldsymbol{r})} \sum_{a,b}^{N/2} \varphi_a^*(\boldsymbol{r}) \varphi_b(\boldsymbol{r}) \left\langle \varphi_b \left| c \right| \varphi_a \right\rangle \\
&= \upsilon_{x,\mathrm{2nd}}^{\mathrm{loc}}(\boldsymbol{r}) + c \frac{2}{n(\boldsymbol{r})} \sum_{a}^{N/2} \varphi_a^*(\boldsymbol{r}) \varphi_a(\boldsymbol{r}) \\
&= \upsilon_{x,\mathrm{2nd}}^{\mathrm{loc}}(\boldsymbol{r}) + c
\end{aligned}
\tag{1.220}
$$

Thus, both sides of Eq. (1.219) undergo the same constant shift c, which cancel each other. In the discussion below, let r be the approximate distance between the reference electron and the system of interest. The local exchange potential obtained by the LHF method should be shifted by a constant value so that it realizes the correct asymptotic behavior of $-1/r$ as was shown in Fig. 1.12. In the work [50] by Sala and Görling, a well-defined approach is proposed to determine the constant c. They impose a condition that

$$
\left\langle \varphi_{N/2} \left| \hat{\upsilon}_x^{\mathrm{loc}} - \hat{\upsilon}_x^{\mathrm{nloc}} \right| \varphi_{N/2} \right\rangle = 0 \tag{1.221}
$$

where $\varphi_{N/2}$ is the HOMO of the system. When the above condition is fulfilled, it is easy to see that the second and the third terms in Eq. (1.219) asymptotically vanish in the large r region. This is because the density $n(\boldsymbol{r})$ is dominated by the density of the HOMO in the region. And hence, $\varphi_a(\boldsymbol{r})\varphi_b(\boldsymbol{r})/n(\boldsymbol{r})$ $(a, b \neq N/2)$ vanishes exponentially as r increases. Consequently, the localized exchange potential $\upsilon_x^{\mathrm{loc}}(\boldsymbol{r})$ in the outer region is dominated by the first term in Eq. (1.219), i.e. Slater's exchange potential, which asymptotically behaves as $-1/r$.

Since the target potential is included in both sides of Eq. (1.219), the equation must be solved self-consistently. The procedure to solve the LHF equation will be summarized as follows.

1. First of all, the SCF calculation for ordinary HF method is performed to construct the nonlocal potential $\hat{\upsilon}_x^{\mathrm{nloc}} := \upsilon_x(\boldsymbol{r}, \boldsymbol{r}')$ in Eq. (1.219).
2. The Hartree potential $\upsilon_{\mathrm{H}}(\boldsymbol{r})$ is constructed in terms of the density provided by the process (1).
3. Using the temporary local potential $\hat{\upsilon}_x^{\mathrm{loc}}$, the KS Hamiltonian is diagonalized once to yield the temporary wave functions $\{\varphi_a\}$ in Eq. (1.219).

4. The matrix elements $\langle \varphi_b | \hat{\upsilon}_x^{\text{loc}} | \varphi_a \rangle$ and $\langle \varphi_b | \hat{\upsilon}_x^{\text{nloc}} | \varphi_a \rangle$ in Eq. (1.219) are calculated and the local potential $\hat{\upsilon}_x^{\text{loc}}$ is updated.

Processes (3) and (4) are iterated until the local potential $\hat{\upsilon}_x^{\text{loc}}$ will achieve convergence. In Ref. [50], a method to obtain $\hat{\upsilon}_x^{\text{loc}}$ without SCF procedure is also presented. However, the review of it will not be in the scope of this book. We refer readers to the reference in more details.

1.4.4 Performance of OEP

Since the success of OEP implies the proof-of-principle for KS-DFT, the assessment of the performance of the OEP method is of critical importance. Table 1.2 shows the total energies of five atoms computed by the OEP for the HF exchange energy [26, 27]. The notation E_{OEP}^A in the table indicates the value obtained by Ivanov et al. using Eq. (1.193), while E_{OEP}^B is for the result by Yang *et al.* using Eq. (1.199). It is shown that the OEP energies are always larger than the HF energies. However, the differences were found to be quite small, which suggests the accuracy of the OEP method. The percentage of error to the total energy is also shown in the last column. It is found that the errors are within 4×10^{-3} % regardless of the magnitude of the total energy. Thus, the mean absolute errors (MAEs) of the energies E_{OEP}^A and E_{OEP}^B are evaluated as 1.75 and 1.78 mHartree, respectively. In the OEP calculations, we note that the potentials as well as the orbitals were represented with Gaussian-type orbitals which were solely designed for the description of wave functions. Thus, the errors in the OEP calculation includes those due to the representation of the potential in addition to the problem associated with the principle itself (if it exists). Anyway,

Table 1.2 Total energies of atoms computed by the Hartree-Fock (HF) and the OEP methods. The OEP values; E_{OEP}^A and E_{OEP}^B are those presented in Refs. [26, 27], respectively, which utilized Eqs. (1.193) and (1.199). In both of these calculations, Partridge-3 basis set [51] was employed to represent both wave functions and potentials. The notation 'ratio' shows the percentage of the error in the OEP result and defined as $\left| \frac{E_{\text{OEP}}^B - E_{\text{HF}}}{E_{\text{HF}}} \right| \times 100$. The integer number S represents the spin multiplicity. MAE stands for mean absolute error

Atom	E_{HF} (Hartree)	$E_{\text{OEP}}^A - E_{\text{HF}}$ (mHartree)	$E_{\text{OEP}}^B - E_{\text{HF}}$ (mHartree)	ratio (10^{-3} %)
Li ($S = 2$)	–7.43275	0.24	0.26	3.4
Be ($S = 1$)	–14.57302	0.58	0.58	3.9
N ($S = 4$)	–54.40454	1.14	1.15	2.1
Ne ($S = 1$)	–128.5471	1.7	1.7	1.3
Ar ($S = 1$)	–526.8175	5.1	5.2	1.0
MAE	–	1.75	1.78	–

the results of the OEP methods are quite encouraging at least for the applications to atoms.

In Ref. [27], the OEP method was also applied to calculate the HF total energies of the small 14 molecules including 2 anionic molecules. The results for the 12 neutral molecules were compared with those obtained in Ref. [26]. The mean absolute error (MAE) of the energy E^A_{OEP} provided in Ref. [26] was evaluated as 3.45 mHartree, while that of E^B_{OEP} in Ref. [27] was 3.47 mHartree. Thus, the error in the energy E^B_{OEP} is slightly larger than that in E^A_{OEP} although the same basis set was used in both of these calculations. The same trend is also found in the calculations for atoms. The authors of Ref. [27] speculated that the difference is due to the fact that no constraint was imposed in the calculation of energy E^A_{OEP} to ensure that the exchange potential exhibits the correct asymptotic behavior. It is also worth noting in the table that the MAE for the molecules is larger than that for atoms although the minimum total energy in the 12 molecular systems is only -198.75244 Hartree for F_2 molecule. This is possibly because the basis set used is not appropriate to represent the potential. Although the basis sets for atoms might work better in describing the exchange potentials for atom systems, it is not so for the systems with chemical bonds. Anyway, these calculations revealed that the local exchange potential yielded by the OEP approach is capable of realizing the nonlocal effect of the HF exchange potential.

The localized Hartree-Fock (LHF) method, as described above, differs from the OEP in the point that it is based on the assumption that the wave function (Slater determinant) optimized on an optimized local potential becomes identical to that given by the HF method. Provided that the assumption is correct, it is readily understood that the wave function of LHF is also identical to that of OEP when the exchange only OEP potential faithfully realizes the HF total energy. In Ref. [50], Sala and Görling provided the values of the overlap $\langle \Psi^{\mathrm{LHF}}_{\mathrm{SD}} | \Psi^{\mathrm{HF}}_{\mathrm{SD}} \rangle$ between the LHF wave function $\Psi^{\mathrm{LHF}}_{\mathrm{SD}}$ and the HF Slater determinant $\Psi^{\mathrm{HF}}_{\mathrm{SD}}$ to examine the validity of the assumption for the method of LHF. For the calculations, they developed a Gaussian-type basis set, referred to as GEO (geometric series basis sets). It was found that the deviations of the overlaps from unity are very small for 13 systems including 3 atoms (Ne, Ar, and Kr). Li_8 among the systems showed the maximum deviation of $\sim 50 \times 10^{-4}$, while it was found that the deviations of the other systems stay within $\sim 10 \times 10^{-4}$. For the three atoms, the deviations were found to be within 0.2×10^{-4}. Also in the LHF calculations, the deviations for the atoms were found to be much smaller than those for the molecular systems. The reason for this might be the same as that for the OEP calculations. We also make a remark on the total energies of molecules obtained by the LHF method. The deviation of the LHF energy from the HF energy for H_2O was computed as 3.31 mHartree using the GEO basis, which is found to be comparable to the OEP result of 2.31 mHartree given by Yang et al. [27] employing the different basis set. Similarly, the deviation of the LHF energy for CO molecule was obtained as 7.72 mHartree [50], while it was computed as 5.12 mHartree by the OEP approach [27]. Thus, it was demonstrated in Ref. [50] that the LHF approach is very accurate and almost comparable to the OEP in reproducing the HF wave function on the local potential.

So far we focused our attention only on the calculation of the total OEP energies of atoms and molecules. However, it is quite difficult with these calculations to elucidate how the OEP approach makes it possible to embed the nonlocal HF exchange potential into a local potential. A possible approach to the issue is to analyze how the OEP exchange potential changes in response to the electric field applied to a molecule. Actually, the density response to the applied field has a technological relevance with the nonlinear optical (NLO) properties of materials. It has long been known that the LDA or GGA functional seriously overestimates the polarizabilities and hyperpolarizabilities of the quasilinear extended systems [52]. On the other hand, the HF and the OEP approaches are able to avoid the overpolarization because of the nonlocal term that counteracts the applied field [53]. In Ref. [54], Gisbergen et al. elucidated the mechanism how the OEP exchange potential counteracts the applied field E through the nonlocality embedded in the potential. As was described above, it is well known that the approximation by Krieger et al. [46] to the OEP is efficient regardless of its simple form. Actually, the exchange potential $\upsilon_x^{\mathrm{KLI}}(\boldsymbol{r})$ by the KLI approximation is represented by a simple form:

$$\upsilon_x^{\mathrm{KLI}}(\boldsymbol{r}) = \upsilon_x^{\mathrm{Slater}}(\boldsymbol{r}) + \sum_i^{occ} w_i \frac{n_i(\boldsymbol{r})}{n(\boldsymbol{r})} \tag{1.222}$$

where $\upsilon_x^{\mathrm{Slater}}$ is Slater's local exchange potential defined by Eq. (1.201) and n_i is the density of the ith orbital. w_i is a constant with the dimension of the energy and is dependent on the orbital i and the potential $\upsilon_x^{\mathrm{KLI}}$ itself. As described below, Gisbergen *et al.* utilized Eq. (1.222) in the discussion [54] to illustrate a rough mechanism of the emergence of the counteracting linear potential in $\upsilon_x^{\mathrm{KLI}}$ in response to an electric field E applied along the linear chain of H_{18} molecule. On the left in Fig. 1.13, the bonding φ_{b} and anti-bonding φ_{a} orbitals in the center of the molecular chain is drawn. These two orbitals are mixed by the application of the field E that generates the potential $V_E(\boldsymbol{r}) = -Ez$ along the chain (on the right in the figure). A positive mixing of the fraction of φ_{a} to φ_{b} leads to the energetically lower orbital φ_{b}' with larger electron population on the right in the wave function. And the negative mixing of φ_{b} to φ_{a} yields the orbital φ_{a}' with larger population on the left. Thus, it is anticipated that the resulting density $n'(\boldsymbol{r})$ is almost the same as the original density $n(\boldsymbol{r})$. Therefore, it cannot be expected that the counteracting potential arises in the exchange potential provided by the LDA or GGA functionals. In contrast, in the OEP potential approximated by KLI, the orbital density n_{b}' contributes to create the counteracting potential since the energy w_i for the lower lying orbitals are anticipated to be larger. These discussion clearly shows how the nonlocal effect of the potential in response to the applied field is embedded in the multiplicative KLI potential. Actually, such a construction can be made possible by means of the energy w_i and the corresponding orbital density n_i.

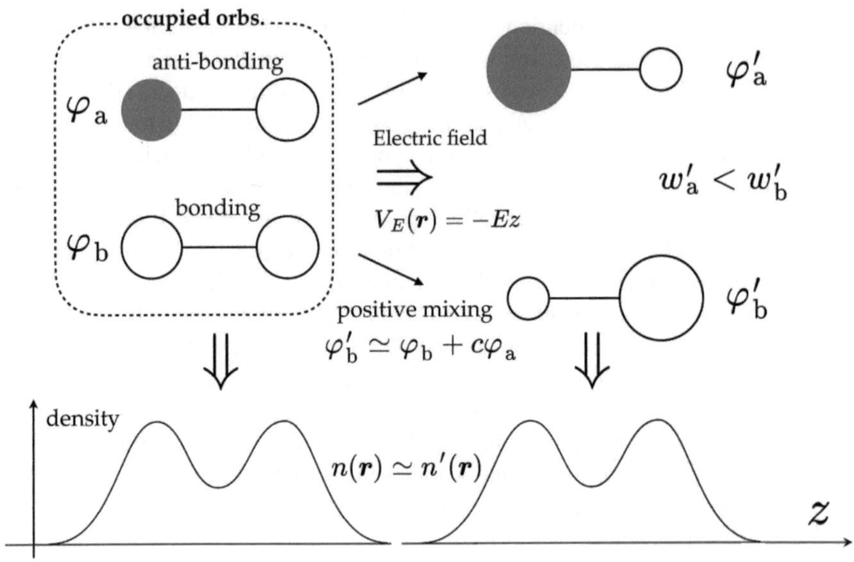

Fig. 1.13 Illustration of the bonding φ_b and anti-bonding φ_a orbitals in the occupied orbital space in the center of the linear chain molecule H_{18} investigated in Ref. [54]. The energy gap between the orbitals increases after the application of the electric field E (on the right). The mixing parameter c is a small positive real value. It is supposed that the electric field is applied along the molecular chain placed on the z axis. The electron densities n and n' are given by $n = |\varphi_b|^2 + |\varphi_a|^2$ and $n' = |\varphi'_b|^2 + |\varphi'_a|^2$, respectively. w'_a denotes the weight w in Eq. (1.222)

1.4.5 Another Equivalent Approach: Inverse Kohn-Sham Method

The HF-OEP or LHF method, of which solution requires rather complicated procedures as described above, can be equivalently solved by another much simpler approach. When the HF-OEP potential is successfully constructed, it will yield a set of wave functions $\{\varphi_i\}$ which provides exactly the energy E_0^{HF} and the density $n_0^{\mathrm{HF}}(\boldsymbol{r})$ of the HF ground state. This issue has a relevance with the representability of n_0^{HF}. Suppose that a ground state density $n_0^{\mathrm{HF}}(\boldsymbol{r})$ is non-interacting v-representable, then the corresponding potential $v^{\mathrm{HF}}(\boldsymbol{r})$ can be determined uniquely due to the one-to-one correspondence as depicted in Fig. 1.11. Thus, the HF-OEP and the LHF methods discussed so far can be regarded as solutions to obtain the local potential $v^{\mathrm{HF}}(\boldsymbol{r})$ that corresponds to $n_0^{\mathrm{HF}}(\boldsymbol{r})$. Provided that $n_0^{\mathrm{HF}}(\boldsymbol{r})$ is non-interacting v-representable, the potential $v^{\mathrm{HF}}(\boldsymbol{r})$ can be obtained numerically by means of the 'inverse Kohn-Sham' (inv-KS) method [31, 55–58]. Actually, it is possible to prove that the HF-OEP is equivalent to the inv-KS method under a condition as will be shown later.

At first, we discuss the OEP method within the framework of the potential-functional theory (PFT) formulated by Yang et al. [59] In this paragraph, we review the PF formalism and understand the HF-OEP in the context of the potential functional. In Appendix C, we provide discussions on the PFT specific to the HF energy in more details. The point of the PF is to switch the fundamental variable in the density-functional theory from the electron density $n(\boldsymbol{r})$ to the external potential $w_{\text{ext}}(\boldsymbol{r})$. The replacement can be made possible, of course, by virtue of the principle of one-to-one correspondence between the v-rep. density and the external potential. In constructing the PF formalism, the Hohenberg-Kohn (HK) functional $F^{\text{HK}}[n]$ defined in Eq. (1.76) in Sect. 1.3.2 plays a central role. In the following, it is always assumed that the ground state is non-degenerate. The potential functional for the system with an external potential v_{ext} can be formally defined by

$$E_{\text{PF},v_{\text{ext}}}\left[w_{\text{ext}}\right]=F^{\text{HK}}\left[n_{w_{\text{ext}}}\right]+\int d\boldsymbol{r}\; n_{w_{\text{ext}}}\left(\boldsymbol{r}\right)v_{\text{ext}}\left(\boldsymbol{r}\right) \tag{1.223}$$

where it is assumed that the electron density $n_{w_{\text{ext}}}$ is derived from the ground state wave function $\Psi_0[n_{w_{\text{ext}}}]$ corresponding to the external potential w_{ext}. Note that $n_{w_{\text{ext}}}$ can be uniquely determined from the potential w_{ext}. Then the ground state energy $E_{0,v_{\text{ext}}}$ can be obtained by the minimization of the potential functional

$$E_{0,v_{\text{ext}}}=\min_{w_{\text{ext}}}E_{\text{PF},v_{\text{ext}}}\left[w_{\text{ext}}\right] \tag{1.224}$$

The minimum of the energy can be achieved when the potential w_{ext} coincides with the external potential v_{ext}. The PF approach can be extended to the non-interacting (KS) systems. The potential functional $E_{\text{PF0},w_{\text{KS}}}\left[w_{\text{KS}}\right]$ is defined as

$$E_{\text{PF0},v_{\text{ext}}}\left[w_{\text{KS}}\right]=F^{\text{HK}}\left[n_{w_{\text{KS}}}\right]+\int d\boldsymbol{r}\; n_{w_{\text{KS}}}\left(\boldsymbol{r}\right)v_{\text{ext}}\left(\boldsymbol{r}\right) \tag{1.225}$$

where w_{KS} is the effective potential of a KS (non-interacting) system and the subscript PF0 stands for potential functional for a non-interacting system. The density $n_{w_{\text{KS}}}$ is assumed to be built from the orbitals $\{\varphi_i\}$ that fulfill the KS equations

$$\left(-\frac{1}{2}\nabla^2+w_{\text{KS}}\left(\boldsymbol{r}\right)\right)\varphi_i\left(\boldsymbol{r}\right)=\epsilon_i\varphi_i\left(\boldsymbol{r}\right) \tag{1.226}$$

Namely, the relation $n_{w_{\text{KS}}}(\boldsymbol{r})=\sum_i^{occ}\varphi_i(\boldsymbol{r})\varphi_i^*(\boldsymbol{r})$ holds. Note that the potential functionals $E_{\text{PF},v_{\text{ext}}}$ is different from $E_{\text{PF0},v_{\text{ext}}}$ because $n_{w_{\text{ext}}}$ is different from $n_{w_{\text{KS}}}$ even when $w_{\text{ext}}(\boldsymbol{r})=w_{\text{KS}}(\boldsymbol{r})$. Anyway, the ground state energy $E_{0,v_{\text{ext}}}$ is obtained by the minimization of the potential functional $E_{\text{PF0},v_{\text{ext}}}$ with respect to the effective potential w_{KS} thus

$$E_{0,v_{\text{ext}}}=\min_{w_{\text{KS}}}E_{\text{PF0},v_{\text{ext}}}\left[w_{\text{KS}}\right] \tag{1.227}$$

It is thus demonstrated that the potential functional is dual to the density functional [59].

So far we have discussed the PF for the ground state energy of a general wave function for Fermi particles. Actually, as shown in Appendix C, it is also possible to prove the principle of one-to-one correspondence between the electron density of the HF ground state and the external potential. As a consequence, the Hohenberg-Kohn functional $F_{\mathrm{HF}}^{\mathrm{HK}}[n]$ for the electron density n of the HF ground state wave function Ψ_{SD} represented with a Slater determinant can be defined as

$$F_{\mathrm{HF}}^{\mathrm{HK}}[n] = \left\langle \Psi_{\mathrm{SD}}[n] \middle| \hat{\mathbf{T}} + \hat{\mathbf{V}}_{\mathrm{ee}} \middle| \Psi_{\mathrm{SD}}[n] \right\rangle \tag{1.228}$$

where $\Psi_0[n]$ is the HF ground state wave function that uniquely corresponds to the argument n. With this HK functional $F_{\mathrm{HF}}^{\mathrm{HK}}[n]$ specific to the HF energy, the potential functional can be defined as follows:

$$E_{\mathrm{PF},\upsilon_{\mathrm{ext}}}^{\mathrm{HF}}[w_{\mathrm{ext}}] = F_{\mathrm{HF}}^{\mathrm{HK}}\left[n_{w_{\mathrm{ext}}}\right] + \int d\boldsymbol{r}\, n_{w_{\mathrm{ext}}}(\boldsymbol{r})\, \upsilon_{\mathrm{ext}}(\boldsymbol{r}) \tag{1.229}$$

where $n_{w_{\mathrm{ext}}}$ is the HF electron density corresponding to the external potential w_{ext}. Note that the functional is completely parallel to Eq. (1.223). Now it is possible to extend the potential functional to that for the non-interacting systems. Namely, the functional $E_{\mathrm{PF0},\upsilon_{\mathrm{ext}}}^{\mathrm{HF}}[w_{\mathrm{KS}}]$ is defined as

$$E_{\mathrm{PF0},\upsilon_{\mathrm{ext}}}^{\mathrm{HF}}[w_{\mathrm{KS}}] = F_{\mathrm{HF}}^{\mathrm{HK}}\left[n_{w_{\mathrm{KS}}}\right] + \int d\boldsymbol{r}\, n_{w_{\mathrm{KS}}}(\boldsymbol{r})\, \upsilon_{\mathrm{ext}}(\boldsymbol{r}) \tag{1.230}$$

With this PF for the HF total energy, the HF ground state energy $E_{0,\upsilon_{\mathrm{ext}}}^{\mathrm{HF}}$ for the system with the external potential υ_{ext} can be obtained through the optimization of the effective potential thus

$$E_{0,\upsilon_{\mathrm{ext}}}^{\mathrm{HF}} = \min_{w_{\mathrm{KS}}} E_{\mathrm{PF0},\upsilon_{\mathrm{ext}}}^{\mathrm{HF}}[w_{\mathrm{KS}}] \tag{1.231}$$

This equation serves as the theoretical foundation for the HF-OEP. Actually, the HF-OEP can be regarded as a numerical realization of Eq. (1.231). More detailed discussion is provided in Appendix C.

Remind that the potential corresponding to a certain non-interacting υ-rep. density can also be obtained through the 'inverse Kohn-Sham' (inv-KS) method [31, 55–58]. It can be readily proved that the KS effective potential obtained by the inv-KS method from the HF ground state density n_0^{HF} is equivalent to that given by the HF-OEP under limited situations. Before we provide the details of the inv-KS, it is useful to discuss the equivalence of the inv-KS with the HF-OEP. Figure 1.14 is presented to help the readers' understandings. The assumption to prove the equivalence is that the set of wave functions $\{\varphi_i^{\mathrm{OEP}}\}$ $(i = 1, 2, \cdots, N)$ on the HF-OEP potential realizes the HF ground state energy E_0^{HF}, that is, $E_{\mathrm{HF}}[\{\varphi_i^{\mathrm{OEP}}\}] = E_0^{\mathrm{HF}}\left[\{\varphi_i^{\mathrm{HF}}\}\right]$. Provided that this equation holds, $\{\varphi_i^{\mathrm{OEP}}\}$ is equivalent to $\{\varphi_i^{\mathrm{HF}}\}$ from the variational principle. As a consequence, the electron density constructed from $\{\varphi_i^{\mathrm{OEP}}\}$ realizes the density $n_0^{\mathrm{HF}}(\boldsymbol{r})$. When n_0^{HF} is non-interacting υ-representable, the corresponding KS potential $\upsilon_{\mathrm{KS}}(\boldsymbol{r})$ can be determined uniquely. On the other hand, n_0^{HF} determines the set of wave

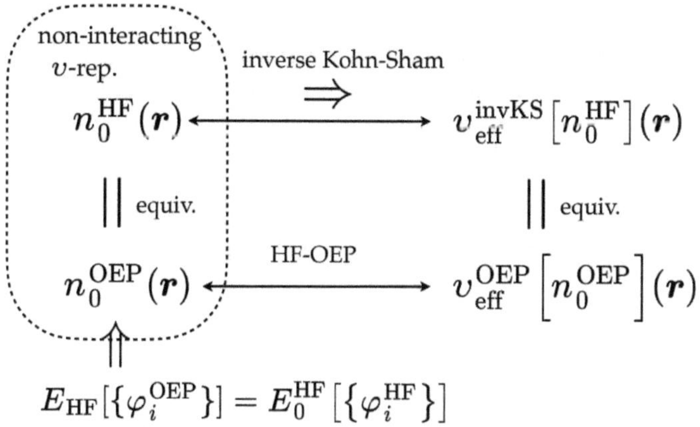

Fig. 1.14 A non-interacting reference system associated with the HF ground state electron density $n_0^{\mathrm{HF}}(\boldsymbol{r})$. The corresponding set of HF wave functions is denoted as $\{\varphi_i^{\mathrm{HF}}\}$. The two lines with arrows at either end show the one-to-one correspondences. When the Kohn-Sham (KS) wave functions $\{\varphi_i^{\mathrm{OEP}}\}$ optimized on a local potential $v_{\mathrm{eff}}^{\mathrm{OEP}}(\boldsymbol{r})$ realize the HF ground state energy E_0^{HF}, the HF-OEP potential $v_{\mathrm{eff}}^{\mathrm{OEP}}(\boldsymbol{r})$ can be directly obtained as $v_{\mathrm{eff}}^{\mathrm{invKS}}(\boldsymbol{r})$ from the HF ground state density $n_0^{\mathrm{HF}}(\boldsymbol{r})$ by means of the inverse Kohn-Sham method

functions $\{\varphi_i^{\mathrm{OEP}}\}$ and the potential $v_{\mathrm{eff}}^{\mathrm{OEP}}(\boldsymbol{r})$ uniquely through the HF-OEP. Thus, we conclude that $v_{\mathrm{KS}} = v_{\mathrm{eff}}^{\mathrm{OEP}}$. Here we have a question whether the assumption on which the proof is based is possible. It can be readily recognized that the condition is fulfilled for the two-electron systems. The homogeneous electron gas is also the system that satisfies the condition. However, it was demonstrated by Ivanov and Levy [60] that $\{\varphi_i^{\mathrm{OEP}}\}$ differs from $\{\varphi_i^{\mathrm{HF}}\}$ for a three-electron hydrogen-like density. Therefore, the equivalence is not guaranteed always. However, as shown in Table 1.2, the HF-OEP is capable of reproducing almost the HF energies of many-electron systems. Thus, it can be expected that the inv-KS for the HF electron density can yield the local potential $v_{\mathrm{eff}}^{\mathrm{invKS}}(\boldsymbol{r})$ almost identical to $v_{\mathrm{eff}}^{\mathrm{OEP}}$ provided by the HF-OEP. One can find more detailed discussions on the equivalency between the inv-KS and the HF-OEP in Ref. [61]. It is also interesting and useful to compare the HF energy $E_{\mathrm{HF}}[\{\varphi_i^{\mathrm{OEP}}\}]$ with $E_{\mathrm{HF}}[\{\varphi_i^{\mathrm{invKS}}\}]$ for various systems. The comparisons were also made in Table I in Ref. [61].

In the inv-KS, the corresponding KS effective potential $v_{\mathrm{KS}}[n](\boldsymbol{r})$ for the density n is generated through an iterative manner. The algorithm proposed by Kadantsev and Stott [31] for inv-KS is based on the variational principle found by Foulkes and Haydock [30]. The proof for the variational principle is worth studying. Suppose that $n_0(\boldsymbol{r})$ is an electron density of an electronic ground state and non-interacting v-representable. We introduce a kinetic energy functional $T[n_0]$ defined by

$$T\left[n_0\right] = \min_{\Psi_{\mathrm{SD}} \to n_0} \left\langle \Psi_{\mathrm{SD}} \middle| \hat{T} \middle| \Psi_{\mathrm{SD}} \right\rangle \tag{1.232}$$

where $\Psi_{\rm SD}$ is a wave function represented by a Slater determinant for the non-interacting N electrons. The minimization is performed within a set of $\Psi_{\rm SD}$ that realizes the density n_0. Actually, the functional is identical to that defined in Eq. (1.112). The constraint minimization of Eq. (1.232) can be expressed in another form by introducing a new functional $\gamma[\Psi_{\rm SD}]$ including Lagrange's multipliers $\lambda(\boldsymbol{r})$ and ϵ_i,

$$\gamma\left[\Psi_{\rm SD}\right] = \sum_i^N \int d\boldsymbol{r}\varphi_i^*\left(\boldsymbol{r}\right)\left(-\frac{1}{2}\nabla_i^2\right)\varphi_i\left(\boldsymbol{r}\right) + \int d\boldsymbol{r}\lambda\left(\boldsymbol{r}\right)\left(n\left(\boldsymbol{r}\right) - n_0\left(\boldsymbol{r}\right)\right)$$
$$-\sum_i^N \epsilon_i\left(\int d\boldsymbol{r}\varphi_i^*\left(\boldsymbol{r}\right)\varphi_i\left(\boldsymbol{r}\right) - 1\right) \quad (1.233)$$

where the set of wave functions $\{\varphi_i\}$ constitutes $\Psi_{\rm SD}$, and $n(\boldsymbol{r})$ is the electron density constructed from $\Psi_{\rm SD}$. The multiplier $\lambda(\boldsymbol{r})$ acts as a potential to ensure the constraint $n(\boldsymbol{r}) = n_0(\boldsymbol{r})$. And ϵ_i are the multipliers to impose the normalization conditions for the wave functions φ_i. When the minimization of Eq. (1.232) is achieved, the functional $\gamma[\Psi_{\rm SD}]$ is stationary with respect to the infinitesimal variation of φ_i, that is,

$$\frac{\delta\gamma\left[\Psi_{\rm SD}\right]}{\delta\varphi_i\left(\boldsymbol{r}\right)} = 0 \quad (i = 1, 2, \cdots, N) \quad (1.234)$$

which directly leads to a Schrödinger equation,

$$\left(-\frac{1}{2}\nabla^2 + \lambda\left(\boldsymbol{r}\right)\right)\varphi_i\left(\boldsymbol{r}\right) = \epsilon_i\varphi_i\left(\boldsymbol{r}\right) \quad (1.235)$$

It is quite important to note that the minimization of Eq. (1.232) or equivalently Eq. (1.233) is to solve the Schrödinger equation with the local potential $\lambda(\boldsymbol{r})$ that realizes the target density n_0. Since there exists one-to-one correspondence between the potential $\lambda(\boldsymbol{r})$ and the wave function $\Psi_{\rm SD}$, the functional $\gamma[\Psi_{\rm SD}]$ can be regarded as a functional $\gamma[\lambda(\boldsymbol{r})]$ of the potential $\lambda(\boldsymbol{r})$. Thus, the minimization of the functional $\gamma[\Psi_{\rm SD}]$ can be equivalently replaced by that of $\gamma[\lambda(\boldsymbol{r})]$, which leads to

$$\frac{\delta\gamma\left[\lambda\left(\boldsymbol{r}\right)\right]}{\delta\lambda\left(\boldsymbol{r}\right)} = \sum_i^N \int d\boldsymbol{r}' \frac{\delta\gamma\left[\lambda\left(\boldsymbol{r}\right)\right]}{\delta\varphi_i\left(\boldsymbol{r}'\right)}\frac{\delta\varphi_i\left(\boldsymbol{r}'\right)}{\delta\lambda\left(\boldsymbol{r}\right)} + n\left(\boldsymbol{r}\right) - n_0\left(\boldsymbol{r}\right) \quad (1.236)$$

Provided that the stationary condition of Eq. (1.234) is fulfilled, the first term of the right-hand side of Eq. (1.236) vanishes. Then, the derivative can be reduced to

$$\frac{\delta\gamma\left[\lambda\left(\boldsymbol{r}\right)\right]}{\delta\lambda\left(\boldsymbol{r}\right)} = n\left(\boldsymbol{r}\right) - n_0\left(\boldsymbol{r}\right) \quad (1.237)$$

Thus it is shown that the minimization is completed when $n(\boldsymbol{r})$ achieves $n_0(\boldsymbol{r})$. This is the variation principle of Foulkes and Haydock [30]. On the basis of this principle the algorithm for the inverse Kohn-Sham(inv-KS) method will be illustrated as follows. First, it is assumed that the target density e.g. $n_0^{\mathrm{HF}}(\boldsymbol{r})$ is already provided. For a potential $v_{\mathrm{eff}}^k(\boldsymbol{r})$ at kth step, one solves the set of the Schrödinger equations for N electrons

$$\left(-\frac{1}{2}\nabla^2 + v_{\mathrm{eff}}^k(\boldsymbol{r})\right)\varphi_i^k(\boldsymbol{r}) = \epsilon_i^k \varphi_i^k(\boldsymbol{r}) \quad (i = 1, 2, \cdots, N) \tag{1.238}$$

Note that $v_{\mathrm{eff}}^k(\boldsymbol{r})$ includes the fixed external potential $v_{\mathrm{ext}}(\boldsymbol{r})$. Then, the ground state density $n^k(\boldsymbol{r})$ for the potential $v_{\mathrm{eff}}^k(\boldsymbol{r})$ is obtained. The $(k+1)$th renewed potential is merely given by

$$v_{\mathrm{eff}}^{k+1}(\boldsymbol{r}) = v_{\mathrm{eff}}^k(\boldsymbol{r}) + \kappa\left(n^k(\boldsymbol{r}) - n_0^{\mathrm{HF}}(\boldsymbol{r})\right) \tag{1.239}$$

with κ a small positive real number. Note, however, that κ can be a function of $\boldsymbol{r}$ to accelerate the convergence since the choice of $\kappa(\boldsymbol{r})$ does not affect the equilibrium condition that $n^k(\boldsymbol{r}) - n_0^{\mathrm{HF}}(\boldsymbol{r}) = 0$. The increment of the potential in Eq. (1.239) is to add locally the potential according to the deviation of the current density n^k from the target one. The exchange potential $v_x^k(\boldsymbol{r})$ during the SCF calculation can be obtained by subtracting the Hartree potential $v_{\mathrm{H}}(\boldsymbol{r})$ and the external potential $v_{\mathrm{ext}}(\boldsymbol{r})$ from $v_{\mathrm{eff}}^k(\boldsymbol{r})$

$$v_x^k\left[n^k\right](\boldsymbol{r}) = v_{\mathrm{eff}}^k\left[n^k\right](\boldsymbol{r}) - v_{\mathrm{H}}\left[n^k\right](\boldsymbol{r}) - v_{\mathrm{ext}}(\boldsymbol{r}) \tag{1.240}$$

It is anticipated that the potential in the outer region of the molecule is very hard to be converged because the density difference $n^k(\boldsymbol{r}) - n_0^{\mathrm{HF}}(\boldsymbol{r})$ is considered to be very small in that region. To expedite the convergence, it is strongly recommended to build the initial guess $v_x^0(\boldsymbol{r})$ for the exchange potential so that it has the correct long-range property. Explicitly, the part of the Fermi-Amaldi (FA) potential can be used as the initial guess, that is,

$$v_x^0(\boldsymbol{r}) = -\frac{1}{N}\int d\boldsymbol{r}'\frac{n_0^{\mathrm{HF}}(\boldsymbol{r}')}{|\boldsymbol{r} - \boldsymbol{r}'|} \tag{1.241}$$

with N being the number of electrons. Another better choice of the guess $v_x^0(\boldsymbol{r})$ might be to take Slater's local exchange potential of Eq. (1.201)

$$v_x^0(\boldsymbol{r}) = v_x^{\mathrm{Slater}}(\boldsymbol{r}) \tag{1.242}$$

Since these potentials have the correct asymptotic behaviors of $-1/r$, only the short range part of the potential is to be optimized through Eq. (1.239).

It is very important to note that the coupled optimization of the potential $v^{\mathrm{HF}}(\boldsymbol{r})$ and the set of wave functions $\{\varphi_i\}$ on the potential automatically minimizes the HF energy $E_{\mathrm{HF}}[\{\varphi_i\}]$ due to the one-to-one correspondence between the density n_0^{HF} and the potential $v^{\mathrm{HF}}(\boldsymbol{r})$, provided that $n_0^{\mathrm{HF}}(\boldsymbol{r})$ is non-interacting v-representable. Thus, the potential $v^{\mathrm{HF}}(\boldsymbol{r})$ given by the inv-KS method is equivalent to the potential provided by the HF-OEP approach. The method using the inv-KS approach can be applied to the ground state densities provided by wave functions based on various types of theories such as CASSCF, coupled cluster method, perturbation theories, $\cdots$. Once a computer program for the inv-KS is obtained, it can be used for those densities without any change of the source code. This can be a great advantage over the OEP method. Furthermore, the inv-KS method is completely free from the difficulties associated with the inversion of the response functions. In the following, comparisons will be made between the OEP and the solution given by inv-KS method [61].

Comparisons between HF-OEP and inverse Kohn-Sham Method

As an example, the exchange potential for H_2O molecule will be considered here. The computations are performed using our code called 'Vmol' based on the real-space grid (RSG) method [62–66]. The overall description of the RSG method is given in Appendix F. We provide here the details specific to the real-space implementations of OEP and inv-KS methods for the HF exchange. To implement the OEP method by the Yang-Wu approach introduced above, the grid itself is regarded as the basis to represent the OEP potential. To define explicitly the basis function $g_p(\boldsymbol{r})$ in Eq. (1.194) in the RSG method, we introduce the cubic fractional region Ω_p^h enclosing the grid point $\boldsymbol{r}_p = (x_p, y_p, z_p)$ in the real-space cell

$$\Omega_p^h := \left(l_p - h/2 \le l \le l_p + h/2\right) \quad (l = x, y, \text{and } z) \tag{1.243}$$

with h the grid width. For all p, it is assumed that the grid function $g_p(\boldsymbol{r})$ at $\boldsymbol{r}_p$ satisfies

$$g_p\left(\boldsymbol{r}\right) = \begin{cases} 1 & \boldsymbol{r} \in \Omega_p^h \\ 0 & \boldsymbol{r} \notin \Omega_p^h \end{cases} \tag{1.244}$$

Thus, the integral $\left\langle \varphi_i \left| g_p \right| \varphi_a \right\rangle$ in Eq. (1.197) can be simply evaluated as $\left\langle \varphi_i \left| g_p \right| \varphi_a \right\rangle = \varphi_i^* \left(\boldsymbol{r}_p\right) \varphi_a \left(\boldsymbol{r}_p\right) h^3$. The coefficient c_p in Eq. (1.197) represents the value of the exchange potential to be optimized at the grid $\boldsymbol{r}_p$. The functional derivative of the HF-exchange energy E_x^{HF} in E_{HF} in Eq. (1.197) with respect to $\varphi_i^*(\boldsymbol{r}')$ is explicitly given by

$$\sum_j \varphi_j\left(\boldsymbol{r}'\right) \int d\boldsymbol{r} \frac{1}{|\boldsymbol{r} - \boldsymbol{r}'|} \varphi_j^*\left(\boldsymbol{r}\right) \varphi_i\left(\boldsymbol{r}\right) \tag{1.245}$$

The major task in the HF-OEP calculation as well as in the HF calculation is the evaluation of the integrand of Eq. (1.245) for all the $i - j$ pairs. That is, to compute the classical Coulomb potential of the pair density $\varphi_j^*\left(\boldsymbol{r}\right) \varphi_i\left(\boldsymbol{r}\right)$. In our implementation, the potential is computed utilizing a Poisson solver based on the parallelized three-dimensional FFT (Fast Fourier Transformation) technique. The details of the method are described in Ref. [67]. For the HF-OEP, it was found that the local exchange potential $v_x^{\mathrm{Slater}}\left(\boldsymbol{r}\right)$ by Slater (Eqs. (1.200)–(1.202)) is appropriate for the initial guess of the potential. This directly corresponds to the fact that the reference potential in

Eq. (1.195) is replaced by

$$v_{\text{ref}}(\boldsymbol{r}) = \int d\boldsymbol{r}' \frac{n_{\text{ref}}(\boldsymbol{r}')}{|\boldsymbol{r}-\boldsymbol{r}'|} + v_x^{\text{Slater}}(\boldsymbol{r}) \tag{1.246}$$

In our implementation, we found that the potential offers better convergence than the Fermi-Amaldi potential as an initial guess of the exchange potential. The update of the coefficient c_p for the grid function g_p in the SCF calculation is simply performed by

$$c_p^{k+1} = c_p^k - \lambda \frac{\partial E_{\text{HF}}[\{\varphi_i\}]}{\partial c_p} \tag{1.247}$$

where the acceleration factor λ is taken to be 5.0 in our calculations. Along with the optimization of the effective potential, the set of 1-electron wave functions $\{\varphi_i\}$ are also to be optimized on the temporal potential. At the convergence of the coefficients $\{c_p\}$, the local HF exchange potential $v_x^{\text{HF}}(\boldsymbol{r})$ is obtained as

$$v_x^{\text{HF}}(\boldsymbol{r}) = \sum_p c_p g_p(\boldsymbol{r}) - \int d\boldsymbol{r}' \frac{n(\boldsymbol{r}') - n_{\text{ref}}(\boldsymbol{r}')}{|\boldsymbol{r}-\boldsymbol{r}'|} + v_x^{\text{Slater}}(\boldsymbol{r}), \tag{1.248}$$

which is completely parallel to Eq. (1.199).

Implementation of the inverse Kohn-Sham (inv-KS) method with the real-space grid approach is rather simple. We directly employed Eq. (1.239) to update the effective local potential $v_{\text{KS}}(\boldsymbol{r})$. The acceleration factor κ in the equation is being fixed at 0.8 irrespective of the position $\boldsymbol{r}$. To insure the $-1/r$ asymptotic behavior of the exchange potential $v_x^{\text{HF}}(\boldsymbol{r})$, we employed Eq. (1.242) as the initial guess. The target ground state density $n_0^{\text{HF}}(\boldsymbol{r})$ in Eq. (1.239) must be obtained beforehand. Of course, the computational settings for the real-space grid (RSG) calculation to construct n_0 should be the same as those used to perform inv-KS. After the convergence of the effective potential v_{KS}, the HF exchange potential v_x^{HF} can be evaluated as

$$v_x^{\text{HF}}(\boldsymbol{r}) = v_{\text{KS}}(\boldsymbol{r}) - v_{\text{ext}}(\boldsymbol{r}) - v_{\text{H}}[n](\boldsymbol{r}) \tag{1.249}$$

where the density n is supposed to be very close to the target density n_0. During the coupled optimizations of the potential v_x^{HF} and the set of one-electron wave functions $\{\varphi_i\}$ on the potential, $\{\varphi_i\}$ is substituted to the HF-energy functional of Eq. (1.27) in each SCF step to monitor the tentative energy.

Now we provide the computational details to perform HF-OEP and inv-KS with the RSG method. The grid width h was set at $h = 0.22214$ a.u. and the number of the uniform grids along each axis was 120. The resultant box size L of the QM cell is 26.65730 a.u. The molecular geometry of a water molecule was determined by CCSD method with the aug-cc-pVTZ basis set, for which Gaussian 16 program package was conducted. The geometrical parameters were computed as R(O-H) = 0.959 Å and ∠HOH = 104.4°.

In Table 1.3, the HF-OEP energy E(OEP) computed by the real-space grid approach is compared with those given by Ivanov et al. [26] and by Yang and Wu (YW) in Ref. [27]. In the work by Ivanov et al., a Gaussian basis set (uncontracted cc-pVTZ) was employed for the potential as well as for the orbitals, while DGauss A2 basis set [68] was used in the YW calculations. It is shown in the table that the OEP energy by the present work (PW) is comparable in accuracy with that by YW. More importantly, it was revealed that the set of wave functions $\{\varphi_i\}$ optimized on the local exchange potential provided by inverse-Kohn-Sham (inv-KS) method gives the better value than the OEP energies. This result is quite striking because the implementation of inv-KS is much simpler than that of OEP in general. In Fig. 1.15, we also show the exchange potentials and the corresponding electron densities along the molecular axis aligned parallel to the z axis. It is seen in the figure that the electron density obtained by inv-KS shows an excellent agreement with the HF density. The

Table 1.3 The energy differences of E(OEP)$-E$(HF) and E(inv-KS)$-E$(HF) in the units of mHartree. Ivanov: Ivanov et al. [26] YW: Yang and Wu [27] PW: present work. The geometrical parameters of H_2O used in the works by Ivanov and YW are R(O-H) = 0.959 Å and ∠HOH = 103.9°. Thus, the angle is slightly different from that of the present work (∠HOH = 104.4°). This parameter is also different from that used in Ref. [61]

Molecule	OEP (Ivanov)	OEP (YW)	OEP (PW)	inv-KS (PW)
H_2O	2.30	2.27	2.36	1.52

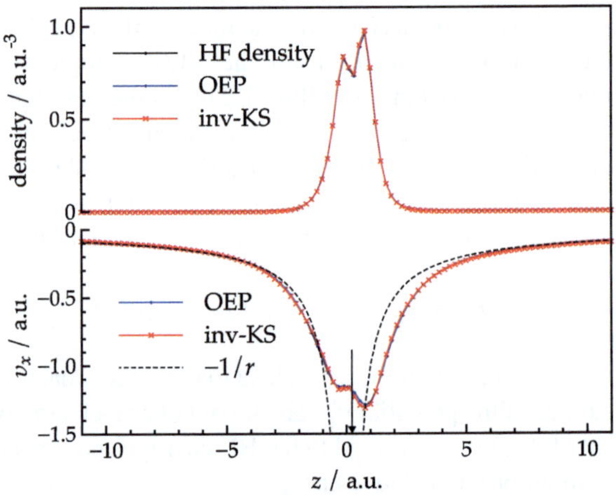

Fig. 1.15 Electron densities n computed by HF, OEP, and inverse-Kohn-Sham (inv-KS) methods are shown in the upper figure, while the local exchange potentials v_x obtained by OEP and inv-KS are presented in the lower figure. n and v_x are plotted as functions of the z coordinate. The arrow indicates the position of the oxygen atom on the z axis, where the z coordinates of the hydrogen atoms are in the negative direction of the axis. The Coulomb potential $-1/r$ is plotted with r being the distance from the origin

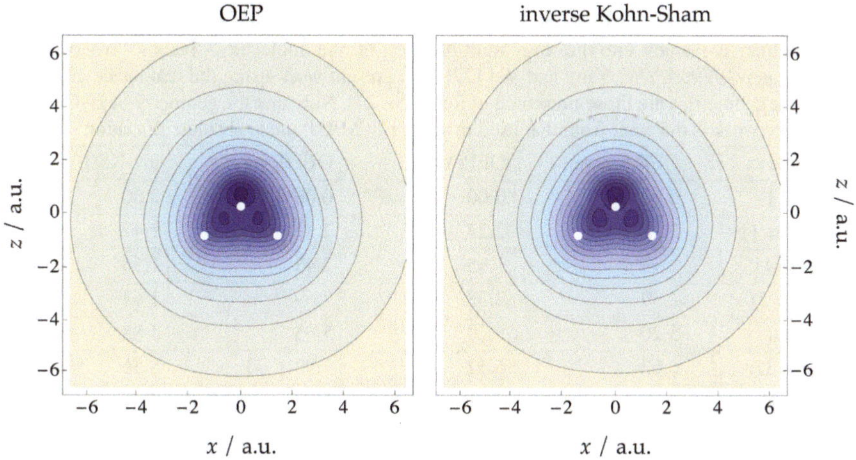

Fig. 1.16 Contour plots of the local HF-exchange potentials $v_x(\boldsymbol{r})$ on the H_2O molecular plane placed on the xz plane(left: OEP, right: inverse Kohn-Sham). The darker region is lower in the energy. The energy range of the plots is from -1.28 a.u. to -0.16 a.u. The interval between the adjacent contours is 0.08. The white bold circles indicate the positions of the oxygen and hydrogen atoms, where the oxygen atom is placed on the z axis ($x = 0$)

density given by OEP also shows a good agreement although a quite small deviation from the HF density is found.

In Fig. 1.16, the contour plot of the exchange potential v_x obtained by OEP on the H_2O molecular plane is compared with that computed by the inv-KS method. It is seen in the figure that the details of the potential near the atomic cores slightly differ from each other. It should be noted that such a small difference in the exchange potential gives rise to the energy difference between the HF-OEP and the inv-KS as shown in Table 1.3. However, the overall behaviors in the OEP exchange potential agrees nicely with the inv-KS potential, suggesting that the two potentials are equivalent in practice although the ways to create these potentials are totally different.

The inv-KS method is also applied to the test set of 12 small molecules in Ref. [26] and the two anions in Ref. [27] to compare the accuracy of the inv-KS with the HF-OEP. The results are summarized in Table 1.4. It is shown in the table that the mean absolute deviation(MAD) of the HF-OEP calculations by Ivanov et al. is 3.45 mHartree. The value is almost the same as the MAD (3.44 mHartree) of the HF-OEP calculations by YW using the different potential basis set. It is also notable that the HF-OEP based on the RSG method is almost comparable in accuracy to the calculations using the gauss functions as the potential basis sets. Importantly, the MAD value (= 2.58 mHartree) of the inv-KS calculation is distinctly smaller than those of the other HF-OEP results. Furthermore, it is shown that the deviation of the inv-KS for each molecule is always smaller than that of the corresponding HF-OEP calculation. At least for the molecular set, it can be concluded that the inv-KS is superior to HF-OEP in providing the local exchange potential $v_x(\boldsymbol{r})$. The

Table 1.4 The energy differences of E(OEP)−E(HF) and E(inv-KS)−E(HF) in the units of mHartree. S values are the spin multiplicities of the molecules. Ivanov: Ivanov et al. citech1Ivanov1999prl YW: Yang and Wu [27] PW: present work using the real-space grids. The molecular geometries are those presented in Refs. [26, 27]. Note that the geometry of H_2O used in the present work is different from that used in Table 1.3. MAD: mean absolute deviation

Molecule	OEP(Ivanov)	OEP(YW)	OEP(PW)	inv-KS(PW)
$H_2(S=1)$	0.00	0.00	0.00	0.00
$H_2O(S=1)$	2.30	2.27	2.37	1.47
$HF(S=1)$	1.99	1.95	2.43	1.50
$OH(S=2)$	2.39	2.38	2.69	1.64
$N_2(S=1)$	5.20	5.20	5.03	3.88
$O_2(S=3)$	6.62	6.52	7.93	5.56
$F_2(S=1)$	8.55	8.51	10.3	7.94
$CH_2(S=1)$	2.89	3.04	3.49	1.81
$CH_2(S=3)$	1.87	1.90	1.45	0.90
$NH_2(S=2)$	2.55	2.64	2.72	1.61
$NH(S=3)$	1.97	2.08	2.35	1.25
$CO(S=1)$	5.08	5.13	5.18	3.84
$CN^-(S=1)$	–	4.41	4.69	3.21
$OH^-(S=1)$	–	2.14	3.53	1.46
MAD	3.45	3.44	3.87	2.58

disadvantage of the HF-OEP arises from the fact that it utilizes the response function $\chi(\boldsymbol{r}, \boldsymbol{r}')$ in constructing the potential $v_x(\boldsymbol{r})$ which necessitates the references to the virtual orbitals within a truncated subspace of orbitals. In the inv-KS approach, on the other hand, the potential can be directly optimized using Eq. (1.237) based on the Foulkes-Haydock variation principle. The direct increment of the fraction of the density difference $n(\boldsymbol{r}) - n_0^{\mathrm{HF}}(\boldsymbol{r})$ to the potential at the current optimization step will be advantageous for the construction of the detailed structure of the exchange potential.

We close this subsection with an application of the inv-KS to an ortho-benzyne radical (C_6H_4) known as a strongly correlated polyatomic system. The molecular geometry of C_6H_4 is taken from Ref. [57]. All the atoms are placed within a xz plane. The computational setup for the RSG method is the same as that given in a paragraph below Eq. (1.249). The diradical character of the molecule is realized with the spin-polarized wave functions in the inv-KS and in the HF calculations. The value of the UHF energy functional $E^{\mathrm{UHF}}[\{\varphi_i^{\mathrm{invKS}}\}]$ is computed as −35.48172 a.u. and the deviation $E^{\mathrm{UHF}}[\{\varphi_i^{\mathrm{invKS}}\}] - E^{\mathrm{UHF}}$ is obtained as 13.53 mHartree. Thus it is found that the deviation from the UHF energy is rather large as compared with the values in Table 1.4. This can be attributed to the system size of the benzyne that includes 28 valence electrons because it is natural to consider that the error is additive with respect to the molecular size. Therefore, the accuracy of the inv-KS for the benzyne

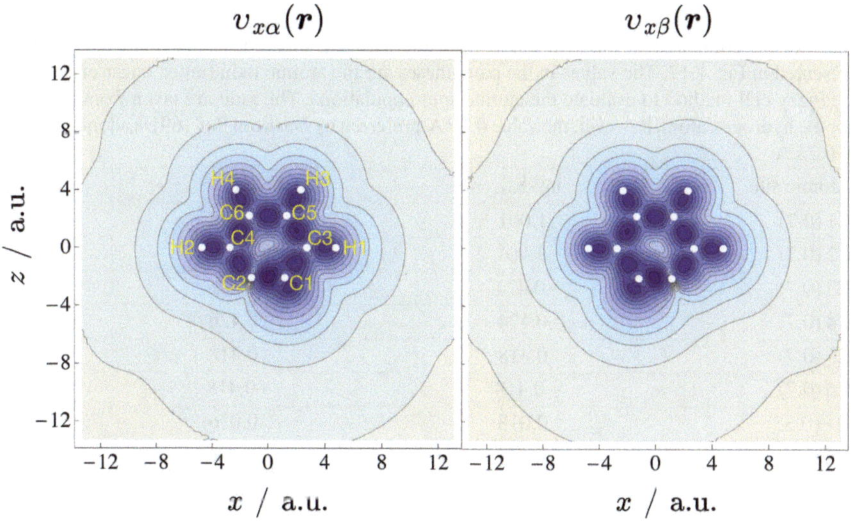

Fig. 1.17 Contour plots of the local HF-exchange potentials $v_{x\sigma}(\boldsymbol{r})(\sigma = \alpha$ and $\beta)$ on the molecular plane of C_6H_4 placed on the xz plane. The potentials are obtained by the inverse KS-DFT method. The darker region is lower in the energy. The energy range of the plots is from -1.04 a.u. to -0.08 a.u. The interval between the adjacent contours is 0.08 a.u. The white bold circles indicate the positions of the atoms

is almost the same as that for the O_2 molecule with 12 valence electrons shown in Table 1.4. The dipole moment d of the spin-polarized electron density is computed as $d = 1.321$ Debye by the inv-KS, which shows nice agreement with $d = 1.331$ Debye in the UHF calculation. In Fig. 1.17, the exchange potentials $v_{x\alpha}(\boldsymbol{r})$ and $v_{x\beta}(\boldsymbol{r})$ optimized by the inv-KS for α and β spins are shown, respectively. It is interesting to note that the large difference between these potentials can be seen only at the atomic sites C1 and C2 where the chemical bonds with the hydrogen atoms are being lost.

To examine the accuracy of the spin density obtained in the inv-KS calculation, the spin populations assigned to the atomic sites are evaluated using the fuzzy cell method[69]. The results are summarized in Table 1.5. As expected, the atomic sites C1 and C2 have the largest spin populations with the opposite direction. In the table, it is revealed that the spin populations are characterized by the α and the β spins emerging alternately along the six-membered ring. It is clearly shown in the table that the atomic spin populations of the UHF wave function is almost faithfully realized in the wave function of the inv-KS method. It is thus demonstrated that the inv-KS can also be applied to a strongly correlated polyatomic system.

Table 1.5 Spin populations assigned to the atomic sites in the ortho-benzyne radical (C_6H_4). Spin densities are computed by the inverse KS-DFT and the UHF methods. Refer to the atom indexes presented in Fig. 1.17. The values in the parentheses are the atomic radii (units: angstrom) used in the fuzzy cell method to evaluate the atomic spin populations. The radii are taken from Ref. [70]. For the hydrogen atom, however, the value 0.35 Å preferred by Becke in Ref. [69] is adopted instead of 0.25 Å

Atomic site	inv-KS	UHF
C1 (0.7)	1.061	1.062
C2 (0.7)	−1.061	−1.062
C3 (0.7)	−0.424	−0.426
C4 (0.7)	0.424	0.426
C5 (0.7)	0.418	0.418
C6 (0.7)	−0.418	−0.418
H1 (0.35)	0.018	0.016
H2 (0.35)	−0.017	−0.016
H3 (0.35)	−0.005	−0.004
H4 (0.35)	0.005	0.004

References

1. A. Szabo, N.S. Ostlund, *Modern Quantum Chemistry* (Macmillan, New York, 1982)
2. R. Mcweeny, *Methods of Molecular Quantum Mechanics* (Academic Press, 2001)
3. F. Jensen, *Introduction to Computational Chemistry* (Wiley, 2009)
4. T. Helgaker, P. Jorgensen, J. Olsen, *Molecular Electronic-Structure Theory* (Wiley, 2012)
5. R.G. Parr, W. Yang, *Density-Functional Theory of Atoms and Molecules* (Oxford University Press, New York, 1989)
6. W. Koch, M.C. Holthausen, *A Chemist's Guide to Density Functional Theory* (Wiley-VCH, 2001)
7. R.M. Martin, *Electronic Structure* (Cambridge University Press, 2008)
8. W. Kohn, L.J. Sham, Phys. Rev. **140**(4A), A1133 (1965)
9. P. Hohenberg, W. Kohn, Phys. Rev. **136**(3B), B864 (1964)
10. M. Levy, Proc. Natl. Acad. Sci. U.S.A. **76**(12), 6062 (1979)
11. J.C. Slater, Phys. Rev. **81**(3), 385 (1951)
12. F. Herman, J.P.V. Dyke, I.B. Ortenburger, Phys. Rev. Lett. **22**(16), 807 (1969)
13. A.D. Becke, Phys. Rev. A **38**(6), 3098 (1988)
14. J.P. Perdew, K. Burke, M. Ernzerhof, Phys. Rev. Lett. **77**(18), 3865 (1996)
15. J.P. Perdew, S. Kurth, Phys. Rev. Lett. **82**(12), 2544 (1999)
16. R.T. Sharp, G.K. Horton, Phys. Rev. **90**, 317 (1953)
17. J.D. Talman, W.F. Shadwick, Phys. Rev. A **14**, 36 (1976)
18. A.D. Becke, J. Chem. Phys. **119**(6), 2972 (2003)
19. H. Takahashi, J. Phys. B: Atomic Mol. Optic. Phys. **51**(5), 055102(11) (2018)
20. N. Matubayasi, M. Nakahara, J. Chem. Phys. **113**(15), 6070 (2000)
21. N. Matubayasi, M. Nakahara, J. Chem. Phys. **117**(8), 3605 (2002)
22. N. Matubayasi, M. Nakahara, J. Chem. Phys. **119**(18), 9686 (2003)
23. R. Pauncz, *The Construction of Spin Eigenfunctions: An Exercise Book* (Springer, 2000)
24. N.D. Mermin, Phys. Rev. **137**(5A), A1441 (1965)
25. J.P. Hansen, I.R. Mcdonald, *Theory of Simple Liquids* (Academic Press, 2013)
26. S. Ivanov, S. Hirata, R.J. Bartlett, Phys. Rev. Lett. **83**(26), 5455 (1999)

27. W. Yang, Q. Wu, Phys. Rev. Lett. **89**, 143002 (2002)
28. W. Kutzelnigg, J. Mol. Struct. **768**, 163 (2006)
29. J.E. Harriman, Phys. Rev. A **24**, 680 (1981)
30. W.M.C. Foulkes, R. Haydock, Phys. Rev. B **39**, 12520 (1989)
31. E.S. Kadantsev, M.J. Stott, Phys. Rev. A **69**, 012502(7) (2004)
32. R. van Leeuwen, E.J. Baerends, Phys. Rev. A **49**(4), 2421 (1994)
33. H. Takahashi, Int. J. Quantum Chem. **122**(20), e26969 (2022)
34. M.A. Buijse, E.J. Baerends, Mol. Phys. **100**(4), 401 (2002)
35. H. Takahashi, R. Kishi, M. Nakano, J. Chem. Theory Comput. **6**, 647 (2010)
36. B.G. Johnson, P.M.W. Gill, J.A. Pople, J. Chem. Phys. **98**(7), 5612 (1993)
37. P. Hohenberg, W. Kohn, Phys. Rev. **136**(3B), B864 (1964)
38. A.D. Becke, Int. J. Quantum Chem. **23**, 1915 (1983)
39. A.D. Becke, M.R. Roussel, Phys. Rev. A **39**(8), 3761 (1989)
40. H. Vosko, L. Wilk, M. Nusair, Can. J. Chem. **58**, 1200 (1980)
41. C. Lee, W. Yang, R.G. Parr, Phys. Rev. B **37**(2), 785 (1988)
42. J.C. Slater, *Quantum Theory of Molecules and Solids: The Self-Consistent Field for Molecules and Solids*, vol. 4 (McGraw-Hill, New York, 1974)
43. L.A. Curtiss, K. Raghavachari, G.W. Trucks, J.A. Pople, J. Chem. Phys. **94**(11), 7221 (1991)
44. A.D. Becke, J. Chem. Phys. **98**(2), 1372 (1993)
45. A.D. Becke, J. Chem. Phys. **98**(7), 5648 (1993)
46. J.B. Krieger, Y. Li, G.J. Iafrate, Phys. Rev. A **45**, 101 (1992)
47. A. Görling, M. Levy, Phys. Rev. A **50**(1), 196 (1994)
48. A. Görling, Phys. Rev. Lett. **83**(26), 5459 (1999)
49. Q. Zhao, R.C. Morrison, R.G. Parr, Phys. Rev. A **50**(3), 2138 (1994)
50. F.D. Sala, A. Görling, J. Chem. Phys. **115**(13), 5718 (2001)
51. H. Partridge, J. Chem. Phys. **87**(11), 6643 (1987)
52. B. Champagne, E.A. Perpète, S.J. van Gisbergen, E.J. Baerends, J.G. Snijders, C. Soubra-Ghaoui, K.A. Robins, B. Kirtman, J. Chem. Phys. **109**(23), 10489 (1998)
53. B. Champagne, F.A. Bulat, W. Yang, S. Bonness, B. Kirtman, J. Chem. Phys. **125**(19), 194114(9) (2006)
54. S.J.A. van Gisbergen, P.R.T. Schipper, O.V. Gritsenko, E.J. Baerends, J.G. Snijders, B. Champagne, B. Kirtman, Phys. Rev. Lett. **83**(4), 694 (1999)
55. Q. Wu, W. Yang, J. Chem. Phys. **118**, 2498 (2003)
56. T. Heaton-Burgess, F.A. Bulat, W. Yang, Phys. Rev. Lett. **98**, 256401(4) (2007)
57. B. Kanungo, P.M. Zimmerman, V. Gavini, Nat. Commun. **10**, 4497 (2019)
58. Y. Shi, A. Wasserman, J. Phys. Chem. Lett. **12**, 5308 (2021)
59. W. Yang, P.W. Ayers, Q. Wu, Phys. Rev. Lett. **92**(14), 146404(4) (2004)
60. S. Ivanov, M. Levy, J. Chem. Phys. **119**(14), 7087 (2003)
61. H. Takahashi, J. Chem. Phys. **161**, 104108(11) (2024)
62. J.R. Chelikowsky, N. Troullier, Y. Saad, Phys. Rev. Lett. **72**(8), 1240 (1994)
63. J.R. Chelikowsky, N. Troullier, K. Wu, Y. Saad, Phys. Rev. B **50**(16), 11355 (1994)
64. H. Takahashi, T. Hori, T. Wakabayashi, T. Nitta, Chem. Lett. **3**, 222 (2000)
65. H. Takahashi, T. Hori, T. Wakabayashi, T. Nitta, J. Phys. Chem. A **105**(17), 4351 (2001)
66. H. Takahashi, T. Hori, H. Hashimoto, T. Nitta, J. Comp. Chem. **22**(12), 1252 (2001)
67. H. Takahashi, S. Sakuraba, A. Morita, J. Chem. Inf. Model. **60**, 1376 (2020)
68. N. Godbout, D.R. Salahub, J. Andzelm, E. Wimmer, Can. J. Chem. **70**(2), 560 (1992)
69. A.D. Becke, J. Chem. Phys. **88**(4), 2547 (1988)
70. J.C. Slater, J. Chem. Phys. **41**(10), 3199 (1964)

Chapter 2
Static Correlation

2.1 Static Correlation in Stretched Chemical Bonds

Dissociation of a chemical bond is, of course, a matter of major interests in the field of chemistry. It is well known, however, that the theoretical description of a chemical bonding is not a trivial task. As seen in Sect. 1.1, the most prevailing approaches to the electronic structure calculations will be categorized into two methods, that is, the molecular orbital theory (MO) and the density-functional theory (DFT). In the former the wave function serves as a fundamental variable in the variational calculation, while the electron density plays a role in the latter. In the following, we first provide a concise review for the methods to describe the bond stretch with these approaches. Then, a novel density-functional theory developed in a recent work will be reviewed, which constitutes the central subject in this chapter.

In the MO theory, the bond dissociation can be described in terms of a multi-configuration wave function. A detailed explanation for the theoretical basis can be found in several textbooks [1, 2]. However, it is quite useful for the following discussions to illustrate how the bond dissociation can be described using a couple of Slater determinants. Stretching the single bond in an H_2 molecule is the simplest model of the bond dissociation. Importantly, the description of the outline of the method illustrates the essence of the electronic structure changes associated with the bond cleavages.

The one-electron wave function for the H_2 molecule can be minimally represented by assigning one atomic orbital on each hydrogen atom. We, thus, approximate the bonding orbital φ_b

$$\varphi_b = \frac{1}{\sqrt{2}} (\phi_A + \phi_B) \tag{2.1}$$

in terms of the normalized atomic orbitals ϕ_A and ϕ_B placed at the atomic centers A and B, respectively. Hereafter, we ignore the overlap between the atomic orbitals in this subsection for the sake of simplicity. In the mean-field approximation, that is,

H. Takahashi, *Density-Functional Theory for Electrons*,
https://doi.org/10.1007/978-981-95-1229-4_2

the Hartree-Fock (HF) approximation, the ground state wave function $\Psi_{GS}(\boldsymbol{x}_1, \boldsymbol{x}_2)$ of H_2 is given by a Slater determinant

$$\Psi_{GS}(\boldsymbol{x}_1, \boldsymbol{x}_2) = \frac{1}{\sqrt{2}}\varphi_b^*(\boldsymbol{r}_1)\varphi_b(\boldsymbol{r}_2)\{\alpha(\omega_1)\beta(\omega_2) - \alpha(\omega_2)\beta(\omega_1)\} \tag{2.2}$$

where $\boldsymbol{x}_1$ stands for the set of the position $\boldsymbol{r}_1$ and the spin ω_1 of electron 1 and the same is true for electron 2. It should be noted in Eq. (2.2) that the bonding orbital φ_b is occupied by two electrons and $\Psi_{GS}(\boldsymbol{x}_1, \boldsymbol{x}_2)$ has the singlet spin symmetry. Throughout this chapter, we only consider the wave functions with correct spin symmetries. The electronic configuration of Eq. (2.2) provides a qualitatively correct potential energy of H_2 in the bonding region. However, it shows a disastrous behavior when the bond is stretched to a dissociative region. Actually, it is well known that the potential energy produced by Eq. (2.2) seriously overshoots the energy of the correct dissociation limit. The mechanism underlying the notorious artifact can be readily elucidated by consulting Eq. (2.2) with the atomic orbital basis. By substituting Eq. (2.1) into Eq. (2.2), we obtain the wave function

$$\begin{aligned}\Psi_{GS}(\boldsymbol{x}_1, \boldsymbol{x}_2) = \frac{1}{2\sqrt{2}}[&\phi_A^*(1)\phi_B(2) + \phi_B^*(1)\phi_A(2) \\ &+ \phi_A^*(1)\phi_A(2) + \phi_B^*(1)\phi_B(2)] \times spin.\end{aligned} \tag{2.3}$$

where the notations '1' and '2' represent $\boldsymbol{r}_1$ and $\boldsymbol{r}_2$, respectively. The first two terms in the bracket in Eq. (2.3) describe the situation that the two electrons are located in the individual atoms, which is referred to as the valence bond (VB) electronic structure. The last two terms in Eq. (2.3), on the other hand, represent the ionic structures where an atom has a positive elementary charge and the other has a negative one. At the dissociation limit, the VB terms offer the correct electron correlation that the two electrons are infinitely separated, while the ionic terms give rise to the erroneous destabilization in energy. It is readily recognized that any augmentation of the basis set does not alleviate such an unfavorable situation associated with the bond dissociation. A mathematical consideration tells us that the single determinant of Eq. (2.2) constitutes only a subset of the anti-symmetric wave functions for two electrons. Thus, the source of the failure arises from the deficiency of the wave function that it is not able to realize the strong electron correlation associated with the bond dissociation. However, as described below, the ionic terms in Eq. (2.3) can be legitimately excluded by mixing another electronic configuration to Eq. (2.2). To this end we introduce the normalized anti-bonding one-electron wave function φ_a

$$\varphi_a = \frac{1}{\sqrt{2}}(\phi_A - \phi_B). \tag{2.4}$$

We, then, construct the doubly excited electronic configuration

$$\Psi_{\mathrm{DE}}\left(\boldsymbol{x}_1, \boldsymbol{x}_2\right) = \frac{1}{\sqrt{2}} \varphi_{\mathrm{a}}{}^*(\boldsymbol{r}_1) \varphi_{\mathrm{a}}(\boldsymbol{r}_2) \left\{\alpha\left(\omega_1\right) \beta\left(\omega_2\right) - \alpha\left(\omega_2\right) \beta\left(\omega_1\right)\right\} \tag{2.5}$$

which reads

$$\begin{aligned} \Psi_{\mathrm{DE}}\left(\boldsymbol{x}_1, \boldsymbol{x}_2\right) = & \frac{1}{2\sqrt{2}} [-\phi_{\mathrm{A}}{}^*(1) \phi_{\mathrm{B}}(2) - \phi_{\mathrm{B}}{}^*(1) \phi_{\mathrm{A}}(2) \\ & + \phi_{\mathrm{A}}{}^*(1) \phi_{\mathrm{A}}(2) + \phi_{\mathrm{B}}{}^*(1) \phi_{\mathrm{B}}(2)] \times spin \end{aligned} \tag{2.6}$$

in terms of the atomic orbitals. By taking the equally weighted sum of $\Psi_{\mathrm{GS}}\left(\boldsymbol{x}_1, \boldsymbol{x}_2\right)$ and $\Psi_{\mathrm{DE}}\left(\boldsymbol{x}_1, \boldsymbol{x}_2\right)$ with opposite signs, we obtain a multi-configurational wave function $\widetilde{\Psi}^{\mathrm{MC}}\left(\boldsymbol{x}_1, \boldsymbol{x}_2\right)$, thus,

$$\widetilde{\Psi}_{\mathrm{MC}}\left(\boldsymbol{x}_1, \boldsymbol{x}_2\right) = \frac{1}{\sqrt{2}} \left(\Psi_{\mathrm{GS}}\left(\boldsymbol{x}_1, \boldsymbol{x}_2\right) - \Psi_{\mathrm{DE}}\left(\boldsymbol{x}_1, \boldsymbol{x}_2\right)\right). \tag{2.7}$$

Interestingly, it leads to the spin-adapted VB wave function

$$\widetilde{\Psi}_{\mathrm{MC}}\left(\boldsymbol{x}_1, \boldsymbol{x}_2\right) = \Psi_{\mathrm{VB}}\left(\boldsymbol{x}_1, \boldsymbol{x}_2\right) = \frac{1}{2} \left(\phi_{\mathrm{A}}{}^*(1) \phi_{\mathrm{B}}(2) + \phi_{\mathrm{B}}{}^*(1) \phi_{\mathrm{A}}(2)\right) \times spin. \tag{2.8}$$

It is, thus, demonstrated that the superposition of the two configurations with distinct characters yields the appropriate electronic structure with the correct spin symmetry for the strongly correlated electrons at the dissociation limit of H_2. The general wave function Ψ_{MC} that can also be used to describe the bonding region of H_2 is expressed by

$$\Psi_{\mathrm{MC}}\left(\boldsymbol{x}_1, \boldsymbol{x}_2\right) = c_{\mathrm{GS}} \Psi_{\mathrm{GS}}\left(\boldsymbol{x}_1, \boldsymbol{x}_2\right) + c_{\mathrm{DE}} \Psi_{\mathrm{DE}}\left(\boldsymbol{x}_1, \boldsymbol{x}_2\right) \tag{2.9}$$

where the expansion coefficients c_{GS} and c_{DE} are to be optimized according to the variational principle under the normalization condition of $|c_{\mathrm{GS}}|^2 + |c_{\mathrm{DE}}|^2 = 1$.

The above discussion can be cast into the analyses in terms of the exchange-correlation hole n_{xc}. The concept of the hole n_{xc} also plays an important role in constructing the theoretical basis of DFT and in the functional development. To introduce the exchange-correlation hole n_{xc}, we first define the second-order reduced density matrix

$$\begin{aligned} n_2\left(\boldsymbol{r}_1, \boldsymbol{r}_2; \boldsymbol{r}_1', \boldsymbol{r}_2'\right) = & \frac{N(N-1)}{2} \int d\omega_1 d\omega_2 d\boldsymbol{x}_3 \cdots d\boldsymbol{x}_N \\ & \Psi\left(\boldsymbol{r}_1\omega_1, \boldsymbol{r}_2\omega_2, \boldsymbol{x}_3 \cdots \boldsymbol{x}_N\right) \Psi^*\left(\boldsymbol{r}_1'\omega_1, \boldsymbol{r}_2'\omega_2, \boldsymbol{x}_3 \cdots \boldsymbol{x}_N\right). \end{aligned} \tag{2.10}$$

The diagonal elements of the matrix $n_2\left(\boldsymbol{r}_1, \boldsymbol{r}_2; \boldsymbol{r}_1, \boldsymbol{r}_2\right) = n_2\left(\boldsymbol{r}_1, \boldsymbol{r}_2\right)$ represent the probability density that one finds two electrons at $\boldsymbol{r}_1$ and $\boldsymbol{r}_2$. Then, the electronic repulsion energy V_{ee} of the system is given by

$$V_{ee} = \int d\boldsymbol{r}_1 d\boldsymbol{r}_2 \frac{1}{r_{12}} n_2(\boldsymbol{r}_1, \boldsymbol{r}_2). \tag{2.11}$$

We subtract the classical Coulomb term from the density matrix $n_2(\boldsymbol{r}_1, \boldsymbol{r}_2)$ to extract the correlation function $h_{xc}(\boldsymbol{r}_1, \boldsymbol{r}_2)$ originating from the quantum mechanical interactions among the electrons, i.e. exchange and correlation energy, thus,

$$V_{ee} = \frac{1}{2} \int d\boldsymbol{r}_1 d\boldsymbol{r}_2 \frac{1}{r_{12}} n(\boldsymbol{r}_1) n(\boldsymbol{r}_2) (1 + h_{xc}(\boldsymbol{r}_1, \boldsymbol{r}_2)). \tag{2.12}$$

From Eqs. (2.11) and (2.12), we obtain

$$n_2(\boldsymbol{r}_1, \boldsymbol{r}_2) = n(\boldsymbol{r}_1) n(\boldsymbol{r}_2) (1 + h_{xc}(\boldsymbol{r}_1, \boldsymbol{r}_2)) \tag{2.13}$$

It is, thus, understood that $h_{xc}(\boldsymbol{r}_1, \boldsymbol{r}_2)$ represents the non-classical correlation between electrons at $\boldsymbol{r}_1$ and $\boldsymbol{r}_2$. In Eq. (2.12), the term $n(\boldsymbol{r}_2) h_{xc}(\boldsymbol{r}_1, \boldsymbol{r}_2)$ is referred to as the exchange-correlation hole for the electron placed at $\boldsymbol{r}_1$ and denoted as $n_{xc}(\boldsymbol{r}_1, \boldsymbol{r}_2)$, thus,

$$n_{xc}(\boldsymbol{r}_1, \boldsymbol{r}_2) = n(\boldsymbol{r}_2) h_{xc}(\boldsymbol{r}_1, \boldsymbol{r}_2). \tag{2.14}$$

It is important to notice that V_{ee} can be obtained by integrating the classical Coulomb interaction between the electron density $n(\boldsymbol{r}_1)$ and the hole $n_{xc}(\boldsymbol{r}_1, \boldsymbol{r}_2)$ over the coordinates $\boldsymbol{r}_1$ and $\boldsymbol{r}_2$. It can be readily recognized that the hole function $n_{xc}(\boldsymbol{r}_1, \boldsymbol{r}_2)$ satisfies the following characteristic property referred to as 'sum rule':

$$\int d\boldsymbol{r}_2\, n_{xc}(\boldsymbol{r}_1, \boldsymbol{r}_2) = -1. \tag{2.15}$$

Here, we introduce the exchange hole function $n_x(\boldsymbol{r}_1, \boldsymbol{r}_2)$ that can be defined in terms of the first-order reduced density matrix for one electron $n_1(\boldsymbol{r}_1, \boldsymbol{r}_2)$,

$$n_x(\boldsymbol{r}_1, \boldsymbol{r}_2) = -\frac{1}{2n(\boldsymbol{r}_1)} |n_1(\boldsymbol{r}_1, \boldsymbol{r}_2)|^2 \tag{2.16}$$

where the density matrix $n_1(\boldsymbol{r}_1, \boldsymbol{r}_2)$ is given by

$$n_1\left(\boldsymbol{r}_1, \boldsymbol{r}_1'\right) = N \int d\omega_1 d\boldsymbol{x}_2 d\boldsymbol{x}_3 \cdots d\boldsymbol{x}_N$$
$$\Psi(\boldsymbol{r}_1\omega_1, \boldsymbol{x}_2, \boldsymbol{x}_3 \cdots \boldsymbol{x}_N) \Psi^*\left(\boldsymbol{r}_1'\omega_1, \boldsymbol{x}_2, \boldsymbol{x}_3 \cdots \boldsymbol{x}_N\right). \tag{2.17}$$

In a special case that $\Psi(\boldsymbol{x}_1, \boldsymbol{x}_2, \boldsymbol{x}_3 \cdots \boldsymbol{x}_N)$ is represented by a single Slater determinant with a closed shell structure, it is readily proved that $n_1\left(\boldsymbol{r}_1, \boldsymbol{r}_1'\right)$ can be expressed in terms of $N/2$ one-electron wave functions, thus,

$$n_1^{\mathrm{SD}}\left(\boldsymbol{r}_1, \boldsymbol{r}_1'\right) = 2\sum_{i=1}^{N/2} \varphi_i\left(\boldsymbol{r}_1\right) \varphi_i^*\left(\boldsymbol{r}_1'\right) \tag{2.18}$$

where N is the number of electrons. It is useful to recall that we already introduced the first-order density matrix in Eq. (1.124) for the HF wave function in Chap. 1. Anyway, the exchange energy E_x for a spin is expressed with the density matrix

$$E_x = -\frac{1}{4}\int d\boldsymbol{r}_1 d\boldsymbol{r}_2 \, \frac{1}{r_{12}} \left|n_1\left(\boldsymbol{r}_1, \boldsymbol{r}_2\right)\right|^2. \tag{2.19}$$

With the definition of Eq. (2.16), E_x can also be described in terms of the exchange hole $n_x\left(\boldsymbol{r}_1, \boldsymbol{r}_2\right)$, thus,

$$E_x = \frac{1}{2}\int d\boldsymbol{r}_1 d\boldsymbol{r}_2 \, \frac{1}{r_{12}} n(\boldsymbol{r}_1) n_x\left(\boldsymbol{r}_1, \boldsymbol{r}_2\right). \tag{2.20}$$

It is possible to prove that the exchange hole $n_x\left(\boldsymbol{r}_1, \boldsymbol{r}_2\right)$ satisfies the sum rule,

$$\int d\boldsymbol{r}_2 \, n_x\left(\boldsymbol{r}_1, \boldsymbol{r}_2\right) \; = \; -1. \tag{2.21}$$

Hence, from Eq. (2.15), the correlation hole $n_c\left(\boldsymbol{r}_1, \boldsymbol{r}_2\right)$ defined by

$$n_{xc}\left(\boldsymbol{r}_1, \boldsymbol{r}_2\right) = n_x\left(\boldsymbol{r}_1, \boldsymbol{r}_2\right) + n_c\left(\boldsymbol{r}_1, \boldsymbol{r}_2\right) \tag{2.22}$$

satisfies another sum rule,

$$\int d\boldsymbol{r}_2 \, n_c\left(\boldsymbol{r}_1, \boldsymbol{r}_2\right) = 0. \tag{2.23}$$

With the discussion noted above, we are now ready to analyze the electron correlation energy associated with the bond dissociation in terms of the hole functions. First, we construct the exchange-correlation hole $n_{xc}\left(\boldsymbol{r}_1, \boldsymbol{r}_2\right)$ for the multi-configuration wave function $\Psi_{\mathrm{MC}}\left(\boldsymbol{r}_1, \boldsymbol{r}_2\right)$ defined by Eq. (2.8) which realizes the appropriate electronic energy at the dissociation limit. From Eqs. (2.12) and (2.14), $n_{xc}\left(\boldsymbol{r}_1, \boldsymbol{r}_2\right)$ is described in terms of the second-order reduced density matrix $n_2\left(\boldsymbol{r}_1, \boldsymbol{r}_2\right)$, thus,

$$n_{xc}\left(\boldsymbol{r}_1, \boldsymbol{r}_2\right) = \frac{2n_2\left(\boldsymbol{r}_1, \boldsymbol{r}_2\right)}{n\left(\boldsymbol{r}_1\right)} - n\left(\boldsymbol{r}_2\right). \tag{2.24}$$

In the following, we assume that the bond distance R is being infinitely stretched to make the discussion simple without the loss of essence. By substituting Eq. (2.8) into the diagonal element of the density matrix of Eq. (2.10), we obtain

$$2n_2\left(\boldsymbol{r}_1, \boldsymbol{r}_2\right) = \left|\phi_{\mathrm{A}}\left(\boldsymbol{r}_1\right)\right|^2 \left|\phi_{\mathrm{B}}\left(\boldsymbol{r}_2\right)\right|^2. \tag{2.25}$$

In deriving Eq. (2.25), it is assumed that the position $\boldsymbol{r}_1$ of the reference electron is in the vicinity of the atomic center A and, therefore, $\phi_B(\boldsymbol{r}_1)$ is supposed to be almost zero since the interatomic distance $R = \infty$ is assumed. Similarly, the density $n(\boldsymbol{r}_1) = \phi_A(\boldsymbol{r}_1)^2 + \phi_B(\boldsymbol{r}_1)^2$ is dominated by $\phi_A(\boldsymbol{r}_1)^2$. Then, it is readily recognized from Eqs. (2.24) and (2.25) that the resultant exchange-correlation hole for the reference electron placed on the atomic site A becomes

$$n_{xc}(\boldsymbol{r}_2)|_{R=\infty,\ \boldsymbol{r}_1 \in A} = -|\phi_A(\boldsymbol{r}_2)|^2 . \tag{2.26}$$

It is worth noting that the hole is completely localized upon dissociation around the reference electron due to the electron correlation. Thus, the two electrons come apart to reside separately in the individual atomic sites when the interatomic distance is being stretched. We, next, discuss the exchange hole distribution $n_x(\boldsymbol{r}_1, \boldsymbol{r}_2)$ constructed from the single determinant wave function (i.e. HF wave function) expressed by Eq. (2.3). Utilizing Eq. (2.18), it is possible to show that

$$n_1(\boldsymbol{r}_1, \boldsymbol{r}_2) = \phi_A(\boldsymbol{r}_1)(\phi_A(\boldsymbol{r}_2) + \phi_B(\boldsymbol{r}_2))^* \tag{2.27}$$

assuming $R = \infty$ and the reference point $\boldsymbol{r}_1$ is placed in the vicinity of the site A. Since the density $n(\boldsymbol{r}_1)$ is dominated by $\phi_A(\boldsymbol{r}_1)^2$ at $\boldsymbol{r}_1$ the exchange hole $n_x(\boldsymbol{r}_1, \boldsymbol{r}_2)$ becomes

$$n_x(\boldsymbol{r}_2)|_{R=\infty,\ \boldsymbol{r}_1 \in A} = -\frac{1}{2}\left(|\phi_A(\boldsymbol{r}_2)|^2 + |\phi_B(\boldsymbol{r}_2)|^2\right) . \tag{2.28}$$

It is, thus, shown that the exchange hole function is delocalized over the two atomic sites separated with an infinite distance and it makes a clear contrast to the localized exchange-correlation function expressed by Eq. (2.25). Now, it is also possible to evaluate the correlation hole distribution $n_c(\boldsymbol{r}_1, \boldsymbol{r}_2)$ from Eq. (2.21), thus,

$$n_c(\boldsymbol{r}_2)|_{R=\infty,\ \boldsymbol{r}_1 \in A} = -\frac{1}{2}|\phi_A(\boldsymbol{r}_2)|^2 + \frac{1}{2}|\phi_B(\boldsymbol{r}_2)|^2 . \tag{2.29}$$

The behaviors of these hole functions [3, 4] are summarized in Fig. 2.1 for the help of readers' understandings. For the later discussions, it is of essential importance to note that both of the exchange and the correlation holes are delocalized over the two atomic sites. However, the superposition of these holes yields an exchange-correlation hole localized around the reference electron as shown in the figure. The electron correlation that localizes the exchange-correlation hole on a single site at the dissociation limit of a chemical bond is often referred to as the static correlation energy E_{sc} or the left-right correlation and it plays a decisive role in providing the correct dissociation. In other words, the lack of the energy E_{sc} (static correlation error) gives rise to an erroneous behavior in the potential energy for the bond dissociation. It is often stated in literatures that the whole of the correlation energy E_c can be roughly given by the sum of E_{sc} and the rest of it, that is, the dynamical correlation E_{dyn} ($E_c = E_{sc} + E_{dyn}$). Note, however, that there are no individual rig-

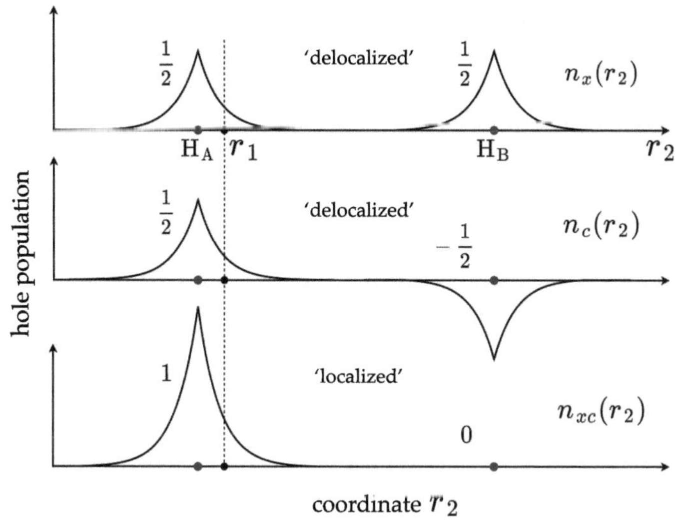

Fig. 2.1 Illustration of the populations of the exchange hole n_x (upper), the correlation hole n_c (middle), and their sum n_{xc} (lower) as functions of the coordinate $\boldsymbol{r}_2$ in the infinitely stretched H_2 molecule [3]. The fractional population assigned to each atomic site is presented. In the constructions of these figures, it is assumed that the reference point $\boldsymbol{r}_1$ is placed in the vicinity of atom H_A

orous definitions for these energies nor discriminations between them since these correlation energies are expressed with the same mathematical form in terms of the one-electron wave functions. However, in the description of the energy E_{sc}, only a few electronic configurations, which characterize the whole electronic structure, play a dominant role. With these discussions in mind, we consider below the exchange and the correlation holes provided within the framework of the DFT approach.

2.2 Treatment of Static Correlation in Wave-Function Theory

2.2.1 Multi-configuration Approach

As demonstrated in the previous Sect. 2.1, the static correlation plays a decisive role in describing dissociations of covalent bonds theoretically. Actually, the static correlation energy becomes quite significant even in the dissociation of the simplest chemical bond, i.e. the bond in a hydrogen molecule. It was demonstrated that the electronic structure of the dissociation limit can be soundly described by a linear combination of only two characteristic electronic configurations. In this subsection, we take a closer look at this approach for later references. For this purpose, we consider here a general molecular system with N electrons. For simplicity, we consider closed shell systems with singlet spin state. Recall that the electronic configuration which represents the electron transition from the bonding orbital to the corresponding anti-bonding orbital should be included in the total wave function Ψ to describe the dissociation of a single σ bond as discussed in Sect. 2.1. This approach can also be extended to a system with a double bond such as the ethene molecule (C_2H_4). In this case, the electronic configuration due to the electron transition from the bonding π orbital to the anti-bonding π orbital ($\pi - \pi^*$ transition) should also be included in the total wave function besides the $\sigma - \sigma^*$ transition. Of course, the similar manner can also be used to describe the dissociation of a triple bond. Thus, the total wave function, which serves to describe the bond dissociation, shall be given by the linear combination of the several spin-adapted electronic configurations. Such a multi-configuration (MC) wave function Ψ_{MC} can be expressed in terms of the configuration state functions (CSF) $\{\Psi_i^{\text{CSF}}\}$,

$$\Psi_{\text{MC}} = \sum_i C_i \Psi_i^{\text{CSF}} \tag{2.30}$$

where the coefficients C_i are supposed to satisfy the normalization condition $\sum_i |C_i|^2 = 1$. As mentioned in Sect. 1.2.4, each Ψ_i^{CSF} is an eigenstate of the spin operator $\mathbf{S}^2$ and supposed to be normalized. The expansion coefficient C_i represents the weight of Ψ_i^{CSF} in the total wave function Ψ^{MC}. In the approach termed MCSCF method [1, 2], not only the expansion coefficients $\{C_i\}$ but also the set of one-electron wave functions $\{\varphi_k\}$ which serve as building blocks of $\{\Psi_i^{\text{CSF}}\}$, are also optimized self-consistently. The MCSCF approach is a classical wave-function theory which offers a robust theoretical description for the bond dissociations within a reasonable computational cost although its numerical implementation is not straightforward as shown below. In a mathematical representation, the ground state energy E_0^{MCSCF} by the MCSCF method can be written as

$$E_0^{\text{MCSCF}} = \min_{\{\varphi_k\}} \left\{ \min_{\{C_i\}} \langle \Psi_{\text{MC}} | \hat{T} + \hat{V}_{ee} + \hat{v}_{\text{ext}} | \Psi_{\text{MC}} \rangle \right\} \tag{2.31}$$

For the inner minimization, only the expansion coefficients C_i are to be optimized with the set of wave functions $\{\varphi_k\}$ being frozen. Then, the inner minimization can be performed in parallel to the configuration interaction (CI) method discussed in Sect. 1.2.4. Remind that the minimization of the expectation value $\langle \Psi_{\text{MC}} | \hat{H} | \Psi_{\text{MC}} \rangle$ under the normalization condition can be reduced to the diagonalization of the CI matrix $\mathbf{H}$ with matrix elements $H_{ij} = \left\langle \Psi_i^{\text{CSF}} \right| \hat{H} \left| \Psi_j^{\text{CSF}} \right\rangle$, where the Hamiltonian $\hat{H}$ is defined as $\hat{H} = \hat{T} + \hat{V}_{ee} + \hat{v}_{\text{ext}}$. The unitary matrix $\mathbf{U}$ that diagonalizes the matrix $\mathbf{H}$ constitutes the set of the expansion coefficients $C_i^{(p)}$ $(p = 0, 1, \cdots)$ for the electronic states indexed with p including the excited states. The eigenvalues are to be obtained as the matrix elements of the diagonalized CI matrix. The outer minimization is, on the other hand, performed with respect to the wave functions φ_i with the coefficients C_i being fixed. To do this, we consider the energy E_{MCSCF} defined as

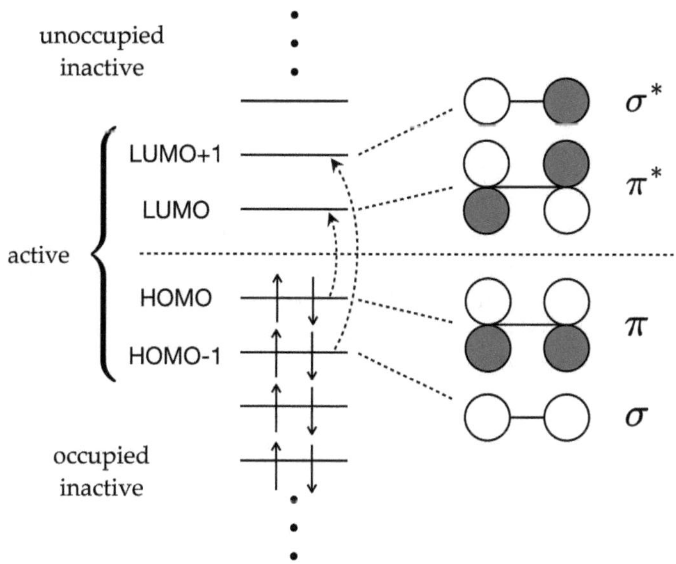

Fig. 2.2 Illustration of the active and inactive orbitals in a covalent double bond that consists of σ and π orbitals. In a multi-configuration self-consistent field (MCSCF) calculation to describe the bond dissociation, the electronic configurations representing at least the HOMO-LUMO and (HOMO-1) to (LUMO+1) electronic transitions should be included in the total wave function

$$E_{\text{MCSCF}} = \langle \Psi_{\text{MC}} | \hat{H} | \Psi_{\text{MC}} \rangle = \left\langle \sum_i C_i \Psi_i^{\text{CSF}} \middle| \hat{H} \middle| \sum_j C_j \Psi_j^{\text{CSF}} \right\rangle \tag{2.32}$$

To formulate the energy, we decompose the set $\{\varphi_k\}$ into the 'active' and 'inactive' orbitals. As shown in Fig. 2.2, which illustrates the electronic configuration of a system with a double bond, the orbitals related to the bond dissociation are treated as the active orbitals and Ψ_i^{CSF} are created by distributing electrons within these orbitals. On the other hand, the inactive orbitals are those fully occupied or completely vacant in all Ψ_i^{CSF}. The orbital space consisting of the active orbitals is often referred to as active space, while that of the inactive orbitals is called inactive space.

In the following, we illustrate a method to perform the MCSCF calculation for later references. To this end, we consider the simplest MCSCF wave function Ψ_{MC} that consists of only two configuration state functions Ψ_i^{CSF}, that is, the ground state configuration Ψ_0^{CSF} and the doubly excited configurations Ψ_1^{CSF} with closed shell structure to avoid complications. The doubly excited configuration Ψ_1^{CSF} is created by the HOMO-LUMO transition as in the description of the H_2 dissociation. Thus, each configuration $\Psi_i^{\text{CSF}} (i = 0, 1)$ consists of a single Slater determinant. To be precise, the wave function Ψ_{MC} is written as

$$\Psi_{\text{MC}} = C_0 \Psi_0^{\text{CSF}} + C_1 \Psi_1^{\text{CSF}} \tag{2.33}$$

Then, the electron density $n(\boldsymbol{r})$ of the system with $\Psi_{\rm MC}$ can be obtained by using Eq. (1.66), thus,

$$n(\boldsymbol{r}) = 2\sum_{k=1}^{N/2-1} \varphi_k(\boldsymbol{r})\,\varphi_k^*(\boldsymbol{r}) + 2\sum_{i}^{\rm act} |C_i|^2\,\overline{\varphi}_i(\boldsymbol{r})\,\overline{\varphi}_i^*(\boldsymbol{r}) \tag{2.34}$$

In Eq. (2.34), the index k is for the inactive orbitals and runs from 1 to $N/2-1$, while the index i is for the active orbitals and it runs from 0 to 1. In the present case, $\overline{\varphi}_0$ and $\overline{\varphi}_1$ correspond to HOMO and LUMO, respectively. In other words, $\overline{\varphi}_0 = \varphi_{N/2}$ and $\overline{\varphi}_1 = \varphi_{N/2+1}$. When we use the occupation numbers $\{p_i\}$ for the active orbitals, Eq. (2.34) can be equivalently rewritten as

$$n(\boldsymbol{r}) = 2\sum_{k=1}^{N/2-1} \varphi_k(\boldsymbol{r})\,\varphi_k^*(\boldsymbol{r}) + \sum_{i=N/2}^{N/2+1} p_i\varphi_i(\boldsymbol{r})\,\varphi_i^*(\boldsymbol{r}) \tag{2.35}$$

where $p_{N/2} = 2\,|C_0|^2$ and $p_{N/2+1} = 2\,|C_1|^2$. For the inactive orbitals, the occupation numbers are, of course, $p_k = 2\ (k = 1, 2, \cdots, N/2-1)$.

The energy expectation value of Eq. (2.32) for the wave function in Eq. (2.33) can be decomposed into the contributions from the one-electron and two-electron Hamiltonians as in the Hartree-Fock energy calculation presented in Sect. 1.2.2. Explicitly, $E_{\rm MCSCF}$ is expressed as

$$E_{\rm MCSCF} = E_1^{\rm MCSCF} + E_2^{\rm MCSCF} \tag{2.36}$$

where $E_1^{\rm MCSCF}$ and $E_2^{\rm MCSCF}$ are, respectively, the expectation values for one-electron and two-electron Hamiltonians. In the following, we discuss the individual terms. The expectation value $E_1^{\rm MCSCF}$ is given by

$$E_1^{\rm MCSCF} = N \times \langle \Psi_{\rm MC}|\, \mathbf{h}_1(1)\, |\Psi_{\rm MC}\rangle \tag{2.37}$$

in parallel to Eq. (1.14). The one-electron operator $\mathbf{h}_1(1)$ for electron 1 is explicitly defined as

$$\mathbf{h}_1(1) := -\frac{1}{2}\nabla_1^2 + v_{\rm ext}(\boldsymbol{r}_1) \tag{2.38}$$

In the calculation of Eq. (2.37), it should be noticed that the energy elements $\langle \Psi_0^{\rm CSF}|\, \mathbf{h}_1(1)\, |\Psi_1^{\rm CSF}\rangle$ between the different CSFs become zero in principle since these two CSFs differ from each other by two or more orbitals. Therefore, only the diagonal elements $\langle \Psi_0^{\rm CSF}|\, \mathbf{h}_1(1)\, |\Psi_0^{\rm CSF}\rangle$ and $\langle \Psi_1^{\rm CSF}|\, \mathbf{h}_1(1)\, |\Psi_1^{\rm CSF}\rangle$ remain in $E_1^{\rm MCSCF}$. These considerations lead to

$$E_1^{\rm MCSCF} = 2\sum_{k=1}^{N/2-1} \langle \varphi_k|\, \mathbf{h}_1(1)\, |\varphi_k\rangle + 2\sum_{i}^{\rm act} |C_i|^2 \langle \overline{\varphi}_i|\, \mathbf{h}_1(1)\, |\overline{\varphi}_i\rangle \tag{2.39}$$

where the index i is for the active orbitals (HOMO and LUMO) and $i = 0, 1$. We next consider the expectation value E_2^{MCSCF} for the two-electron operator in E_{MCSCF}. E_2^{MCSCF} is given by

$$E_2^{\mathrm{MCSCF}} = \frac{N(N-1)}{2} \times \langle \Psi_{\mathrm{MC}} | \frac{1}{r_{12}} | \Psi_{\mathrm{MC}} \rangle \tag{2.40}$$

where r_{12} is the distance between electrons 1 and 2. By substituting Eqs. (2.33)–(2.40), E_2^{MCSCF} can be expanded as

$$\begin{aligned} E_2^{\mathrm{MCSCF}} = \frac{N(N-1)}{2} \Bigg(& |C_0|^2 \left\langle \Psi_0^{\mathrm{CSF}} \right| \frac{1}{r_{12}} \left| \Psi_0^{\mathrm{CSF}} \right\rangle + |C_1|^2 \left\langle \Psi_1^{\mathrm{CSF}} \right| \frac{1}{r_{12}} \left| \Psi_1^{\mathrm{CSF}} \right\rangle \\ & + \left(C_0 C_1^* \left\langle \Psi_1^{\mathrm{CSF}} \right| \frac{1}{r_{12}} \left| \Psi_0^{\mathrm{CSF}} \right\rangle + c.c. \right) \Bigg) \end{aligned} \tag{2.41}$$

E_2^{MCSCF} reads

$$\begin{aligned} E_2^{\mathrm{MCSCF}} = & \sum_{k,l=1}^{N/2-1} 2 (\varphi_k \varphi_k | \varphi_l \varphi_l) - (\varphi_k \varphi_l | \varphi_l \varphi_k) \\ & + 2 \sum_i^{\mathrm{act}} C_i^2 \left(\sum_{k=1}^{N/2-1} 2 (\overline{\varphi}_i \overline{\varphi}_i | \varphi_k \varphi_k) - (\overline{\varphi}_i \varphi_k | \varphi_k \overline{\varphi}_i) \right) \\ & + \sum_{i,j}^{\mathrm{act}} C_i C_j (\overline{\varphi}_i \overline{\varphi}_j | \overline{\varphi}_i \overline{\varphi}_j) \end{aligned} \tag{2.42}$$

Remind that the indices i and j are for the two active orbitals in the present case, and $i, j = 0, 1$. In Eq. (2.42), we use the shorthand notation $(\varphi_a \varphi_b | \varphi_c \varphi_d)$ to make the equation concise. It is explicitly defined as

$$(\varphi_a \varphi_b | \varphi_c \varphi_d) := \int d\boldsymbol{r}_1 d\boldsymbol{r}_2 \varphi_a(\boldsymbol{r}_1) \varphi_b(\boldsymbol{r}_1) \frac{1}{r_{12}} \varphi_c(\boldsymbol{r}_2) \varphi_d(\boldsymbol{r}_2) \tag{2.43}$$

It should also be noted that orbitals are supposed to be real in deriving Eq. (2.42) for simplicity. E_2^{MCSCF} includes the classical Coulomb energy $E_{\mathrm{H}}^{\mathrm{MCSCF}}$ constructed from the density $n(\boldsymbol{r})$ given by Eq. (2.34). Using this equation, $E_{\mathrm{H}}^{\mathrm{MCSCF}}$ can be written as

$$\begin{aligned} E_{\mathrm{H}}^{\mathrm{MCSCF}} = & \sum_{k,l=1}^{N/2-1} 2 (\varphi_k \varphi_k | \varphi_l \varphi_l) + 2 \sum_i^{\mathrm{act}} C_i^2 \sum_{k=1}^{N/2-1} 2 (\overline{\varphi}_i \overline{\varphi}_i | \varphi_k \varphi_k) \\ & + 2 \sum_{i,j}^{\mathrm{act}} C_i^2 C_j^2 (\overline{\varphi}_i \overline{\varphi}_i | \overline{\varphi}_j \overline{\varphi}_j) \end{aligned} \tag{2.44}$$

By subtracting $E_{\mathrm{H}}^{\mathrm{MCSCF}}$ of Eq. (2.44) from E_2^{MCSCF} of Eq. (2.42), the static correlation energy $E_{\mathrm{SC}}^{\mathrm{MCSCF}}$ can be extracted, thus,

$$E_{\mathrm{SC}}^{\mathrm{MCSCF}} = -\sum_{k,l=1}^{N/2-1} (\varphi_k\varphi_l|\,\varphi_l\varphi_k) - 2\sum_{i}^{\mathrm{act}} C_i^2 \sum_{k=1}^{N/2-1} (\overline{\varphi}_i\varphi_k|\,\varphi_k\overline{\varphi}_i)$$
$$+\sum_{i,j}^{\mathrm{act}} C_i C_j\,(\overline{\varphi}_i\overline{\varphi}_j|\,\overline{\varphi}_i\overline{\varphi}_j) - 2\sum_{i,j}^{\mathrm{act}} C_i^2 C_j^2\,(\overline{\varphi}_i\overline{\varphi}_i|\,\overline{\varphi}_j\overline{\varphi}_j) \tag{2.45}$$

In the above discussion, we formulated the MCSCF energy E_{MCSCF} in terms of the one-electron orbitals $\{\varphi_k\}$ $(k = 1, 2, \cdots, N/2-1)$ and $\{\varphi_i\}$ $(i = N/2, N/2+1)$ and the expansion coefficients $\{C_i\}$ $(i = 0, 1)$ for the minimal MC wave function Ψ_{MC} representing a dissociation of a single bond. Now it is possible to evaluate the functional derivatives needed to perform the outer minimization in Eq. (2.31). At the equilibrium, we obtain the relations for the derivatives,

$$\frac{\partial E_{\mathrm{MCSCF}}}{\partial \varphi_j(\boldsymbol{r})} = 0 \quad (j = 1, 2, \cdots, N/2+1) \tag{2.46}$$

with the fixed coefficients C_i. As noted above, C_i can be obtained by diagonalizing the CI matrix. The self-consistent minimization of E_{MCSCF} with respect to the set $\{\varphi_i\}$ $(i = N/2, N/2+1)$ must be performed under the orthonormalization conditions $\langle\varphi_i|\,\varphi_j\rangle = \delta_{ij}$ which leads to the set of Schrödinger equations in the canonical form for the one-electron wave functions,

$$\hat{f}_{\mathrm{MCSCF}}\varphi_i = \epsilon_i\varphi_i \quad (i = 1, 2, \cdots, N/2+1) \tag{2.47}$$

in completely parallel to the derivation of the HF equation of Eq. (1.49). Only the difference is the contents of the operator $\hat{f}_{\mathrm{MCSCF}}$. To be more precise, the major difference comes from the differentiation of E_2^{MCSCF} in Eq. (2.42) with respect to φ_i. Fortunately, the derivative can be readily obtained as in the evaluation of the operator $\hat{f}$ in the HF equation. However, it should be noticed that the mathematical form of the mean field applied to the active orbitals is different from that to the inactive orbitals. The double minimization in Eq. (2.31), that is for the set of wave functions $\{\varphi_i\}$ and the CI coefficients $\{C_j\}$, is thus made possible. The set $\{\varphi_i\}$ shall be minimized with the fixed CI coefficients, and the coefficients are minimized under the frozen set of wave functions. Thus the two coupled minimizations are performed in a self-consistent manner in a numerical implementation of the MCSCF procedure.

In closing this paragraph, we discuss the construction of the active space (see Fig. 2.2) in the MCSCF calculations. For the description of the single σ bond in H_2 or the double bond in ethene, how to select the orbitals to be placed within the active space is apparent and straightforward. Actually, however, there are a lot of molecular systems where the proper orbitals to be selected are quite ambiguous. Unfortunately, the properties of the system of interest depends seriously on the choice of the active orbitals. It is written in a certain textbook that the orbital selection should be undertaken by specialists. However, 'specialists' is neither a scientific word nor a well-defined technical term. If a specialist knows how to select the proper active orbitals on the basis of some *scientific* reasons, the method can be clearly described so that any *non*-specialist can recognize the validity of the selection. Furthermore, it will be made possible to cast the method into a robust procedure and implement it as a computer algorithm. To the best of the author's knowledge, no such method was never developed so far. A naive solution to the problem of the orbital selection is, of course, to increase the number of active orbitals and the number of electrons to be correlated. However, the energy convergence is quite slow and the number of electronic configurations included in the MCSCF wave function increases drastically for the increase in size of the active space, which makes the MCSCF calculations intractable. Although the MCSCF approach is robust from the theoretical viewpoint, it suffers from the problem of the orbital selections in describing a lot of actual systems of interest.

2.3 Treatment of Static Correlation in Density-functional Theory

2.3.1 General Descriptions

The treatment of the static correlation with DFT approach shows a clear contrast to that in the MO theory. We illustrate below how the static correlation is being realized though only partly in the DFT approach. To this end, we consider again the dissociation of an H_2 molecule. The theoretical framework developed by Kohn and Sham (KS-DFT) is obviously a one and only workhorse in DFT to analyze chemical events and properties of interest adequately. The detailed explanation for the construction the KS equation was provided in Sect. 1.3.5. Below, we make a concise review of the equation for later discussions. The KS equation for one-electron wave function φ_i is represented as

$$\left(-\frac{1}{2}\nabla_1^2 + v_{\mathrm{ext}}(\boldsymbol{r}_1) + \int \frac{n(\boldsymbol{r}_2)}{|\boldsymbol{r}_1 - \boldsymbol{r}_2|} d\boldsymbol{r}_2 + \overline{v}_{xc}[n](\boldsymbol{r}_1)\right) \varphi_i(\boldsymbol{r}_1) = \epsilon_i \varphi_i(\boldsymbol{r}_1). \tag{2.48}$$

The first term in the parenthesis of the left hand side of Eq. (2.48) is the kinetic energy operator, the second one stands for the external potential applied to the electrons, the third is the classical Coulomb potential formed by electron density n. The exchange-correlation potential $\overline{v}_{xc}[n](\boldsymbol{r}_1)$ in Eq. (2.48) plays a decisive role in the KS-DFT and it is defined as a functional derivative of the exchange-correlation energy $\overline{E}_{xc}[n]$ with respect to the density at $\boldsymbol{r}$

$$\overline{v}_{xc}[n](\boldsymbol{r}) = \frac{\delta \overline{E}_{xc}[n]}{\delta n(\boldsymbol{r})}. \tag{2.49}$$

Note that the notation of $\overline{E}_{xc}$ indicates that the functional includes the portion of the kinetic energy of the interacting system in addition to the exchange-correlation energy E_{xc} itself. In the following, however, we are not concerned with the notational difference between $\overline{E}_{xc}$ and E_{xc} because a lot of approximate functionals do not include the kinetic energy of the interacting system. Equation (2.49) indicates that the potential $\overline{v}_{xc}[n](\boldsymbol{r})$ is local in the real space. That is, the effect of the nonlocal correlation among the electrons is projected onto the local potential. However, note that it doesn't imply the limitation of the KS equation since the KS theory is formulated so that it is formally exact. As mentioned in Sect. 1.1, in principle, the functional derivative of Eq. (2.49) can implicitly realize the nonlocal response of the exchange-correlation energy $\overline{E}_{xc}[n]$ to the variation of the density at $\boldsymbol{r}$ although the explicit form of the exact functional is not yet available. In practice, however, only the local or semi-local response to the density can be considered since the functional is represented in terms of the density itself in the local-density approximation (LDA) and its correction is described merely with the gradient $\nabla n(\boldsymbol{r})$ of the density (GGA). Importantly, within the LDA approximation the exchange hole is modeled by that of the homogeneous electron gas and, hence, it is approximated even in the real system that a spherically symmetric hole is being located around the reference point $\boldsymbol{r}_1$. When the interatomic distance R is being stretched infinitely, the wave function with spin σ is split into two equivalent fragments placed on the two atomic centers H_A and H_B. In the treatment with the HF theory the exchange hole for the electron with a spin σ behaves similarly as described in Eq. (2.28). However, in the LDA-based approach, the hole, of which population is normalized to 1, is always localized at the reference point and this feature makes a distinct difference from the HF theory in the evaluation of the exchange energy in the dissociation. In Fig. 2.3 the LDA exchange hole is compared with the holes of HF method and that given by multi-configurational approach.

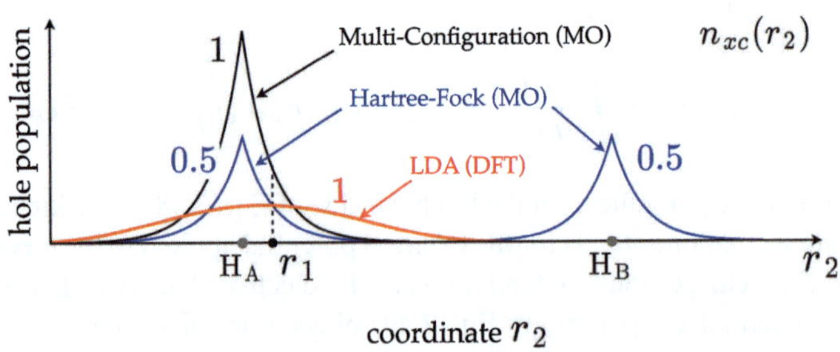

Fig. 2.3 Comparison of the LDA (local-density approximation) exchange hole with those of the HF (Hartree-Fock) and the multi-configurational method for the H_2 molecule assuming the infinite stretch of the chemical bond. Each quantity near the hole function represents the fractional population assigned to the atomic sites

As shown in Eq. (2.26) the exchange-correlation hole in the method with the multi-configuration wave function is localized at the position of the reference electron. Thus, the exchange hole simulated in LDA is more advantageous than the HF method. However, the depth of the hole in LDA is estimated as just half of the true hole since the on-top depth of the exchange hole, that is the depth at $\boldsymbol{r}_1 = \boldsymbol{r}_1'$, is $n_x(\boldsymbol{r}_1, \boldsymbol{r}_1) = \frac{1}{2}n(\boldsymbol{r}_1, \boldsymbol{r}_1) = n_\sigma(\boldsymbol{r}_1, \boldsymbol{r}_1)$. Thus, within the framework of the LDA approach, the exchange hole is appropriately localized upon dissociation around the reference point although its depth becomes half of the true hole. Consequently, the Coulomb interaction between the reference electron and the exchange hole is weakened in LDA as compared with the true hole though it is much larger than that in the HF method where the hole contains a half an electron in the vicinity of each nuclear. We show in Fig. 2.4 the schematic behaviors of the resultant potential energy curves of a dissociating H_2 molecule provided by the LDA and the HF methods in comparison with the description using the proper exchange-correlation hole. The corresponding structures of the holes are also depicted. The LDA-based DFT is, thus, superior to the HF method and, hence it might be possible to mention that DFT can *simulate* the static correlation to a certain extent. The problem is that it seems quite difficult to incorporate the remaining static correlation into an approximate functional within the conventional framework of DFT. We also refer readers to the discussions in, e.g. Refs. [5, 6].

A remedy for the static correlation error in DFT is to employ the multi-configuration approach often referred to as MC-DFT [7–11]. In this method, the static correlation is taken into considerations by means of the wave functions given in the form of Eq. (2.9), while the remaining dynamical correlation is being evaluated through a density functional. Such a straightforward approach, however, suffers from two major problems. The first one is that the MC-DFT calculation is rather time-consuming and hence, the foremost advantage in DFT will be lost. The

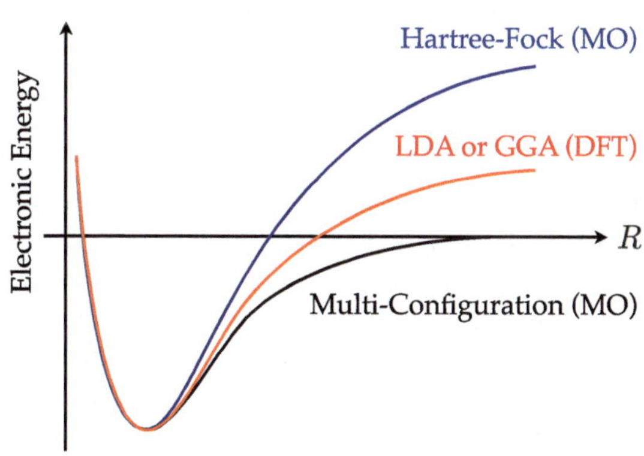

Fig. 2.4 Schematic illustration of the potential energy curves for H_2 dissociations described with Hartree-Fock method, multi-configuration approach in molecular orbital theory (MO) and the density-functional theory (DFT) with local-density approximation (LDA) or generalized-gradient approximation (GGA). The potential curves are shifted so that their bottoms are matched

second is that some double counting of the electron correlation energy inevitably arises since there is no clear difference between the static and dynamical correlations. In excluding the double count in the electron correlation, however, there are subtle problems. Assuming that the complete elimination of the double count of the correlation energy is made possible, it directly indicates that the static correlation itself can be explicitly represented in terms of a certain density functional. Then, the introduction of the MC wave function is no longer needed. Thus, an attempt to develop a density functional that can be used for the MC-DFT leads to a contradiction. Resolving the problem of the static correlation error (SCE) in DFT is regarded as the toughest challenge in the development of the exchange-correlation functional. In the next subsection we review a rare DFT functional for the static correlation.

2.3.2 *Approach of the Becke03 Functional*

Becke-Roussel Exchange Hole Model

The pure DFT functional to incorporate the static correlation was first developed in 2003 by A. D. Becke (Becke03) [5] who continuously revised the functional in the subsequent works [12, 13]. We describe here the outline of the development to make comparisons with our approach to be introduced later as a major issue in this chapter. As a notable feature in the development of the Becke03 functional, the exchange hole model proposed by Becke and Roussel (BR) [14] plays a crucial role. More explicitly, a generalized form of the BR machinery is utilized to measure the delocalization of the exchange hole in a system of interest. In the following, we first present a concise review of the original BR functional. In the BR functional, the exchange hole associated with a reference point in a system of interest is modeled by the electron density of a hydrogenic atom instead of the homogeneous electron gas. Explicitly, the model exchange hole for the electron with a spin σ has the form of the Slater function

$$n_{x\sigma}^{\mathrm{BR}}(\zeta, r) = A(\zeta)\exp(-\zeta r) \tag{2.50}$$

where $A(\zeta)$ is the normalization factor and satisfies $A(\zeta) = \frac{\zeta^3}{8\pi}$ to fulfill the sum rule of Eq. (2.15). The variables ζ and r represent the width of the hole function and the distance from the center of the hole and the reference point $\boldsymbol{r}$, respectively. $(\zeta,\ r)$ are the parameters to be determined from the coefficients of the Taylor expansion of the average exchange hole $\overline{n}_{x\sigma}(\boldsymbol{r}, s)$ of a real system. Explicitly, $\overline{n}_{x\sigma}$ stands for the exchange hole $n_{x\sigma}$ averaged over the sphere Ω centered at $\boldsymbol{r}$ with radius s. As derived in Eq. (1.146), the Taylor expansion of $\overline{n}_{x\sigma}(\boldsymbol{r}, s)$ for spin σ is given by

$$\overline{n}_{x\sigma}(\boldsymbol{r}, s) = n_\sigma(\boldsymbol{r}) + Q_\sigma(\boldsymbol{r})s^2 + \cdots \tag{2.51}$$

where the coefficient $Q_\sigma(\boldsymbol{r})$ of the quadratic term is expressed by

$$Q_\sigma(\boldsymbol{r}) = \frac{1}{6}\left[\nabla^2 n_\sigma(\boldsymbol{r}) - 2\tau_\sigma(\boldsymbol{r}) + \frac{(\nabla n_\sigma(\boldsymbol{r}))^2}{2n_\sigma(\boldsymbol{r})}\right]. \tag{2.52}$$

In Eq. (2.52), τ_σ is the kinetic energy density and defined in terms of the one-electron wave functions, thus,

$$\tau_\sigma(\boldsymbol{r}) = \sum_i^{occ} |\nabla\varphi_{i\sigma}(\boldsymbol{r})|^2 . \tag{2.53}$$

The BR parameters $(\zeta,\ r)$ are determined by equating the coefficients of the zeroth- and second-order terms in the expansion of Eq. (2.51) of the BR reference system to those of the real system. Then, from the zeroth-order term, we obtain

$$\zeta^3 \exp(-\zeta r) = 8\pi n_\sigma(\boldsymbol{r}) \tag{2.54}$$

while the second-order term gives

$$\zeta^2 r - 2\zeta = 6r\frac{Q_\sigma(\boldsymbol{r})}{n_\sigma(\boldsymbol{r})}. \tag{2.55}$$

It is important to note that the parameters $(\zeta,\ r)$ can be uniquely determined for any given electron density n and its curvature Q_σ at a reference point $\boldsymbol{r}$. Once the BR parameters are obtained, the exchange energy density $U_{x\sigma}^{\mathrm{BR}}(\boldsymbol{r})$ is readily evaluated as

$$U_{x\sigma}^{\mathrm{BR}}(\boldsymbol{r}) = -\frac{1}{r}\left(1 - \exp(-\zeta r) - \frac{1}{2}\zeta \exp(-\zeta r)\right). \tag{2.56}$$

Then, the exchange energy of the system can be obtained as

$$E_{x\sigma}^{\mathrm{BR}} = \frac{1}{2}\int d\boldsymbol{r}\ n_\sigma(\boldsymbol{r})U_{x\sigma}^{\mathrm{BR}}(\boldsymbol{r}). \tag{2.57}$$

It is, thus, possible to evaluate the exchange energy for a given electron density n without introducing any empirical parameters. The BR approach shows a clear contrast to the series of the development of the exchange functionals based on LDA since it employs an *inhomogeneous* system to build an exchange hole. It should also be noted that the BR functional can be categorized into the meta-GGA functional since it includes the terms of second derivative of the density as well as the kinetic energy density in its construction.

Generalized BR Model

In the Becke-Roussel exchange hole model of Eq. (2.50), the hole function $n_{x\sigma}^{\mathrm{BR}}$ is always normalized to one to guarantee the sum rule of Eq. (2.21). In the construction of the Becke03 functional [5], however, the normalization is not necessarily evaluated to be one, but considered as a flexible parameter. Namely, in the generalized BR exchange hole model, the normalization of the hole serves as a parameter to measure the degree of delocalization of the exchange hole. In this paragraph, we illustrate how the parameter will be determined besides the quantities (ζ, r) in Eq. (2.50). The generalized BR exchange hole function $n_{x\sigma}^{\mathrm{GBR}}$ has the form similar to Eq. (2.50),

$$n_{x\sigma}^{\mathrm{GBR}}(A, \zeta, r) = A \exp(-\zeta r) \quad . \tag{2.58}$$

However, the parameter A is not given by the normalization condition, but determined to realize the exact exchange energy $U_{x\sigma}^{\mathrm{exact}}$ of the real system. $U_{x\sigma}^{\mathrm{exact}}$ at $\boldsymbol{r}$ for a spin σ can be expressed as

$$U_{x\sigma}^{\mathrm{exact}}(\boldsymbol{r}) = \frac{1}{2} \int d\boldsymbol{r}' \frac{1}{|\boldsymbol{r} - \boldsymbol{r}'|} n_{x\sigma}\left(\boldsymbol{r}, \boldsymbol{r}'\right) \tag{2.59}$$

where the exchange hole function $n_{x\sigma}$ is defined with Eqs. (2.16) and (2.18). On the other hand, by referring to Eq. (2.56) the exchange energy density $U_{x\sigma}^{\mathrm{GBR}}(\boldsymbol{r})$ due to the model hole $n_{x\sigma}^{\mathrm{GBR}}$ is given by

$$U_{x\sigma}^{\mathrm{GBR}}(\boldsymbol{r}) = \frac{4\pi A}{\zeta^3 r} \left(2 - 2\exp(-\zeta r) - \zeta \exp(-\zeta r)\right) \tag{2.60}$$

because the normalization of $n_{x\sigma}^{\mathrm{GBR}}$ is $\frac{8\pi A}{\zeta^3}$. Then, by equating the exchange energy density $U_{x\sigma}^{\mathrm{GBR}}(\boldsymbol{r})$ with the exact energy $U_{x\sigma}^{\mathrm{exact}}$, i.e. $U_{x\sigma}^{\mathrm{GBR}}(\boldsymbol{r}) = U_{x\sigma}^{\mathrm{exact}}$, we have

$$\frac{4\pi A}{\zeta^3 r} \left(2 - 2\exp(-\zeta r) - \zeta \exp(-\zeta r)\right) = -U_{x\sigma}^{\mathrm{exact}}(\boldsymbol{r}). \tag{2.61}$$

With Eq. (2.61), in addition to the conditions Eqs. (2.54) and (2.55), the parameter A can be determined through a certain numerical method. Then, the normalization $N_{x\sigma}^{\mathrm{GBR}}$ of the generalized BR exchange hole at $\boldsymbol{r}$ is given as

$$N_{x\sigma}^{\mathrm{GBR}}(\boldsymbol{r}) = \frac{8\pi A}{\zeta^3}. \tag{2.62}$$

Due to the imposition of the equality of Eq. (2.61) on $n_{x\sigma}^{\mathrm{GBR}}$, the delocalized exchange hole in a stretched bond of a real system will be projected onto a fractional hole localized at the reference point $\boldsymbol{r}$. Thus, $N_{x\sigma}^{\mathrm{GBR}}$ is always less or equal to one. The concept of the projected exchange hole plays a central role in the generalized BR approach. For understanding the projected hole, we again consider an extreme situation that the interatomic distance R in H_2 is infinitely stretched. As was shown in Eq. (2.28), the exchange hole is split into two fragments with the population of half an electron located around the two nuclei. Therefore, $N_{x\sigma}^{\mathrm{GBR}}$ for a spin is estimated to be 0.5 irrespective of the position $\boldsymbol{r}$ of the reference electron. When the normalization $N_{x\sigma}^{\mathrm{GBR}}$ is much less than one, it implies that the exchange hole of the real system is seriously delocalized. Thus, the generalized BR model offers a flexibility to accommodate the exchange hole in a stretched chemical bond.

Static Correlation Functional

The key idea in the Becke03 functional [5] that simulates the static (nondynamical) correlation in a bond dissociation is to deepen the exchange hole for the electron with a spin σ at $\boldsymbol{r}$ by mixing the hole for the electron with the opposite spin at the same position. Namely, the total exchange hole n_x^{static} for spin α involving the static correlation is expressed by

$$n_{x\alpha}^{\mathrm{static}}(\boldsymbol{r}) = n_{x\alpha}^{\mathrm{GBR}}(\boldsymbol{r}) + f_{\alpha\beta} n_{x\beta}^{\mathrm{GBR}}(\boldsymbol{r}) \tag{2.63}$$

where $n_{x\alpha}^{\mathrm{GBR}}$ and $n_{x\beta}^{\mathrm{GBR}}$ stand for the generalized BR holes in the form of Eq. (2.58) for α and β spins, respectively. The factor $f_{\alpha\beta}$ specifies the degree of the mixing of the hole for the opposite spin. It can be readily shown that the factor is symmetric, i.e. $f_{\alpha\beta} = f_{\beta\alpha}$, and hence, we can use the same notation $f(= f_{\alpha\beta} = f_{\beta\alpha})$ for different spins. Corresponding to Eq. (2.63), it is straightforward to derive the relation for the normalizations,

$$N_{xc\alpha}^{\mathrm{static}} = N_{x\alpha}^{\mathrm{GBR}} + f N_{x\beta}^{\mathrm{GBR}} \quad . \tag{2.64}$$

Since the population of the total exchange hole should be always less than one, a condition: $0 \leq N_{x\alpha}^{\mathrm{static}} \leq 1$, must be imposed. The correlation factor $f \ (0 \leq f \leq 1)$ in Eq. (2.64) is determined so as to satisfy this condition, of which details were described in the original paper. In Fig. 2.5, the outline of the procedure to construct the generalized BR exchange hole is illustrated for the infinitely stretched chemical bond in H_2. It is shown that the exchange hole $n_{x\alpha}^{\mathrm{static}}$ that incorporates the static correlation is localized around the reference point $\boldsymbol{r}_1$ and its population is normalized to one as in the exchange-correlation hole given by the multi-configurational approach (see Fig. 2.3). Thus, the construction of the hole utilizing the generalized BR model offers a localized hole with the correct population.

The hole distribution of Eq. (2.63) directly leads to the exchange energy density $U_{x\alpha}^{\mathrm{static}}$ for α spin with static correlation

$$U_{x\alpha}^{\mathrm{static}}(\boldsymbol{r}) = U_{x\alpha}^{\mathrm{exact}}(\boldsymbol{r}) + f U_{x\beta}^{\mathrm{exact}}(\boldsymbol{r}) \tag{2.65}$$

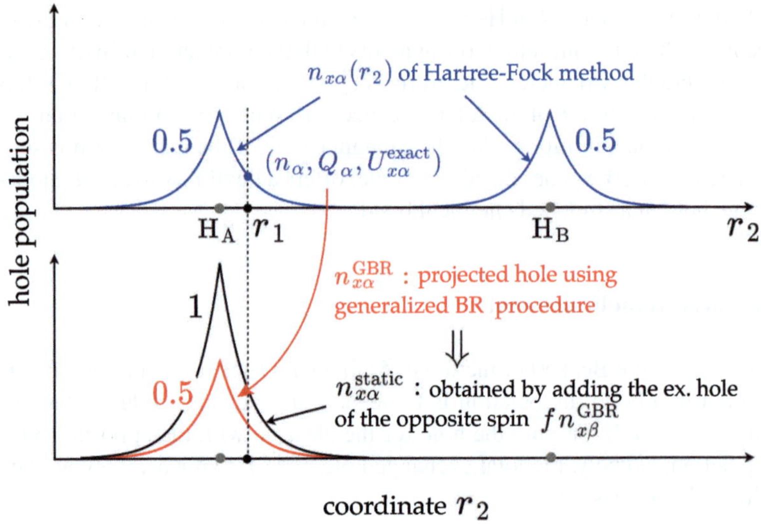

Fig. 2.5 Illustration of the generalized Becke-Roussel exchange hole $n_{x\alpha}^{\text{GBR}}$ and of the engineering to construct the exchange hole $n_{x\alpha}^{\text{static}}$ involving the static correlation for the H_2 molecule assuming the infinite stretch of the chemical bond. Each quantity near the hole function represents the fractional population assigned to the atomic sites

where we use the relation $U_{x\sigma}^{\text{GBR}}(\boldsymbol{r}) = U_{x\sigma}^{\text{exact}}(\boldsymbol{r})$. Then, the exchange energy $E_{x\alpha}^{\text{static}}$ becomes

$$E_{x\alpha}^{\text{static}} = E_{x\alpha}^{\text{exact}} + \frac{1}{2}\int d\boldsymbol{r}\; n_\alpha(\boldsymbol{r}) f U_{x\beta}^{\text{exact}}(\boldsymbol{r}) \tag{2.66}$$

and similar for the electrons with β spin. The leading term in the right hand side of Eq. (2.66) is actually the exact Hartree-Fock exchange energy, and the second term is the energy due to the static correlation. Therefore, the formulation of Eq. (2.66) is referred to as 'post Hartree-Fock' method in the original paper. In the subsequent developments, some extensions were made on the Becke03 functional to include the dynamical correlation and dispersion interactions, which led to Becke07 [12] and Becke13 [13] functionals. The Becke03 functional provided a great improvement by realizing the static correlation within the framework of DFT, although there is no physical justification of the method as noted in Ref. [5]. In the next chapter, we develop a new theoretical framework of the density-functional theory, which affords a natural description of the exchange hole including the static correlation. Before that, we discuss another approach to realize the static correlation using the implicit density functional, referred to as optimized effective potential (OEP) introduced in Sect. 1.4.

2.4 Multi-configuration Optimized Effective Potential

The previous section describes Becke's construction of an explicit density functional (Becke03) [5] which realizes the static correlation arising in a bond dissociation. A notable feature of the exchange functional for the density n_σ with spin σ is to incorporate the part of the electron density of the *opposite* spin according to the degree of the delocalization of n_σ at each reference point $\boldsymbol{r}$. Such a manipulation in Becke03 is needed to take into consideration the serious nonlocality of the exchange hole in a stretched system as depicted in Fig. 2.1. In Sect. 1.4, it was demonstrated that the Hartree-Fock (HF) nonlocal exchange potential can be reduced faithfully to a local potential through the optimized effective potential (OEP) or equivalently by the inverse Kohn-Sham method. In the present section, we will examine whether the OEP method can also be applied to the stretched systems with the separated exchange holes. This study is important because it provides a theoretical justification for the development of the static correlation functional. It is readily anticipated that the effective multiplicative potential including the static correlation can be constructed by applying the OEP method to the energy expectation value of the multi-configuration (MC) wave function introduced in Sect. 2.2.1. However, it should be stressed that the approach, called MCOEP, differs from the HF-OEP method by the fact that the density realized in MCOEP is constructed from a MC wave function and not a single determinant. The electron density obtained in the HF-OEP procedure resides within the subset of the non-interacting υ-rep. densities. However, the MCOEP approach offers another subset of densities called MCOEP υ-representable densities. In this sense, the MCOEP approach cannot be regarded as a sort of KS-DFT. Of course, the subset of MCOEP υ-rep. densities includes the subset of non-interacting υ-rep. densities. Whether the subset of non-interacting υ-rep. densities covers that of the MCOEP υ-rep. densities will be a difficult issue to resolve. The representability problem constitutes the heart of the theoretical basis for developing the static correlation functional within the framework of KS-DFT. It should also be noticed that the inverse KS-DFT (inv-KS) method introduced in Sect. 1.4.5 can also be applied to the electron density for the MCSCF wave function as long as the density is υ-representable. However, it seems impossible to evaluate the corresponding total energy since the wave function yielded in the inv-KS is represented with a single determinant in principle. In this respect, the inv-KS for the density of the MC wave function is not equivalent to the MCOEP method. With these kept in mind, in the following, we formulate the MCOEP method, along the method proposed by Weimer et al. [15].

To be concise in the formulation, we consider the simplest MCSCF wave function as discussed in Sect. 2.2.1. The system under consideration consists of N electrons with N being an even number. In the ground state configuration Ψ_0^{CSF}, the one-electron wave functions $(\varphi_1, \varphi_2, \cdots, \varphi_{N/2})$, arrayed in the ascending order in the energy, are supposed to be doubly occupied to form a closed shell structure. And the excited configuration Ψ_1^{CSF} is obtained by the transitions of the two electrons on the highest occupied molecular orbital (HOMO) $= \varphi_{N/2}$ to the lowest unoccupied molecular orbital (LUMO) $= \varphi_{N/2+1}$. It is of essential importance to note that the

HOMO= $\varphi_{N/2}$ is supposed to be a bonding orbital pertaining to the bond dissociation of interest and the LUMO= $\varphi_{N/2+1}$ is the corresponding anti-bonding orbital. Thus, the HOMO and the LUMO constitute the active space of the MC wave function Ψ_{MC} and the two electrons are correlated within the active space. The configurations Ψ_1^{CSF} as well as Ψ_0^{CSF} are automatically an eigenfunction of the spin operator $\mathbf{S}^2$. The MCSCF total wave function Ψ_{MC} shall be expanded in terms of the two configuration state functions as shown in Eq. (2.33). We show again the equation for convenience,

$$\Psi_{\text{MC}} = C_0 \Psi_0^{\text{CSF}} + C_1 \Psi_1^{\text{CSF}} \tag{2.67}$$

where C_n $(n = 0, 1)$ are the expansion coefficients. In the MCOEP approach, as in the HF-OEP method described in Sect. 1.4, the local multiplicative potential $\upsilon_{\text{eff}}(\boldsymbol{r})$ is optimized so that the eigenfunctions $\varphi_i (i = 1, 2, \cdots, N/2 + 1)$ on the potential $\upsilon_{\text{eff}}(\boldsymbol{r})$ minimizes the energy expectation value $E_{\text{MCSCF}} = \langle \Psi_{\text{MC}} | \hat{H} | \Psi_{\text{MC}} \rangle$. $\hat{H} = \hat{T} + \hat{V}_{ee} + \hat{\upsilon}_{\text{ext}}$ is the full Hamiltonian of the N electrons under the influence of the external potential $\hat{\upsilon}_{\text{ext}}$. The components E_1^{MCSCF} and E_2^{MCSCF} of the energy $E_{\text{MCSCF}} (= E_1^{\text{MCSCF}} + E_2^{\text{MCSCF}})$ are explicitly given by Eqs. (2.39) and (2.42), which are, respectively, the expectation values for the one-electron and the two-electron Hamiltonians. In the MCOEP approach, the one-electron wave functions (or, equivalently, orbitals) φ_i are the eigenfunctions of the KS equation,

$$\left(-\frac{1}{2}\nabla^2 + \upsilon_{\text{eff}}(\boldsymbol{r}) \right) \varphi_i(\boldsymbol{r}) = \epsilon_i \varphi_i(\boldsymbol{r}) \quad (i = 1, 2, \cdots, N/2 + 1) \tag{2.68}$$

In parallel to Eq. (1.179) for the HF-OEP, we consider the functional derivative of the energy E_{MCSCF} with respect to the potential $\upsilon_{\text{eff}}(\boldsymbol{r})$. When the minimization with respect to the potential is achieved, the derivative becomes zero,

$$\frac{\delta E_{\text{MCSCF}}[\{\varphi_i\}]}{\delta \upsilon_{\text{eff}}(\boldsymbol{r})} = 0 \tag{2.69}$$

The energy E_0^{MCOEP} computed by MCOEP can be expressed by a minimization procedure,

$$\begin{aligned} E_0^{\text{MCOEP}} &= \min_{\upsilon_{\text{eff}}} \left\{ \min_{\{C_n\}} \left\langle \Psi^{\text{MC}} \right| \hat{T} + \hat{V}_{ee} + \hat{\upsilon}_{\text{ext}} \left| \Psi^{\text{MC}} \right\rangle \right\} \\ &= \min_{\upsilon_{\text{eff}}} \left\{ \min_{\{C_n\}} \left\langle \sum_n C_n \Psi_n^{\text{CSF}}[\upsilon_{\text{eff}}] \right| \hat{T} + \hat{V}_{ee} + \hat{\upsilon}_{\text{ext}} \left| \sum_n C_n \Psi_n^{\text{CSF}}[\upsilon_{\text{eff}}] \right\rangle \right\} \end{aligned} \tag{2.70}$$

where the expansion of Eq. (2.67) is used. Note that each configuration state function (CSF) Ψ_n^{CSF} in Eq. (2.70) is composed of the wave functions φ_i obtained as the eigenfunctions of Eq. (2.68). Equation (2.70) corresponds to Eq. (1.175) for the minimization in HF-OEP. Through the outer minimization of Eq. (2.70), Ψ_n^{CSF} itself is optimized, while in the inner minimization the CI expansion coefficients C_n for the CSFs are determined to minimize the energy expectation value.

To formulate the derivative of Eq. (2.69), we first decompose the energy E^{MCSCF} into the contributions,

$$E_{\mathrm{MCSCF}} = E_{\mathrm{kin}}^{\mathrm{MCSCF}} + E_{\mathrm{H}}^{\mathrm{MCSCF}} + E_{\mathrm{SC}}^{\mathrm{MCSCF}} + E_{\mathrm{ext}}^{\mathrm{MCSCF}} \tag{2.71}$$

The Hartree energy $E_{\mathrm{H}}^{\mathrm{MCSCF}}$ for the MCSCF wave function Ψ^{MC} is provided in Eq. (2.44) in Sect. 2.2, while $E_{\mathrm{SC}}^{\mathrm{MCSCF}}$ for the static correlation energy is given in Eq. (2.45). The energies $E_{\mathrm{kin}}^{\mathrm{MCSCF}}$ and $E_{\mathrm{ext}}^{\mathrm{MCSCF}}$ are, respectively, the kinetic energy, and the interaction between the external potential and the electron density $n(\boldsymbol{r})$ expressed by Eq. (2.34). Hereafter, the superscript 'MCSCF' on the energy components in the right hand side of Eq. (2.71) will be omitted for the sake of simplicity. Using the chain rule for the functional differentiation, Eq. (2.69) reads

$$\frac{\delta E_{\mathrm{MCSCF}}\left[\{\varphi_i\}\right]}{\delta v_{\mathrm{eff}}(\boldsymbol{r})} = \sum_n \frac{\partial E_{\mathrm{MCSCF}}}{\partial C_n}\frac{\delta C_n}{\delta v_{\mathrm{eff}}(\boldsymbol{r})} + \sum_i^{N/2+1} \int d\boldsymbol{r}' \frac{\delta E_{\mathrm{MCSCF}}}{\delta \varphi_i^*(\boldsymbol{r}')}\frac{\delta \varphi_i^*(\boldsymbol{r}')}{\delta v_{\mathrm{eff}}(\boldsymbol{r})} \tag{2.72}$$

The first term in the right hand side is zero since the relation $\frac{\partial E_{\mathrm{MCSCF}}}{\partial C_n} = 0$ is fulfilled when the inner minimization of Eq. (2.70) is achieved. Thus we consider the derivative $\frac{\delta E_{\mathrm{MCSCF}}}{\delta \varphi_i(\boldsymbol{r})}$ in the second term. By substituting the decomposition of Eq. (2.71), we have

$$\frac{\delta E_{\mathrm{MCSCF}}}{\delta \varphi_i^*(\boldsymbol{r})} = \left(\frac{\delta E_{\mathrm{kin}}}{\delta \varphi_i^*(\boldsymbol{r})} + \frac{\delta E_{\mathrm{H}}}{\delta \varphi_i^*(\boldsymbol{r})} + \frac{\delta E_{\mathrm{SC}}}{\delta \varphi_i^*(\boldsymbol{r})} + \frac{\delta E_{\mathrm{ext}}}{\delta \varphi_i^*(\boldsymbol{r})}\right) \tag{2.73}$$

Remind that the wave functions φ_i hold the relation of Eq. (2.68). The variation δE_{kin} in the kinetic energy for the variation $\delta\varphi_i$ in the wave function satisfies the relation,

$$\left\langle \delta\varphi_i \left| \hat{T} \right| \varphi_i \right\rangle + \left\langle \delta\varphi_i \left| \hat{v}_{\mathrm{eff}} \right| \varphi_i \right\rangle + c.c. = \epsilon_i \left\langle \delta\varphi_i \left| \varphi_i \right. \right\rangle + c.c. \tag{2.74}$$

By virtue of the orthonormalization condition, $\left\langle \delta\varphi_i \left| \varphi_i \right. \right\rangle = 0$ holds. This directly leads to

$$\frac{\delta E_{\mathrm{kin}}}{\delta \varphi_i^*(\boldsymbol{r})} = -v_{\mathrm{eff}}(\boldsymbol{r})\,\varphi_i(\boldsymbol{r}) \tag{2.75}$$

Then it is possible to extract the local potential $v_{\mathrm{SC}}(\boldsymbol{r})$ corresponding to the static correlation energy E_{SC} as

$$\begin{aligned}\frac{\delta E_{\mathrm{kin}}}{\delta \varphi_i^*(\boldsymbol{r})} + \frac{\delta E_{\mathrm{H}}}{\delta \varphi_i^*(\boldsymbol{r})} + \frac{\delta E_{\mathrm{ext}}}{\delta \varphi_i^*(\boldsymbol{r})} &= \left(-v_{\mathrm{eff}}(\boldsymbol{r}) + v_{\mathrm{H}}(\boldsymbol{r}) + v_{\mathrm{ext}}(\boldsymbol{r})\right)\varphi_i(\boldsymbol{r}) \\ &= -v_{\mathrm{SC}}(\boldsymbol{r})\,\varphi_i(\boldsymbol{r})\end{aligned} \tag{2.76}$$

Note that $\upsilon_{SC}(\boldsymbol{r})$ is the potential of target to be optimized in the MCOEP procedure. At the stationary state, since the relation Eq. (2.69) is fulfilled, one obtains the MCOEP equation to specify $\upsilon_{SC}(\boldsymbol{r})$, thus,

$$\int d\boldsymbol{r}' \chi_{MC}(\boldsymbol{r}, \boldsymbol{r}') \upsilon_{SC}(\boldsymbol{r}') = \sum_{i}^{N/2+1} \int d\boldsymbol{r}' \frac{\delta E_{SC}}{\delta \varphi_i^*(\boldsymbol{r}')} \frac{\delta \varphi_i^*(\boldsymbol{r}')}{\delta \upsilon_{\text{eff}}(\boldsymbol{r})} + c.c. \tag{2.77}$$

where $\chi_{MC}(\boldsymbol{r}, \boldsymbol{r}')$ is the response function of Ψ_{MC} and is defined by

$$\chi_{MC}(\boldsymbol{r}, \boldsymbol{r}') = \frac{\delta n(\boldsymbol{r})}{\delta \upsilon_{\text{eff}}(\boldsymbol{r}')} \tag{2.78}$$

The density $n(\boldsymbol{r})$ for the wave function Ψ_{MC} is given in Eq. (2.34) in Sect. 2.2. In deriving Eq. (2.77), the relation Eq. (1.184) in Sect. 1.4.1 is also used. Readers might be noticed that the way to derive the MCOEP equation represented in Eq. (2.77) is almost parallel to that for the HF-OEP equation in Eq. (1.182). The response function $\chi_{MC}(\boldsymbol{r}, \boldsymbol{r}')$ can be obtained through the perturbation theory, thus,

$$\begin{aligned} \chi_{MC}(\boldsymbol{r}, \boldsymbol{r}') &= \frac{\sum_i^{N/2+1} p_i \varphi_i(\boldsymbol{r}') \delta \varphi_i^*(\boldsymbol{r}')}{\delta \upsilon_{\text{eff}}(\boldsymbol{r})} + c.c. \\ &= -2 \sum_i^{N/2+1} p_i \sum_a^{vir} \frac{1}{\epsilon_a - \epsilon_i} \varphi_i(\boldsymbol{r}') \varphi_a^*(\boldsymbol{r}') \varphi_a(\boldsymbol{r}) \varphi_i^*(\boldsymbol{r}) + c.c. \end{aligned} \tag{2.79}$$

where p_i is the occupation number for the wave function $\varphi_i (i = 1, 2, \cdots, N/2 + 1)$. In the present case, for the inactive orbitals; $p_i = 2\ (1 \leq i \leq N/2 - 1)$, and for the active orbitals; $p_{N/2} = 2|C_0|^2$ and $p_{N/2+1} = 2|C_1|^2$.

In closing this section, we present results of the MCOEP calculation for the ground state H_2 molecule performed by Weimer et al. [15] In their approach, the active space for the MCOEP calculation composed of the bonding $1\sigma_g$ orbital (HOMO) and the anti-bonding $1\sigma_u$ orbital (LUMO). Thus the configuration of the active space corresponds exactly to the case $N = 2$ without inactive orbitals in the above discussion. As a notable feature of their calculation, the static correlation potential υ_{SC} was constructed within the framework of the localized Hartree-Fock (LHF) method [16] illustrated in Sect. 1.4.3. The augmented U2T basis set was used for the calculation. The MC wave function is explicitly expressed as

$$|\Psi_{MC}\rangle = c_g \left|1\sigma_g^2\right\rangle + c_u \left|1\sigma_u^2\right\rangle \tag{2.80}$$

At the dissociation limit of H_2 molecule, it should be noticed that there is no dynamical correlation energy $E_{\text{dyn}}\ (= E_c - E_{SC})$. The computed potential energy curves are presented in Fig. 2.6, where the LHF, MCSCF, and full CI calculations are also provided to make comparisons. The LHF or OEP calculation for H_2 molecule is equivalent to the HF calculation in principle. Thus the potential energy curve com-

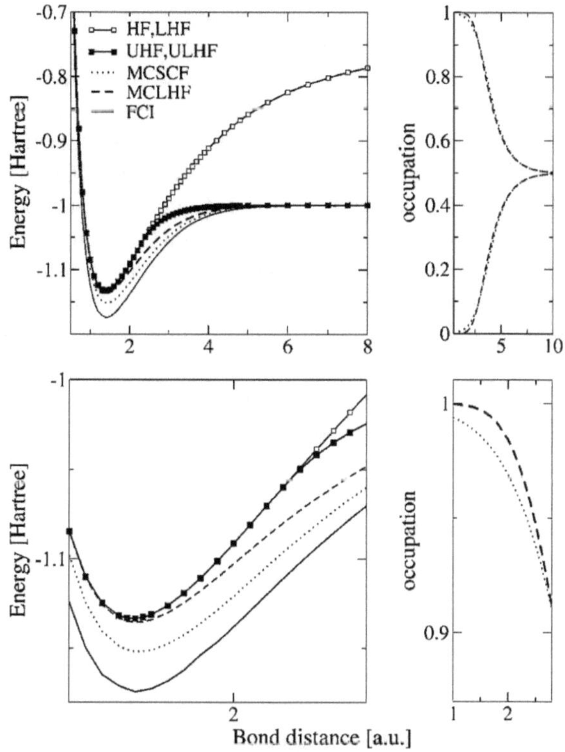

Fig. 2.6 Left: Potential energy curves of H_2 molecule in the electronic ground state. Right: Occupations of the bonding ($1\sigma_g$) and the anti-bonding ($1\sigma_u$) orbitals in the multi-configuration wave function as functions of the bond distance. Lower panels show the magnified potential energy curves(left) and the occupations(right). Reprinted with permission by AIP publishing for Fig. 2 in Ref. [15]

puted by the LHF method faithfully reproduces that provided by the HF calculation. These curves, however, overshoot seriously the energy of the correct dissociation limit due to the lack of the static correlation. On the other hand, spin unrestricted HF and LHF calculations (UHF and ULHF) [17] provide the correct dissociation limit although the spin symmetry of the wave functions are broken. A full CI (FCI) calculation is also performed using a cc-pV5Z basis set, which serves as a reference. The MCSCF calculation shows reasonable agreement with the FCI calculation over the whole region of the bond distance. It is seen, however, the energy at the bottom of the MCSCF curve is higher than that of FCI by ~ 0.022 Hartree. This is, of course, due to the lack of the dynamical correlation E_{dyn} in the MCSCF calculation. The curve by the MCLHF (=MCOEP) calculation gives slightly higher energy in the bonding region as compared to the MCSCF calculation. The discrepancy can be attributed to the fact that the one-electron wave functions as building blocks were optimized on a local multiplicative potential in the MCOEP calculation. Provided that the electron density of H_2 is MCOEP-υ representable, the MCOEP curve will coincide with that given by the MCSCF *in principle*. However, the MCOEP-υ rep. condition does not guarantees the coincidence in practice, because the representability is also related to the basis set used to describe the potential. Anyway it was demonstrated in their work that the MCOEP method provides a correct dissociation curve of H_2 showing a reasonable agreement with the MCSCF calculation. However, it should also be stressed that the success of MCOEP does *not* mean that it provides the justification for developing the static correlation functional within the framework of KS-DFT because it is possible that an MCOEP-υ rep. density does not belong to the subset of non-interacting υ-rep. densities as discussed above. The non-interacting υ-representability of the density yielded by MCOEP or MCSCF method might be directly examined by the inverse Kohn-Sham method.

Lastly, we make a brief comment on the occupation number presented in the right panels in Fig. 2.6. In their work, the occupation $\eta_{1\sigma_g}$ of the orbital $1\sigma_g$ is defined as c_g^2 in Eq. (2.80). And the same applies to the occupation $\eta_{1\sigma_u}$ of the orbital $1\sigma_u$, that is, $\eta_{1\sigma_u} = c_u^2$. In the bonding region, it is shown that the occupation $\eta_{1\sigma_g}$ of the bonding orbital is almost 1.0, while $\eta_{1\sigma_u}$ of the anti-bonding orbital is 0.0 both in the MCSCF and the MCOEP(=MCLHF) calculations. As the bond distance increases, $\eta_{1\sigma_g}$ decreases and $\eta_{1\sigma_u}$ increases monotonically. At the dissociation limit, these occupations both converge to 0.5 in consistent with the discussion provided in Sect. 2.1. It is also worth noting that the behavior of the occupation of MCSCF with the increase in the bond distance is almost reproduced in the MCOEP calculation.

References

1. R. Mcweeny, *Methods of Molecular Quantum Mechanics* (Academic Press, 2001)
2. T. Helgaker, P. Jorgensen, J. Olsen, *Molecular Electronic-Structure Theory* (Wiley, 2012)
3. E.J. Baerends, O.V. Gritsenko, J. Phys. Chem. A **101**(30), 5383 (1997)
4. W. Koch, M.C. Holthausen, *A Chemist's Guide to Density Functional Theory* (Wiley-VCH, 2001)
5. A.D. Becke, J. Chem. Phys. **119**(6), 2972 (2003)
6. A.J. Cohen, P. Mori-Sanchez, W. Yang, Science **321**, 792 (2008)
7. F. Moscardo, E. San-Fabian, Phys. Rev. A **44**, 1549 (1991)
8. B. Miehlich, H. Stoll, A. Savin, Mol. Phys. **91**(3), 527 (1997)
9. N.O. Malcolm, J.J. McDouall, Chem. Phys. Lett. **282**(9), 121 (1998)
10. J. Gräfenstein, D. Cremer, Chem. Phys. Lett. **316**(21), 569 (2000)
11. R. Takeda, S. Yamanaka, K. Yamaguchi, Chem. Phys. Lett. **366**, 321 (2002)
12. A.D. Becke, E.R. Johnson, J. Chem. Phys. **127**(12), 124108(8) (2007)
13. A.D. Becke, J. Chem. Phys. **138**(7), 074109(10) (2013)
14. A.D. Becke, M.R. Roussel, Phys. Rev. A **39**(8), 3761 (1989)
15. M. Weimer, F. Sala, A. Görling, J. Chem. Phys. **128**(14), 144109 (2008)
16. F.D. Sala, A. Görling, J. Chem. Phys. **115**(13), 5718 (2001)
17. F.D. Sala, A. Görling, J. Chem. Phys. **118**(23), 10439 (2003)

Chapter 3
Foundation of the Density-Functional Theory for the Energy Electron Density

3.1 Introduction of Energy Electron Density

We introduce the electron distribution $n^e(\epsilon)$ defined on the energy coordinate ϵ, which serves as a fundamental variable in the construction of the new functional [1] discussed as a major issue in this chapter. To this end, we provide the definition of the energy coordinate ϵ given individually for the system of interest. The system is explicitly specified by the external potential $\upsilon_{\text{ext}}(\boldsymbol{r})$ with $\boldsymbol{r}$ being the position vector of electrons. In most cases of our concern, $\upsilon_{\text{ext}}(\boldsymbol{r})$ is given by the Coulomb potential due to the nuclei in a molecule, and it serves to define the coordinate ϵ. In such a case, υ_{ext} can be explicitly expressed as

$$\upsilon_{\text{ext}}(\boldsymbol{r}) = -\sum_a \frac{Z_a}{|\boldsymbol{r} - \boldsymbol{R}_a|} \tag{3.1}$$

where $\boldsymbol{R}_a$ and Z_a are the position vector and the charge in the atomic unit (a.u.) of nucleus a, respectively. We note, however, that in the following discussion, $\upsilon_{\text{ext}}(\boldsymbol{r})$ in general need not necessarily be the Coulomb potential due to the nuclei. Using the potential υ_{ext} as a defining potential υ_{def} $(= \upsilon_{\text{ext}})$, the electron density $n_\sigma(\boldsymbol{r})$ for spin σ in the molecule is projected onto the energy coordinate ϵ, thus,

$$n_\sigma^e(\epsilon) = \int d\boldsymbol{r}\; n_\sigma(\boldsymbol{r})\delta\left(\epsilon - \upsilon_{\text{def}}(\boldsymbol{r})\right) \tag{3.2}$$

It should be stressed that the electron density $n_\sigma^e(\epsilon)$ on the energy coordinate ϵ is dependent on the choice of the defining potential υ_{def} since ϵ itself depends on the potential. For the help of intuitive understandings of Eq. (3.2) we present in Fig. 3.1 the isosurfaces of υ_{def} for a diatomic molecule. The integration of the electron density $n_\sigma(\boldsymbol{r})$ in Eq. (3.2) is carried out over the isosurface with the value ϵ of the defining potential. With the projection of Eq. (3.2) the electron distribution in the 3-dimensional space is, of course, reduced to 1-dimensional distribution, which

H. Takahashi, *Density-Functional Theory for Electrons*,
https://doi.org/10.1007/978-981-95-1229-4_3

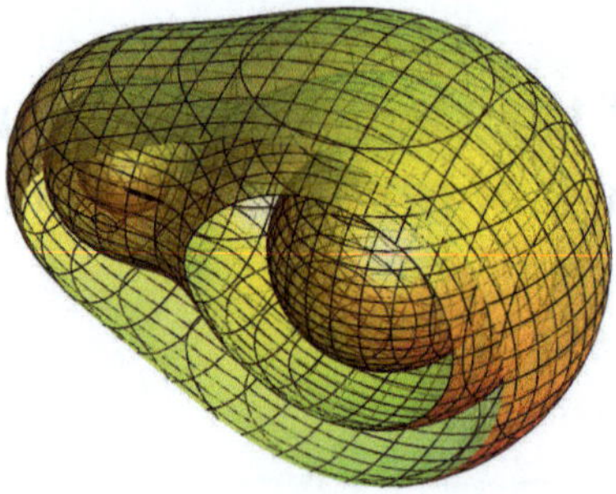

Fig. 3.1 Three-dimensional representation of the contour surfaces of a Coulomb potential due to nuclei as an example of $\upsilon_{\rm ext}(\boldsymbol{r})$. Isosurfaces with the values of −3.5 (outer), −4.5 (inner), and −6.5 (innermost) a.u. are shown for an H-Li molecule with the interatomic distance $R = 1.4$ a.u.

inevitably gives rise to the degradation of the information contents of the original electron density. However, it causes no disadvantage in the special case that the electron density is exactly constant over the isosurface of the potential $\upsilon_{\rm nuc}$ because the faithful reconstruction of $n(\boldsymbol{r})$ from $n^e(\epsilon)$ is possible, through

$$\eta_\sigma(\boldsymbol{r})|_{\epsilon=\upsilon_{\rm def}(\boldsymbol{r})} = n^e_\sigma(\epsilon)\Omega(\epsilon)^{-1} \tag{3.3}$$

where $\Omega(\epsilon)$ is the volume of the spatial region with the energy coordinate ϵ and defined by

$$\Omega(\epsilon) = \int d\boldsymbol{r}\ \delta\left(\epsilon - \upsilon_{\rm def}(\boldsymbol{r})\right). \tag{3.4}$$

We note such a situation is actually realized, for instance, in a hydrogenic system, because the electron density $n(\boldsymbol{r})$ is spherically symmetric and strictly constant on isosurfaces of the energy coordinate. Here, we consider the LDA exchange functional $E^{\rm LDA}_{x\sigma}[n_\sigma]$ for the conventional electron density n_σ. The explicit form of $E^{\rm LDA}_{x\sigma}[n_\sigma]$ is given as

$$E^{\rm LDA}_{x\sigma}[n_\sigma] = -2^{1/3}C^{\rm LDA}_x \int d\boldsymbol{r}\ n_\sigma(\boldsymbol{r})^{4/3} \tag{3.5}$$

where $C^{\rm LDA}_x$ is the coefficient of the LDA exchange functional and given by $C^{\rm LDA}_x = \frac{3}{4}\left(\frac{3}{\pi}\right)^{1/3}$ for each spin as described in Eq. (1.133) and in Appendix A. For the function $\widetilde{n}^e_\sigma$ defined as the right hand side of Eq. (3.3), namely,

$$\widetilde{n}^e_\sigma(\epsilon) = n^e_\sigma(\epsilon)\Omega(\epsilon)^{-1} \tag{3.6}$$

Note that the dimension of $\widetilde{n}^e_\sigma$ is the same as the conventional electron density $n_\sigma(\boldsymbol{r})$. Actually, $\widetilde{n}^e_\sigma(\epsilon)$ can be considered as the average electron density of the region with the energy coordinate ϵ. Then it is straightforward to adopt the conventional LDA functional, thus,

$$E_{x\sigma}^{e,\text{LDA}}[\widetilde{n}_\sigma^e] = -2^{1/3}C_x^{\text{LDA}} \int d\epsilon \, \widetilde{n}_\sigma^e(\epsilon)^{4/3}\Omega(\epsilon) \tag{3.7}$$

In this equation, dependence of the functional $E_{x\sigma}^{e,\text{LDA}}[\widetilde{n}_\sigma^e]$ on $\Omega(\epsilon)$ is not shown explicitly. Actually, the volume $\Omega(\epsilon)$ is determined solely when the defining potential is introduced. Thus the dependence of the functional on $\Omega(\epsilon)$ is implicitly assumed. This also applies to the functionals which appear below. It is possible to show that the functional of Eq. (3.7) is exactly equivalent to the conventional one. Actually, it can be transformed as follows:

$$\begin{aligned}
E_{x\sigma}^{e,\text{LDA}}[\widetilde{n}_\sigma^e] &= -2^{1/3}C_x^{\text{LDA}} \int d\epsilon \, \widetilde{n}_\sigma^e(\epsilon)^{4/3}\Omega(\epsilon) \\
&= -2^{1/3}C_x^{\text{LDA}} \int d\epsilon \, n_\sigma(\boldsymbol{r})^{\frac{4}{3}}\Big|_{\epsilon=\upsilon(\boldsymbol{r})} \int d\boldsymbol{r} \, \delta(\epsilon - \upsilon_{\text{def}}(\boldsymbol{r})) \\
&= -2^{1/3}C_x^{\text{LDA}} \int d\boldsymbol{r} \, n_\sigma(\boldsymbol{r})^{\frac{4}{3}} \\
&= E_{x\sigma}^{\text{LDA}}[n_\sigma].
\end{aligned} \tag{3.8}$$

It should be reminded, however, that Eq. (3.8) holds only when the electron density $n_\sigma(\boldsymbol{r})$ is constant over the isosurface of the potential energy υ_{def}. We also make a remark on a notational issue to avoid confusions. There will be two ways to express the functional e.g. for the exchange energy E_x^e based on the energy electron density $n^e(\epsilon)$, namely, $\overline{E}_x^e[n^e]$ and $E_x^e[\widetilde{n}^e]$. Of course, these two functionals are different because the the arguments are different. However, they provide the same value when $n^e(\epsilon)$ and $\widetilde{n}^e(\epsilon)$ are built from the same electron density $n(\boldsymbol{r})$. This can be readily understood thus

$$\begin{aligned}
\overline{E}_x^{e,\text{LDA}}\left[n^e\right] &= -C_x^{\text{LDA}} \int d\epsilon \left(\frac{n^e(\epsilon)}{\Omega(\epsilon)}\right)^{\frac{4}{3}} \Omega(\epsilon) \\
&= -C_x^{\text{LDA}} \int d\epsilon \, \widetilde{n}^e(\epsilon)^{\frac{4}{3}} \, \Omega(\epsilon) \\
&= E_x^{e,\text{LDA}}\left[\widetilde{n}^e\right]
\end{aligned} \tag{3.9}$$

The confusion can be avoided by specifying the argument ($n^e(\epsilon)$ or $\widetilde{n}^e(\epsilon)$) explicitly for the functional. However, in the formulations below, we will mostly use the simple notation $E_{xc}^e[n^e]$ to show that E_{xc}^e is simply a functional of $n^e(\epsilon)$. The specific notation $\overline{E}_x^e[n^e]$ or $E_x^e[\widetilde{n}^e]$ will be used only when it is considered to be useful for the sake of clarity.

The introduction of the energy electron density $n^e(\epsilon)$ of Eq. (3.2) has its origin in the physical insight given by Parr and Berk [2], who advocated a hypothesis that the contours of the electron densities in molecules are more or less parallel to those of the bare-nuclear potentials υ_{nuc}. Their hypothesis was examined by computing the potential energy curve of the hydrogen molecule ion (H_2^+), where the condition

$\varphi(\boldsymbol{r}) = \varphi(\upsilon_{\text{nuc}}(\boldsymbol{r}))$ (note that φ is a *function* of $\upsilon_{\text{nuc}}(\boldsymbol{r})$ and not the functional of υ_{nuc}) is imposed in the construction of the electronic wave function φ of the ground state. The result was found to be rather favorable producing almost the same binding energy obtained by the LCAO approach. The development of the DFT on the basis of the energy electron density was also motivated by the novel density-functional theory of solutions, referred to as the theory of energy representation (ER) [3]. In the ER theory the distribution function $\rho(\varepsilon)$ of the solute-solvent interaction energy ε plays a fundamental role in describing the solvation free energy of a solute of interest. Actually, it can be readily proved that there exists one-to-one correspondence between the solute-solvent interaction potentials and the resultant energy distribution functions [3], which constitutes the theoretical basis for the development of the approximate free energy functional in terms of the distribution functions [4]. The outline of the DFT for solution systems is described in Appendix G. Importantly, one-to-one correspondence is also established between the *subset* of the energy electron density $\{n^e(\epsilon)\}$ and the *subset* of the external potential $\{\upsilon_{\text{def}}(\boldsymbol{r})\}$ defined on the energy coordinate, as will be shown later. This means that the Hohenberg-Kohn theorem also applies to the subset of the external potentials and the energy distribution functions [1]. Thus, it is possible to formulate the density-functional theory in terms of the function $n^e(\epsilon)$.

3.2 Hohenberg-Kohn Theorem for the Energy Electron Density

3.2.1 Partitioning of Set of External Potentials

Now we discuss the Hohenberg-Kohn (HK) theorem for the DFT in terms of the energy electron density $n^e(\epsilon)$ defined in Eq. (3.2) [1]. To this end we partition the whole set of the external potentials $\{\upsilon(\boldsymbol{r})\}$ into the subsets $\{\upsilon(\boldsymbol{r})\}_I (I = 1, 2, \cdots)$. Within each subset $\{\upsilon(\boldsymbol{r})\}_I$ it is assumed that the isosurfaces of any two different potentials $\upsilon_{\text{A}}(\boldsymbol{r})$ and $\upsilon_{\text{B}}(\boldsymbol{r})$ are mutually parallel and do not intersect with each other. Precisely, it can be stated as follows. Provided that the following condition is fulfilled, υ_{A} and υ_{B} shall belong to the same subset $(\upsilon_{\text{A}}, \upsilon_{\text{B}} \in \{\upsilon(\boldsymbol{r})\}_I)$. That is, for *all* $\boldsymbol{r}$ and $\boldsymbol{r}'$ $(\boldsymbol{r} \neq \boldsymbol{r}')$ satisfying

$$\upsilon_{\text{A}}(\boldsymbol{r}) = \upsilon_{\text{A}}\left(\boldsymbol{r}'\right) \tag{3.10}$$

υ_{B} also satisfies the same relation

$$\upsilon_{\text{B}}(\boldsymbol{r}) = \upsilon_{\text{B}}\left(\boldsymbol{r}'\right) \tag{3.11}$$

In the above notation, it is assumed that $\boldsymbol{r}$ and $\boldsymbol{r}'$ are on an isosurface which constitutes a connected or disconnected region in space. Thus, when $\upsilon_{\text{A}}, \upsilon_{\text{B}} \in \{\upsilon(\boldsymbol{r})\}_I$, any isosurface of $\upsilon_{\text{A}}(\boldsymbol{r})$ does not intersect with that of $\upsilon_{\text{B}}(\boldsymbol{r})$. For readers' understandings,

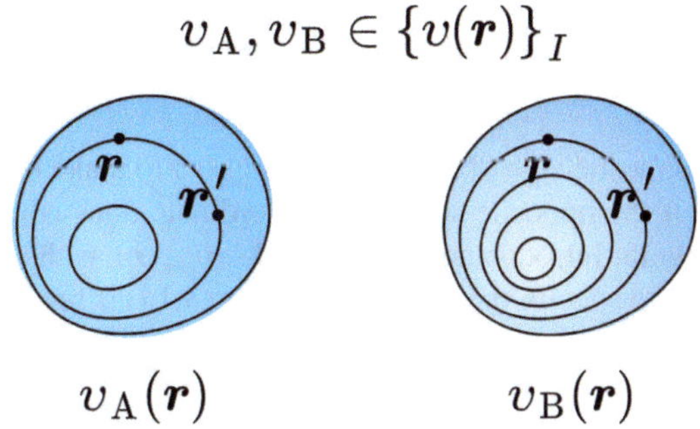

Fig. 3.2 Illustration of the isosurfaces of the potentials v_A and v_B belonging to the same subset $\{v(\boldsymbol{r})\}_I$. Any points $\boldsymbol{r}$ and $\boldsymbol{r}'$ placed on the same isosurface $(v_A(\boldsymbol{r})) = v_A(\boldsymbol{r}'))$ of the potential v_A are also always on an isosurface $(v_B(\boldsymbol{r}) = v_B(\boldsymbol{r}'))$ of the potential v_B. However, $v_A(\boldsymbol{r})$ is not necessarily equal to $v_B(\boldsymbol{r})$

the illustrations of the potentials v_A and v_B belonging to the same subset $\{v(\boldsymbol{r})\}_I$ are presented in Fig. 3.2. Of course, we do not take into consideration the potential differences due to the translational or rotational displacement of the potential.

With this definition of the subset, the entire set of the potential $\{v(\boldsymbol{r})\}$ can be expressed by an infinite sum of the subsets, thus,

$$\{v(\boldsymbol{r})\} = \{v(\boldsymbol{r})\}_1 \cup \{v(\boldsymbol{r})\}_2 \cup \cdots \cup \{v(\boldsymbol{r})\}_I \cup \cdots . \tag{3.12}$$

From the definition, it is obvious that the overlap of any pair of the different subsets($I \neq J$) is the empty set,

$$\{v(\boldsymbol{r})\}_I \cap \{v(\boldsymbol{r})\}_J = \varnothing. \tag{3.13}$$

So far, we have not considered the bounds of the local potential $v(\boldsymbol{r})$ in a mathematically rigorous sense. Actually, in the present book, we are not concerned with the details of the mathematical properties of the potentials. However, the present theory can be mostly applied to the potentials with well behaviors. Each subset of the potentials consists of infinite number of potentials and any potential can be uniquely categorized into one of the subsets. It is quite important to note that any potential $v^I(\boldsymbol{r})$ within the subset I can be equivalently described using a single scalar quantity ϵ instead of the vector $\boldsymbol{r}$. Here ϵ is the energy coordinate as introduced in Eq. (3.2). We consider in more details the coordinate ϵ. Suppose that we are interested in electronic properties of a molecule of which nuclear potential $v_{\text{nuc}}(\boldsymbol{r})$ is described by Eq. (3.1). The potential $v_{\text{nuc}}(\boldsymbol{r})$ can also be categorized into a certain subset $\{v(\boldsymbol{r})\}_I$. Thus the potential is denoted as $v^I_{\text{nuc}}(\boldsymbol{r})$. This potential can be used to define the energy coordinate ϵ for the potentials in subset I. For a given spatial coordinate $\boldsymbol{r}$, the corresponding energy coordinate ϵ is given by

$$\epsilon = -v_{\text{nuc}}^{I}(\boldsymbol{r}) = \sum_{a} \frac{Z_a}{|\boldsymbol{r} - \boldsymbol{R}_a|} \tag{3.14}$$

Here we take the opposite sign of the potential $v_{\text{nuc}}^{I}(\boldsymbol{r})$ to make ϵ to be positive. We note that the use of the opposite sign of the coordinate does not affect the formulation in the following. Introducing the potential $-v_{\text{nuc}}^{I}(\boldsymbol{r})$ as the defining potential $v_{\text{def}}^{I}(\boldsymbol{r}) = -v_{\text{nuc}}^{I}(\boldsymbol{r})$ to specify the coordinate ϵ, Eq. (3.14) is rewritten as

$$\epsilon = v_{\text{def}}^{I}(\boldsymbol{r}) \tag{3.15}$$

It is then possible to specify a spatial region $\mathcal{R}_{\epsilon}^{I}$ on which the energy coordinate ϵ is constant. In a mathematical notion, the region $\mathcal{R}_{\epsilon}^{I}$ can be expressed as,

$$\mathcal{R}_{\epsilon}^{I} := \left\{ \boldsymbol{r} \in \mathbb{R}^3 : v_{\text{def}}^{I}(\boldsymbol{r}) = \epsilon \right\} \tag{3.16}$$

where $0 \leq \epsilon \leq \infty$ is fulfilled by definition. A potential $v^{e,I}(\epsilon)$ on the coordinate ϵ can be uniquely defined for any potential $v^{I}(\boldsymbol{r})$ in the subset I. Explicitly, $v^{e,I}(\epsilon)$ is defined as

$$v^{e,I}(\epsilon) := v^{I}(\boldsymbol{r})\Big|_{\boldsymbol{r} \in \mathcal{R}_{\epsilon}^{I}} \tag{3.17}$$

In Eq. (3.17), it is important to remind that $v^{I}(\boldsymbol{r})$ is constant on $\mathcal{R}_{\epsilon}^{I}$ for all ϵ since v^{I} and v_{def}^{I} belong to the same subset I. Specifically, the potential $v_{\text{def}}^{e,I}(\epsilon)$ corresponding to $v_{\text{def}}^{I}(\boldsymbol{r})$ satisfies the relation

$$v_{\text{def}}^{e,I}(\epsilon) = \epsilon \tag{3.18}$$

by definition. Importantly, any potential $v^{I}(\boldsymbol{r})$ can be faithfully reconstructed from the corresponding potential $v^{e,I}(\epsilon)$ by means of the defining potential v_{def}^{I}. Explicitly, $v^{I}(\boldsymbol{r})$ is obtained by

$$v^{I}(\boldsymbol{r}) = v^{e,I}(\epsilon)\Big|_{\epsilon = v_{\text{def}}^{I}(\boldsymbol{r})} \tag{3.19}$$

Thus the potential $v^{I}(\boldsymbol{r})$ on the coordinate space $\{\boldsymbol{r} \in \mathbb{R}^3\}$ can be reduced to $v^{e,I}(\epsilon)$ on the energy coordinate space $\{\epsilon \in \mathbb{R}\}$ with the help of the single potential $v_{\text{def}}^{I}(\boldsymbol{r})$ without loss of information.

It is important to discuss whether any potential v^{I} in subset I can be used as a defining potential. Actually, the answer to the issue is 'No'. This can be easily examined by the consideration of the subset I_{Ω} that consists of spherically symmetric potentials. Obviously, the single subset I_{Ω} covers the whole of the spherically symmetric potentials. Any potential $v^{I_{\Omega}}$ in the subset I_{Ω} can be represented as a function of the radial distance r from the center of the potential. With this fact in mind, we introduce a defining potential $v_{\text{def}}^{I_{\Omega}}$, thus,

$$v_{\text{def}}^{I_{\Omega}}(r) = \frac{Z}{r} \tag{3.20}$$

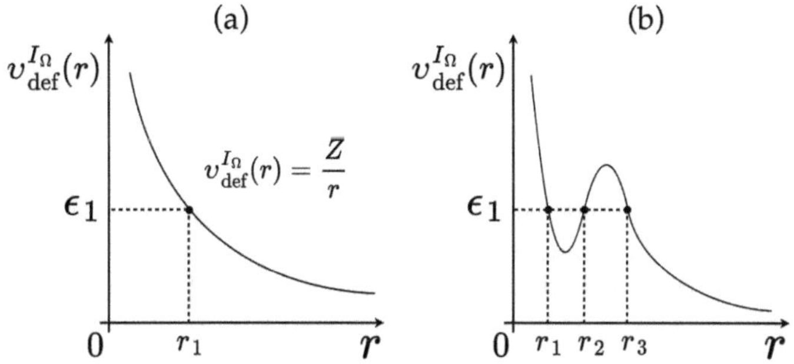

Fig. 3.3 Radial behaviors of the defining potential functions $v_{\text{def}}^{I_\Omega}(r)$ in the subset I_Ω of the spherically symmetric potentials. Figure 3.3a illustrates a monotonically decreasing potential ($v_{\text{def}}^{I_\Omega}(r) = Z/r$), while Fig. 3.3b shows a function with a non-monotonic behavior. The correspondence between r and ϵ is one-to-one in the potential of Fig. 3.3a. As described in the main test, $v_{\text{def}}^{I_\Omega}(r)$ depicted in Fig. 3.3a is appropriate for the defining potential of subset I_Ω

where Z is a positive real number. This is merely a Coulomb potential due to a nuclear charge. Since the correspondence from r to $\epsilon = v_{\text{def}}^{I_\Omega}(r)$ is obviously one-to-one(see Fig. 3.3a), the potential $v^e(\epsilon)$ given as a function of the coordinate ϵ can represent all the potentials in the subset I_Ω. Actually, potential $v^{I_\Omega}(r)$ can be readily transformed to the corresponding function $v^{e,I_\Omega}(\epsilon)$ on the energy coordinate, thus,

$$v^{I_\Omega}(r) = v^{I_\Omega}\left(\frac{Z}{\epsilon}\right) = v^{e,I_\Omega}(\epsilon) \tag{3.21}$$

However, as shown in Fig. 3.3b, some potential cannot be used as a defining potential because there exists three different radial distances, r_1, r_2, and r_3 that correspond to the same energy coordinate ϵ_1. As a consequence, $v^I(r_1)$, $v^I(r_2)$, and $v^I(r_3)$ have inevitably the same value $v^{e,I}(\epsilon_1)$ as shown in Eq. (3.19). Such a defining potential cannot describes the monotonic function in the subset I_Ω. On the other hand, the defining potential e.g. Equation (3.20), which shows a monotonic behavior, can describe non-monotonic functions as well as monotonic ones. Thus, the defining potential must be monotonic with respect to increasing of r. However, when we are interested in molecules, this criterion cannot be applied because the radial distance r cannot be well defined for molecules in general. The Coulomb potential v_{nuc} itself formed by the nuclei in the molecule of interest would be an easy and appropriate choice as the defining potential for the potential subset $\{v(\boldsymbol{r})\}_I$ that includes the potential v_{nuc}.

In a mathematical framework, the decomposition of the whole set of the external potentials as shown in Eq. (3.12) can be regarded as a construction of a 'fiber bundle', where the set of the defining potentials v_{def}^I constitutes the base space of the bundle and the elements belonging to the subset I form the fiber $\mathcal{F}^I$ for each defining potential. The fiber $\mathcal{F}^I$ constitutes a vector space because the potential $v_C(\boldsymbol{r})$ given by the linear combination of the elements $v_A^I(\boldsymbol{r})$ and $v_B^I(\boldsymbol{r})$ in the fiber $\mathcal{F}^I$ is also included in the same fiber. To prove this, we assume that a potential $\upsilon_C(\boldsymbol{r})$ is given by the linear combination,

$$\upsilon_C(\boldsymbol{r}) = a\upsilon_A^I(\boldsymbol{r}) + b\upsilon_B^I(\boldsymbol{r}) \tag{3.22}$$

using the real numbers a and b, from the potentials $\upsilon_A^I(\boldsymbol{r}), \upsilon_B^I(\boldsymbol{r}) \in \mathcal{F}^I$. Since $\upsilon_A^I(\boldsymbol{r})$ and $\upsilon_B^I(\boldsymbol{r})$ belong to the same subset I, or equivalently the same fiber $\mathcal{F}^I$, there exist different $\boldsymbol{r}$ and $\boldsymbol{r}'$ that satisfy the relations in Eqs. (3.10) and (3.11) by definition. For these $\boldsymbol{r}$ and $\boldsymbol{r}'$, it can be readily confirmed that $\upsilon_C(\boldsymbol{r})$ always satisfies

$$\begin{aligned}\upsilon_C(\boldsymbol{r}) &= a\upsilon_A^I(\boldsymbol{r}) + b\upsilon_B^I(\boldsymbol{r}) \\ &= a\upsilon_A^I(\boldsymbol{r}') + b\upsilon_B^I(\boldsymbol{r}') \\ &= \upsilon_C(\boldsymbol{r}')\end{aligned} \tag{3.23}$$

Thus, it is proved that $\upsilon_C(\boldsymbol{r})$ belongs to the fiber $\mathcal{F}^I$, that is,

$$\upsilon_C(\boldsymbol{r}) \in \mathcal{F}^I \tag{3.24}$$

Because of the one-to-one mapping between the set of the potentials $\upsilon(\boldsymbol{r})$ and the set of υ-rep. densities $n(\boldsymbol{r})$, it is also possible to construct the fiber bundle for the υ-rep. densities where the density $n_{\mathrm{def}}^I(\boldsymbol{r})$ corresponding to $\upsilon_{\mathrm{def}}^I(\boldsymbol{r})$ serves as an element of the base space of the fiber bundle. This subject will be revisited at the end of this Section.

The projection of the potential defined on the spatial coordinate to the energy coordinate is illustrated in Fig. 3.4 for the help of readers' understandings. In completely parallel to the partitioning in Eq. (3.12), the potential defined on the energy coordinate can also be partitioned, thus,

$$\{\upsilon^e(\epsilon)\} = \{\upsilon^e(\epsilon)\}_1 \cup \{\upsilon^e(\epsilon)\}_2 \cup \cdots \cup \{\upsilon^e(\epsilon)\}_I \cup \cdots. \tag{3.25}$$

Note, however, that the energy coordinate ϵ in each subset is individually defined using a certain potential $\upsilon_{\mathrm{def}}^I$ within the subset.

Next, let us consider the partitioning of the electron density $n(\boldsymbol{r})$ corresponding to Eq. (3.12). By virtue of the HK theorem, the set of the υ-rep. electron densities of N-electron systems can be uniquely partitioned as

$$\{n(\boldsymbol{r})\}_{\upsilon-\mathrm{rep}} = \{n(\boldsymbol{r})\}_1 \cup \{n(\boldsymbol{r})\}_2 \cup \cdots \cup \{n(\boldsymbol{r})\}_I \cup \cdots. \tag{3.26}$$

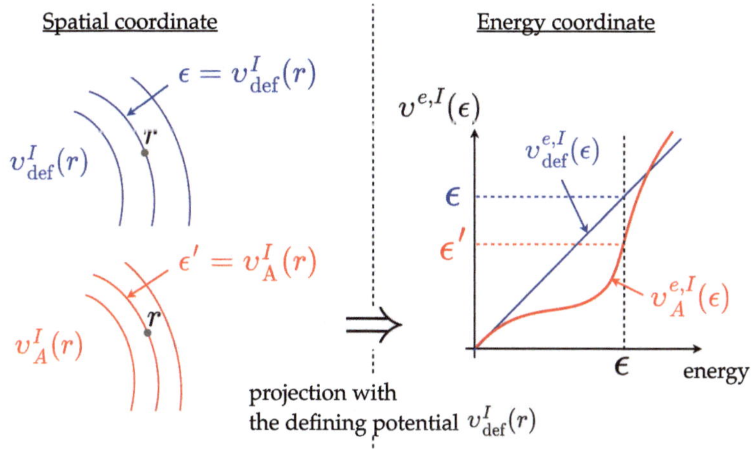

Fig. 3.4 The external potential functions in a subset $\{v(\boldsymbol{r})\}_I$ defined on the spatial region(left) and on the energy coordinate (right). The projection from left to right can be performed employing the defining potential $v^I_{\text{def}}(\boldsymbol{r})$ which can be chosen arbitrarily from the subset I

Then, we obtain the partitioning of the energy electron density $\{n^e(\epsilon)\}$ by projecting the densities in the subsets onto the energy coordinates with the defining potentials v^I_{def},

$$\{n^e(\epsilon)\}_{v-\text{rep}} = \{n^e(\epsilon)\}_1 \cup \{n^e(\epsilon)\}_2 \cup \cdots \cup \{n^e(\epsilon)\}_I \cup \cdots . \tag{3.27}$$

The discussions noted above are summarized in Fig. 3.5. For the v-rep. electron density on the spatial coordinate, the one-to-one correspondence is established within the individual subsets of $\{v(\boldsymbol{r})\}_I$ and $\{n(\boldsymbol{r})\}_I$. In the next Subsection, we examine the one-to-one correspondence between the set of the energy electron densities and that of the potentials defined on the energy coordinate.

3.2.2 Hohenberg-Kohn Theorem for the Novel DFT

Now we have an interesting question, namely whether the one-to-one mapping also applies to the energy electron densities n^e within a subset. In other words, is the one-to-one correspondence established between a subset of the electron densities $\{n(\boldsymbol{r})\}_I$ and the energy electron density $\{n^e(\epsilon)\}_I$? Actually, it is possible to prove the one-to-one correspondence as in the same way provided for the proof of the HK theorem. To start with, we make an assumption that two different external potentials $v(\boldsymbol{r})$ and $v'(\boldsymbol{r})$ in a subset I yield the same energy electron density $n^e(\epsilon)$ within the subset I. Of course, it is premised that a certain defining potential $v^I_{\text{def}}(\boldsymbol{r})$ is chosen beforehand to specify the subset of potential functions. It should also be noted that the potentials on the spatial coordinate can be transformed to those defined on the

Spatial coordinate

$$\{v(\boldsymbol{r})\} = \{v(\boldsymbol{r})\}_1 \cup \{v(\boldsymbol{r})\}_2 \cup \cdots \cup \{v(\boldsymbol{r})\}_I \cup \cdots$$

one-to-one $\updownarrow$

$$\{n(\boldsymbol{r})\}_{v\text{-rep}} = \{n(\boldsymbol{r})\}_1 \cup \{n(\boldsymbol{r})\}_2 \cup \cdots \cup \{n(\boldsymbol{r})\}_I \cup \cdots$$

$\Downarrow$ projection with the defining potential $v^I_{\text{def}}(\boldsymbol{r})$

Energy coordinate

$$\{v^e(\epsilon)\} = \{v^e(\epsilon)\}_1 \cup \{v^e(\epsilon)\}_2 \cup \cdots \cup \{v^e(\epsilon)\}_I \cup \cdots$$

one-to-one ? $\updownarrow$

$$\{n^e(\epsilon)\}_{v\text{-rep}} = \{n^e(\epsilon)\}_1 \cup \{n^e(\epsilon)\}_2 \cup \cdots \cup \{n^e(\epsilon)\}_I \cup \cdots$$

Fig. 3.5 Partitioning of a set of the external potentials v and that of the v-representable electron densities n on the spatial coordinate(upper) and that on the energy coordinate(lower). The subsets are characterized by the individual defining potentials $v^I_{\text{def}}(\boldsymbol{r})$. The one-to-one mapping between the potentials and the densities is established within each subset represented with the spatial coordinate. The question is whether the one-to-one correspondence also applies to the energy electron densities within a subset

energy coordinate. We consider two different Hamiltonians $\hat{H}$ and $\hat{H}'$ for N-electrons systems with the different external potentials, thus,

$$\hat{H} = \hat{T} + \hat{V}_{ee} + \hat{v}^e \tag{3.28}$$

$$\hat{H}' = \hat{T} + \hat{V}_{ee} + \hat{v}^{e\prime} \tag{3.29}$$

Then, we have two different ground state wave functions Ψ_0 and Ψ_0' and the corresponding energies E_0 and E_0' for these Hamiltonians in parallel to Eqs. (1.69) and (1.71). Applying the variational principle to the energy E_0 we have an inequality

$$\begin{aligned} E_0 < \left\langle \Psi_0' | \hat{H} | \Psi_0' \right\rangle &= \left\langle \Psi_0' \middle| \hat{H}' \middle| \Psi_0' \right\rangle + \left\langle \Psi_0' \middle| \hat{H} - \hat{H}' \middle| \Psi_0' \right\rangle \\ &= E_0' + \int d\epsilon \left(v^e(\epsilon) - v^{e\prime}(\epsilon) \right) n^e(\epsilon) \end{aligned} \tag{3.30}$$

The construction of Eq. (3.30) is completely parallel to Eq. (1.72). However, the integration in the last equality is being performed on the energy coordinate ϵ, which differs from Eq. (1.72). Similarly for E_0' we obtain

$$\begin{aligned} E_0' < \left\langle \Psi_0 | \hat{H}' | \Psi_0 \right\rangle &= \left\langle \Psi_0 \middle| \hat{H} \middle| \Psi_0 \right\rangle + \left\langle \Psi_0 \middle| \hat{H}' - \hat{H} \middle| \Psi_0 \right\rangle \\ &= E_0 + \int d\epsilon \left(v^{e\prime}(\epsilon) - v^e(\epsilon) \right) n^e(\epsilon). \end{aligned} \tag{3.31}$$

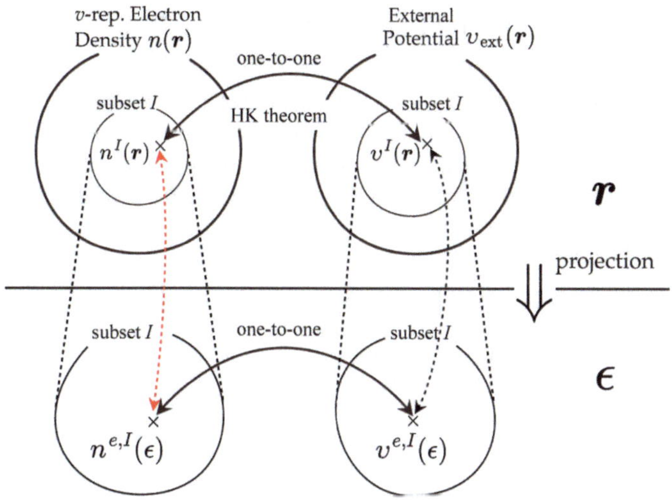

Fig. 3.6 Upper figure shows the one-to-one mapping between the potential subset $\{v(\boldsymbol{r})\}_I$ and the 'v-rep.' electron density subset $\{n(\boldsymbol{r})\}_I$. Lower figure shows the counterpart for the subset $\{v^e(\epsilon)\}_I$ and $\{n^e(\epsilon)\}_I$. The correspondence between $\{n^e(\epsilon)\}_I$ and $\{n(\boldsymbol{r})\}_I$ is also one-to-one (indicated with a red broken line). The curves with arrows on both ends indicate the one-to-one correspondences

Sum of Eqs. (3.30) and (3.31) leads to an contradiction that $E_0 + E_0' < E_0 + E_0'$, which guarantees the one-to-one correspondence between $\{v^e(\epsilon)\}_I$ and $\{n^e(\epsilon)\}_I$. Of course, the proof noted above corresponds to the Hohenberg-Kohn theorem in the conventional DFT. We also note that the potential $v^I(\boldsymbol{r})$ in the subset I uniquely corresponds to $v^{e,I}(\epsilon)$. This result leads to the conclusion that there also exists a one-to-one mapping between a subset of the 'v-rep.' electron densities $\{n(\boldsymbol{r})\}_I$ and the subset of the 'v-rep.' energy electron densities $\{n^e(\epsilon)\}_I$. These results are summarized in Fig. 3.6.

Now it is possible to state that the Hohenberg-Kohn (HK) theorem also applies to the individual subsets of the external potentials specified with the defining potentials. The proof provides us with a theoretical justification to construct the energy functional $E^{e,I}_{\mathrm{HK}}[n^{e,I}]$ using the energy electron density $n^{e,I}(\epsilon)$ as a fundamental variable. The superscript I is attached to E_{HK} since the functional is *not* universal and specific to the subset I associated with a defining potential. In parallel to Eq. (1.75), it is possible to extract the potential energy term from $E^{e,I}_{\mathrm{HK}}$

$$E^{e,I}_{\mathrm{HK}}[n^{e,I}] = F^{e,I}_{\mathrm{HK}}[n^{e,I}] + \int d\epsilon\, n^{e,I}(\epsilon) v^{e,I}_{\mathrm{ext}}(\epsilon) \tag{3.32}$$

In Eq. (3.32), it is supposed that the external potential of interest belongs to the subset I. The HK functional $F^{e,I}_{\mathrm{HK}}[n^{e,I}]$ provides the sum of the kinetic and the electron repulsion energies, and it is defined as

$$F_{\rm HK}^{e,I}[n^{e,I}] = \left\langle \Psi_0[n^{e,I}] \middle| \hat{T} + \hat{V}_{ee} \middle| \Psi_0[n^{e,I}] \right\rangle. \tag{3.33}$$

It should be noticed in Eq. (3.33) that the ground state wave function $\Psi_0[n^{e,I}]$ can be uniquely determined from the energy electron density $n^{e,I}$, since $n^{e,I}$ is assumed to be υ-representable. From Eq. (3.33), it is straightforward to define the functionals $T^{e,I}\left[n^{e,I}\right]$ and $V_{ee}^{e,I}\left[n^{e,I}\right]$, respectively, for the kinetic and electron repulsion energies,

$$T^{e,I}\left[n^{e,I}\right] = \left\langle \Psi_0\left[n^{e,I}\right] \middle| \hat{T} \middle| \Psi_0\left[n^{e,I}\right] \right\rangle \tag{3.34}$$

$$V_{ee}^{e,I}\left[n^{e,I}\right] = \left\langle \Psi_0\left[n^{e,I}\right] \middle| \hat{V}_{ee} \middle| \Psi_0\left[n^{e,I}\right] \right\rangle \tag{3.35}$$

Fiber Bundle

In closing this section, we consider again the mathematical structures of the whole set of the potentials $\upsilon(\boldsymbol{r})$ and that of the υ-rep. electron densities $n(\boldsymbol{r})$. As discussed in Sect. 3.2.1, it is possible to introduce a mathematical structure, i.e. a fiber bundle $\mathcal{V}$, to the set of the potentials $\upsilon(\boldsymbol{r})$ through its partitioning $\{\upsilon(\boldsymbol{r})\}_I$. Let us take a closer look at this fiber bundle. The fiber bundle $\mathcal{V}$ is given by the product of the fiber $\mathcal{F}$ and the base space $\mathcal{B}$, thus, $\mathcal{V} = \mathcal{F} \times \mathcal{B}$. The total space $\mathcal{V}$ constitutes the whole set of the potentials $\upsilon(\boldsymbol{r})$. The base space $\mathcal{B}_\upsilon$ consists of the defining potentials $\upsilon_{\rm def}^I$ and the fiber $\mathcal{F}_\upsilon^I$ is composed of the potentials in the set $\{\upsilon(\boldsymbol{r})\}_I$. As was proved in Sect. 3.2.1, $\mathcal{F}_\upsilon^I$ is a vector space. By virtue of the one-to-one mapping between the set of the potentials and the set of the υ-rep. electron densities, exactly the same mathematical structure can be introduced to the set $\{n(\boldsymbol{r})\}$ as into the potential set. In Fig. 3.7, a schematic illustration is presented to show the concept of the fiber bundles representing the sets of the potentials and of the densities. Basically, the subset $\{\upsilon(\boldsymbol{r})\}_I$ is equivalent to the fiber $\mathcal{F}_\upsilon^I$ where each fiber is supposed to include the zero vector. However, when it is supposed that each subset $\{\upsilon(\boldsymbol{r})\}_I$ includes the potential $\upsilon(\boldsymbol{r}) = 0$, it leads to a contradiction with Eq. (3.13). Thus, the fiber $\mathcal{F}_\upsilon^I$ is slightly different from the subset $\{\upsilon(\boldsymbol{r})\}_I$. The lower part of Fig. 3.7 shows the fibers $\mathcal{F}_\upsilon^{e,I}$ and $\mathcal{F}_n^{e,I}$ for the potential and the corresponding densities, respectively. As shown in Fig. 3.6, the correspondence from the fiber $\mathcal{F}_\upsilon^I$ to $\mathcal{F}_\upsilon^{e,I}$ is one-to-one. Therefore, the fiber $\mathcal{F}_\upsilon^I$ on the real space can be equivalently reduced to the fiber $\mathcal{F}_\upsilon^{e,I}$ on the energy coordinate. And the same relationship also applies to the fibers $\mathcal{F}_n^I$ and $\mathcal{F}_n^{e,I}$ for the densities. Furthermore, the correspondence between the fibers $\mathcal{F}_\upsilon^{e,I}$ and $\mathcal{F}_n^{e,I}$ is also, of course, one-to-one. Thus, the HK functional $F_{\rm HK}^{e,I}\left[n^{e,I}\right]$ is defined for the fiber $\mathcal{F}_n^{e,I}$.

Since the fiber bundle can be regarded as a product of the fiber $\mathcal{F}$ and the base space $\mathcal{B}$, it is possible to interpret the fiber $\mathcal{F}_\upsilon^{e,I}$ as a line perpendicular to the axis of $\mathcal{B}$. A potential $\upsilon^e(\epsilon)$ on the fiber $\mathcal{F}_\upsilon^{e,I}$ is called a section. And a graph drawn

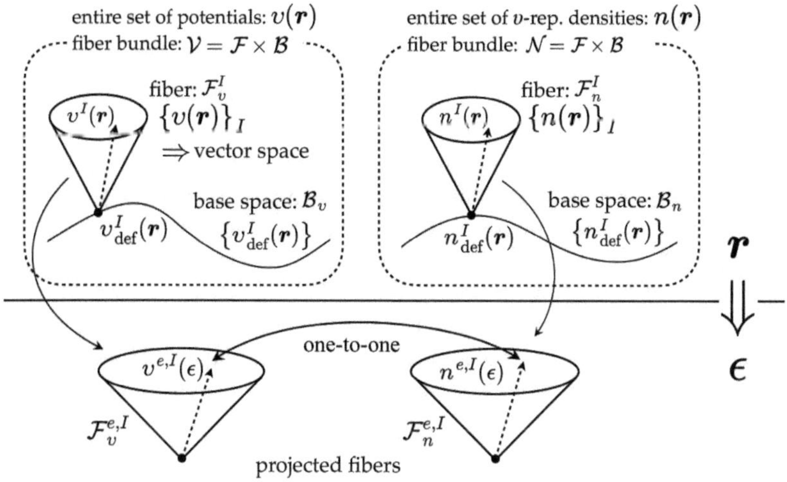

Fig. 3.7 Illustration of the fiber bundles for the potentials $v(\boldsymbol{r})$ and the v-representable electron densities $n(\boldsymbol{r})$. Each inverted cone represents a fiber. The potential fiber bundle $\mathcal{V}$ is given by the product of the fiber $\mathcal{F}$ and the base space $\mathcal{B}$. The lower figure shows the fibers projected onto the energy coordinate

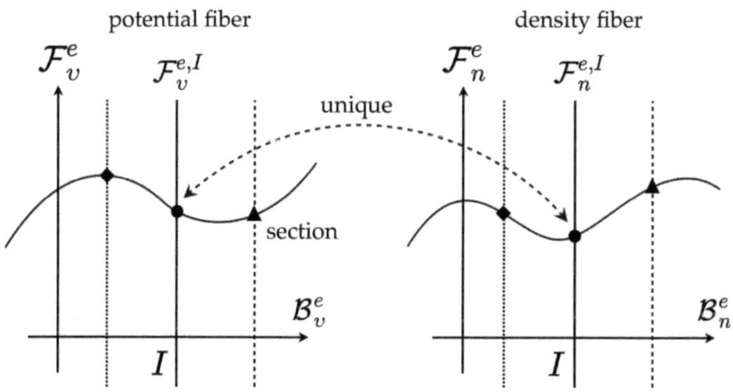

Fig. 3.8 The potential and the corresponding density given as sections on the fiber bundles. The fibers $\mathcal{F}_v^{e,I}$ and $\mathcal{F}_n^{e,I}$ are drawn as the lines perpendicular to the axes representing the base spaces $\mathcal{B}_v^e$ and $\mathcal{B}_n^e$. The potential on the fiber $\mathcal{F}_v^{e,I}$ with index I uniquely corresponds to the density on the fiber $\mathcal{F}_n^{e,I}$ with the same index I

on the $\mathcal{F} \times \mathcal{B}$ plane is considered as the consecutive sections as shown in Fig. 3.8. The section on the line $\mathcal{F}_v^{e,I}$ for the potential uniquely corresponds to the section on the line $\mathcal{F}_n^{e,I}$ for the density. Thus, the concept of the fiber bundle provides a comprehensive picture for the subset of the potential and the corresponding subset of the densities.

3.3 Legendre Transformation

It is also possible to formulate the Hohenberg-Kohn (HK) functional $F_{\mathrm{HK}}^{e,I}[n^{e,I}]$ through the Legendre transformation of the energy functional $E_0^{e,I}[\upsilon_{\mathrm{ext}}^{e,I}]$. Actually, the formulation is completely parallel to that given in Sect. 1.3.3. Only the difference is that the transformation is performed for the potential subset $\{\upsilon^e(\epsilon)\}_I$ and the corresponding density subset $\{n^e(\epsilon)\}_I$. Remind that the concavity of the ground state energy E_0 as a functional of the external potential $\upsilon_{\mathrm{ext}}(\boldsymbol{r})$ is essential to the Legendre transformation of $E_0[\upsilon_{\mathrm{ext}}]$ in the conventional DFT. We discuss whether the concavity of E_0 is satisfied for the potentials within each subset $\{\upsilon^e(\epsilon)\}_I$ $(I = 1, 2, \cdots)$. To this end, we construct the linear combination of the potentials $\upsilon_0^I(\boldsymbol{r})$ and $\upsilon_1^I(\boldsymbol{r})$ in the subset $\{\upsilon(\boldsymbol{r})\}_I$ with the coupling parameter λ $(0 \leq \lambda \leq 1)$ in parallel to Eq. (1.82)

$$\upsilon_\lambda^I(\boldsymbol{r}) = (1-\lambda)\,\upsilon_0^I(\boldsymbol{r}) + \lambda \upsilon_1^I(\boldsymbol{r}) \tag{3.36}$$

It is indicated in the equation that the resultant function υ_λ shall be included in the subset I. This can be readily confirmed as follows. Provided that $\boldsymbol{r}'(\neq \boldsymbol{r})$ satisfies the relation $\upsilon_0^I(\boldsymbol{r}) = \upsilon_0^I(\boldsymbol{r}')$, the relation $\upsilon_1^I(\boldsymbol{r}) = \upsilon_1^I(\boldsymbol{r}')$ also holds from the condition given in Sect. 3.2. Then for these $\boldsymbol{r}$ and $\boldsymbol{r}'$, $\upsilon_\lambda^I(\boldsymbol{r}) = \upsilon_\lambda^I(\boldsymbol{r}')$ is fulfilled. Here we remind that Eq. (1.81), that represents the concavity condition, holds for any pair of υ_0 and υ_1. Therefore, it is found that the concavity of the functional $E_0[\upsilon^I]$ is established within individual potential subset I thus

$$E_0\left[\upsilon_\lambda^I\right] \geq (1-\lambda)\,E_0\left[\upsilon_0^I\right] + \lambda E_0\left[\upsilon_1^I\right] \tag{3.37}$$

As depicted in Fig. 3.6, the correspondence between the subsets $\{\upsilon(\boldsymbol{r})\}_I$ and $\{\upsilon^e(\epsilon)\}_I$ is one-to-one. Hence, Eq. (3.36) can be equivalently transformed to the corresponding counterpart defined on the energy coordinate,

$$\upsilon_\lambda^{e,I}(\epsilon) = (1-\lambda)\,\upsilon_0^{e,I}(\epsilon) + \lambda \upsilon_1^{e,I}(\epsilon) \tag{3.38}$$

For any choice of parameter λ, the potential $\upsilon_\lambda^{e,I}$ is identical to υ_λ^I. Thus, the corresponding ground state energy $E_0^{e,I}\left[\upsilon_\lambda^{e,I}\right]$ is also equivalent to $E_0\left[\upsilon_\lambda^I\right]$. Then we obtain the inequality,

$$E_0^{e,I}\left[\upsilon_\lambda^{e,I}\right] \geq (1-\lambda)\,E_0^{e,I}\left[\upsilon_0^{e,I}\right] + \lambda E_0^{e,I}\left[\upsilon_1^{e,I}\right] \tag{3.39}$$

which shows that the ground state energy E_0 is concave for the potential $\upsilon^{e,I}$ within each subset $\{\upsilon^e(\epsilon)\}_I$. Hereafter in this Subsection, we omit the superscript $\{e, I\}$ on E_0 for simplicity since there is no possibility of confusion.

Next we consider the functional differentiation of the ground state energy $E_0[\upsilon_{\text{ext}}^{e,I}]$ with respect to $\upsilon_{\text{ext}}^{e,I}(\epsilon)$. To do this, we start the discussion with the equation

$$E_0\left[\upsilon_{\text{ext}}^{e,I}\right] = \inf_{n^{e,I}\rightarrow\upsilon-\text{rep.}} \left\{F_{\text{HK}}^{e,I}\left[n^{e,I}\right] + \int d\epsilon\, n^{e,I}(\epsilon)\,\upsilon_{\text{ext}}^{e,I}(\epsilon)\right\} \tag{3.40}$$

This equation itself is a kind of a Legendre transformation that switch $F_{\text{HK}}^{e,I}\left[n^{e,I}\right]$ to $E_0\left[\upsilon_{\text{ext}}^{e,I}\right]$. The quantity in the parenthesis of the right hand side of Eq. (3.40) is the energy functional of Eq. (3.32) and $F_{\text{HK}}^{e,I}$ is the Hohenberg-Kohn functional and defined in Eq. (3.33). As discussed in Sect. 1.3.3, we consider the small variation $\delta\upsilon^{e,I}$ of the potential $\upsilon^{e,I}$. Then the variation δE_0 of the energy up to the first order of $\delta\upsilon^{e,I}$ is given by

$$\delta E_0 = \left\langle\Psi_0\left|\delta\hat{\upsilon}_{\text{ext}}^{e,I}\right|\Psi_0\right\rangle + \left(\left\langle\delta\Psi_0\left|\hat{T} + \hat{V}_{ee} + \hat{\upsilon}_{\text{ext}}^{e,I}\right|\Psi_0\right\rangle + c.c.\right) \tag{3.41}$$

in parallel to Eq. (1.87). The second term in Eq. (3.41) becomes zero because Ψ_0 is supposed to be the ground state wave function for the potential $\hat{\upsilon}_{\text{ext}}^{e,I}$ and it satisfies the normalization condition. Thus, only the first term of the right hand side of Eq. (3.41) remains, which reads

$$\delta E_0 = \int d\epsilon\; \delta n^{e,I}(\epsilon)\,\upsilon_{\text{ext}}^{e,I}(\epsilon) \tag{3.42}$$

Then we obtain the derivative

$$n^{e,I}(\epsilon) = \frac{\delta E_0\left[\upsilon_{\text{ext}}^{e,I}\right]}{\delta\upsilon_{\text{ext}}^{e,I}(\epsilon)} \tag{3.43}$$

as the counterpart of Eq. (1.86). Here it should be noticed that the potential $\upsilon_{\text{ext}}^{e,I}$ and the density $n^{e,I}$ conjugate to the potential are both local on the energy coordinate. Then, we can obtain the functional $F_{\text{HK}}^{e,I}\left[n^{e,I}\right]$ through the backward Legendre transformation of Eq. (3.40), thus,

$$-F_{\text{HK}}^{e,I}\left[n^{e,I}\right] = \inf_{\upsilon_{\text{ext}}^{e,I}} \left\{\int d\epsilon n^{e,I}(\epsilon)\,\upsilon_{\text{ext}}^{e,I}(\epsilon) - E_0\left[\upsilon_{\text{ext}}^{e,I}\right]\right\} \tag{3.44}$$

The representation of the functional $F_{\text{HK}}^{e,I}$ by the Legendre transformation of $E_0\left[\upsilon_{\text{ext}}^{e,I}\right]$ is schematically illustrated in Fig. 3.9. As one can easily recognize, the figure is a counterpart of Fig. 1.5 for the conventional HK functional. Thus the Legendre transformation between the functionals $E_0[\upsilon]$ and $F_{\text{HK}}[n]$ can be replaced in parallel by the transformation between $E_0[\upsilon^e]$ and $F_{\text{HK}}^e[n^e]$.

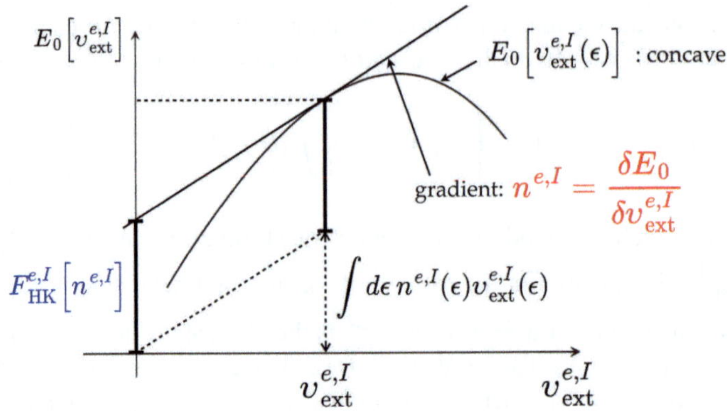

Fig. 3.9 Legendre transformation of the energy functional $E_0[v_{\text{ext}}^{e,I}]$ to the Hohenberg-Kohn functional $F_{\text{HK}}^{e,I}[n^{e,I}]$ given as a functional of the v-representable density $n^{e,I}$. The density $n^{e,I}$ is given as the gradient of E_0 with respect to $v_{\text{ext}}^{e,I}$. Then, the intercept of the tangent at $v_{\text{ext}}^{e,I}$ gives the functional $F_{\text{HK}}^{e,I}[n]$. Note that the transformation is completely parallel to the Legendre transformation given in Sect. 1.3.3 (see also Fig. 1.5)

It is, of course, possible to derive the functional derivative of $F_{\text{HK}}^{e,I}$ with respect to $n^{e,I}$ in parallel to the derivation given in Sect. 1.3.3. In parallel to Eq. (1.92), we express the functional $F_{\text{HK}}^{e,I}$ as

$$F_{\text{HK}}^{e,I}\left[n^{e,I}\right] = E_0\left[n^{e,I}\right] - \int d\epsilon\; n^{e,I}(\epsilon)\, v_{\text{ext}}^{e,I}\left[n^{e,I}\right](\epsilon) \tag{3.45}$$

Note that $v_{\text{ext}}^{e,I}\left[n^{e,I}\right]$ indicates the potential defined on the energy coordinate and corresponds to the v-rep. density $n^{e,I}$. $F_{\text{HK}}^{e,I}\left[n^{e,I}\right]$ can be explicitly written as

$$\begin{aligned} F_{\text{HK}}^{e,I}\left[n^{e,I}\right] &= \left\langle \Psi_0\left[n^{e,I}\right]\right| \hat{T} + \hat{V}_{ee} + \hat{v}_{\text{ext}}^{e,I}\left[n^{e,I}\right] \left|\Psi_0\left[n^{e,I}\right]\right\rangle \\ &\quad - \int d\epsilon\; n^{e,I}(\epsilon)\, v_{\text{ext}}^{e,I}[n](\epsilon) \end{aligned} \tag{3.46}$$

For the small variation $\delta n^{e,I}$ of $n^{e,I}$, the variation $\delta F_{\text{HK}}^{e,I}$ up to the first order of $\delta n^{e,I}$ is given by

$$\begin{aligned} \delta F_{\text{HK}}^{e,I} &= E_0 \langle \delta\Psi_0 | \Psi_0 \rangle + \langle \Psi_0 | \delta \hat{v}_{\text{ext}}^{e,I} | \Psi_0 \rangle \\ &\quad - \int d\epsilon\; \delta n^{e,I}(\epsilon)\, v_{\text{ext}}^{e,I}(\epsilon) - \int d\epsilon\; n^{e,I}(\epsilon)\, \delta v_{\text{ext}}^{e,I}(\epsilon) \end{aligned} \tag{3.47}$$

With the same reason as noted in Sect. 1.3.3, only the third term remains, from which we obtain

$$\frac{\delta F_{\text{HK}}^{e,I}\left[n^{e,I}\right]}{\delta n^{e,I}(\epsilon)} = -v_{\text{ext}}^{e,I}(\epsilon) \tag{3.48}$$

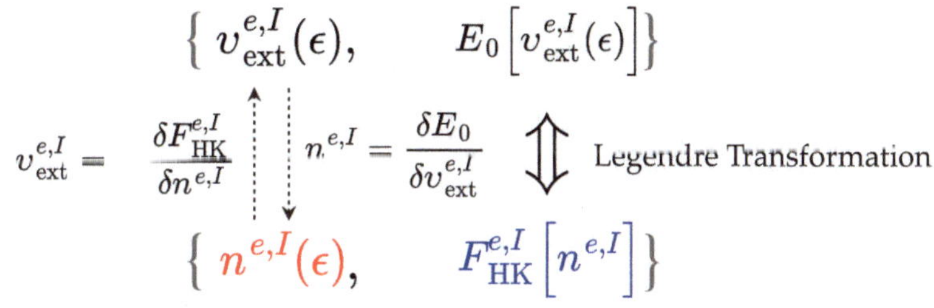

Fig. 3.10 Legendre transformations between the potential functional $E_0[\upsilon_{\mathrm{ext}}^{e,I}]$ and the Hohenberg-Kohn functional $F_{\mathrm{HK}}^{e,I}[n^{e,I}]$. The υ-representable density $n^{e,I}$ and the potential $\upsilon_{\mathrm{ext}}^{e,I}$ are given as the slopes (derivatives) of the functionals E_0 and $F_{\mathrm{HK}}^{e,I}$ with respect to $\upsilon_{\mathrm{ext}}^{e,I}$ and $n^{e,I}$, respectively. The transformation between the functionals defined on the energy coordinate is completely parallel to the scheme for the conventional DFT as shown in Fig. 1.6

In Sect. 3.2, we partitioned the whole set of local potentials $\upsilon(\boldsymbol{r})$ into infinite number of subsets and also the corresponding set of υ-rep. electron densities $n(\boldsymbol{r})$ using the appropriate defining potential assigned to each subset. Then, it was proved that the Hohenberg-Kohn (HK) theorem can be established within each subsets $\{\upsilon\,(\boldsymbol{r})\}_I$ and $\{n\,(\boldsymbol{r})\}_I$. Importantly, these subsets can be equivalently projected onto the space of energy coordinate ϵ by means of the defining potential. Therefore, it was found that the HK theorem is also established within the corresponding subsets $\{\upsilon^e\,(\epsilon)\}_I$ and $\{n^e\,(\epsilon)\}_I$. This leads to construction of the well defined functional, i.e. HK functional $F_{\mathrm{HK}}^{e,I}$ with $n^{e,I}\,(\epsilon)$ as an argument. In the present Section, the HK functional has been reformulated by means of the Legendre transformation within each subset of the potential or density on the energy coordinate ϵ. The essential points in realizing the Legendre transform are the concavity of the ground state energy $E_0[\upsilon^e]$ within the potential subset $\{\upsilon^e\,(\epsilon)\}_I$ and the localities of the potential $\upsilon^{e,I}(\epsilon)$ and the density $n^{e,I}(\epsilon)$.

In Fig. 3.10, the Legendre transformation between $E_0[\upsilon^e]$ and $F_{\mathrm{HK}}^e[n^e]$ is summarized, which again shows a complete parallelism to Fig. 1.6 provided in Sect. 1.3.3 for the conventional DFT. Note that the potential $\upsilon^e(\epsilon)$ is conjugate to the density $n^e(\epsilon)$, which are both local on the energy coordinate.

3.4 Levy's Constraint Search for the Energy Electron Density

Levy's constraint search method guarantees the existence of the density functionals where '$N-$rep.' density serves as a fundamental variable. Here, we extend the method to the novel density functional theory that utilizes the energy electron density $n^e(\epsilon)$ instead of $n(\boldsymbol{r})$. To this end, we introduce the variational search with respect to the '$N-$rep.' energy electron density $n^e(\epsilon)$ in parallel to the original Levy's constraint search, which reads

$$
\begin{aligned}
E_0 &= \min_{n^e \to N_{\mathrm{rep}}} \left\{ \min_{\Psi \to n^e} \left\langle \Psi \left| \hat{T} + \hat{V}_{ee} + \hat{v}_{\mathrm{ext}} \right| \Psi \right\rangle \right\} \\
&= \min_{n^e \to N_{\mathrm{rep}}} \left\{ \min_{\Psi \to n^e} \left\langle \Psi \left| \hat{T} + \hat{V}_{ee} \right| \Psi \right\rangle + \int d\epsilon \; v^e_{\mathrm{ext}}(\epsilon) n^e(\epsilon) \right\}
\end{aligned} \tag{3.49}
$$

In deriving Eq. (3.49), it is assumed that the external potential $v_{\mathrm{ext}}(\boldsymbol{r})$ is always constant on each contour surface of the defining potential $v_{\mathrm{def}}(\boldsymbol{r})$. That is, the external potential v_{ext} and the defining potential v_{def} are supposed to belong to the same potential subset $\{v(\boldsymbol{r})\}_I$ identified with I as discussed in Sect. 3.2. Therefore, it is possible to identify the external potential within the subset I using the coordinate ϵ. Throughout the present paragraph, superscript I for v^e and n^e will be omitted for the sake of simplicity. The interaction between the electrons and the external potential can be described in terms of the energy electron density as shown in the right hand side of the second equality of Eq. (3.49). Thus, the interaction between the electrons and the external potential $v^e_{\mathrm{ext}}(\epsilon)$ can be placed outside the constraint minimization with respect to the wave function Ψ that realizes the distribution n^e. It is quite important to note the complete parallelism of Eq. (3.49) to the conventional approach expressed by Eq. (1.97). Similar to the functional defined by Eq. (1.98), the functional for the kinetic and the electron repulsion energies can be given in terms of the energy electron density, thus,

$$
F^{e,I}_{\mathrm{Levy}}[n^e] = \min_{\Psi \to n^e} \left\langle \Psi \left| \hat{T} + \hat{V}_{ee} \right| \Psi \right\rangle \tag{3.50}
$$

It should be noticed that the functional $F^{e,I}_{\mathrm{Levy}}$ is *not* general and specific to the defining potential v_{def} in the subset I. The argument n^e of the functional $F^{e,I}_{\mathrm{Levy}}$ should be $N-$representable, that is, n^e should be derived from an antisymmetric wave function with N electrons. However, the bound of the subset of the '$N-$rep' electron density n^e for a given defining potential is not yet known. In this sense, the DFT based on the energy electron density has a flaw in its construction. However, it does not matter in the practical applications, because it will be always assumed that n^e is yielded from an '$N-$rep' density $n(\boldsymbol{r})$. The functionals for the kinetic and the electron repulsion energies can also be formulated using n^e in parallel to Eqs. (1.100) and (1.101), thus,

$$
T^e[n^e] = \left\langle \Psi_{\mathrm{min}}[n^e] | \hat{T} | \Psi_{\mathrm{min}}[n^e] \right\rangle \tag{3.51}
$$

and

$$
V^e_{ee}[n^e] = \left\langle \Psi_{\mathrm{min}}[n^e] | \hat{V}_{ee} | \Psi_{\mathrm{min}}[n^e] \right\rangle \tag{3.52}
$$

where $\Psi_{\mathrm{min}}[n^e]$ is the wave function that realizes the distribution n^e and minimizes the expectation value of the operator $\hat{T} + \hat{V}_{ee}$. In the next Section, we formulate a framework of the new Kohn-Sham DFT utilizing the exchange-correlation energy functional in terms of the energy electron density n^e.

3.5 Kohn-Sham Method for the Energy Electron Density

In this Section we formulate the Kohn-Sham method in terms of the energy electron density n^e [1]. It is supposed that the system under consideration includes N electrons. We again introduce the single Slater determinant $\Psi_{\rm SD}$ constructed from a set of orthonormalized N orbitals $\{\varphi_i\}$. We then define the functional $G^e[n^e]$ of the non-interacting reference system, thus,

$$G^e\left[n^e\right] = \min_{n \to n^e}\left\{\min_{\Psi_{\rm SD}\to n}\left\langle \Psi_{\rm SD}|\hat{T}|\Psi_{\rm SD}\right\rangle + E_{\rm H}[n]\right\} \tag{3.53}$$

where n is the electron density that yields the energy electron density n^e and $E_{\rm H}[n]$ is the Hartree energy defined in Eq. (1.108). It is important to note that $G^e\left[n^e\right]$ involves the minimization of the Hartree energy as well as the kinetic energy of the non-interacting system since there will be a lot of electron densities n that give n^e. The kinetic energy functional for n^e is defined as

$$T_s^e\left[n^e\right] = \left\langle \Psi_{\rm SD}^{\rm min}|\hat{T}|\Psi_{\rm SD}^{\rm min}\right\rangle \tag{3.54}$$

and the Hartree energy is represented as

$$E_{\rm H}^e\left[n^e\right] = \frac{1}{2}\int d\boldsymbol{r}_1 d\boldsymbol{r}_2 \frac{n_{\rm min}\left(\boldsymbol{r}_1\right) n_{\rm min}\left(\boldsymbol{r}_2\right)}{|\boldsymbol{r}_1 - \boldsymbol{r}_2|} \tag{3.55}$$

With these functionals it is possible to define the exchange-correlation functional $E_{xc}^e[n^e]$ with argument n^e, thus,

$$E_{xc}^e\left[n^e\right] = T^e\left[n^e\right] - T_s^e\left[n^e\right] + V_{ee}^e\left[n^e\right] - E_{\rm H}^e\left[n^e\right] \tag{3.56}$$

where $T^e\left[n^e\right]$ and $V_{ee}^e\left[n^e\right]$ are defined in Eqs. (3.51) and (3.52), respectively.

Now we are ready to formulate the Kohn-Sham minimization using the functional $E_{xc}[n^e]$,

$$\begin{aligned} E_0 &= \min_{n^e \to N}\left[G^e\left[n^e\right] + E_{xc}^e\left[n^e\right] + \int d\epsilon\, n^e(\epsilon)\, v_{\rm ext}^e(\epsilon)\right] \\ &= \min_{n^e \to N}\left\{T_s^e\left[n^e\right] + E_{\rm H}^e\left[n^e\right] + E_{xc}^e\left[n^e\right] + \int d\epsilon n^e(\epsilon) v_{\rm ext}^e(\epsilon)\right\} \\ &= \min_{\Psi_{\rm SD}}\left\{\left\langle \Psi_{\rm SD}|\hat{T}|\Psi_{\rm SD}\right\rangle + E_{\rm H}^e[n^e] + E_{xc}^e\left[n^e\right] + \int d\epsilon n^e(\epsilon) v_{\rm ext}^e(\epsilon)\right\} \end{aligned} \tag{3.57}$$

In the right-hand side of the first equality, the minimization is performed with respect to the 'N-rep.' energy electron density n^e, while in the second one the single determinant wave function $\Psi_{\rm SD}$, that is, the set of one-electron wave functions $\{\varphi_i\}$, is employed for the variational search. Thus, $n^e(\epsilon)$ is actually constructed from a single determinant wave function, and hence it is automatically N-representable. To realize the Kohn-Sham equation based on the functional $E^e_{xc}[n^e]$, it is necessary to evaluate the functional derivative of $E^e_{xc}[n^e]$ with respect to the one-electron wave function $\varphi_i(\boldsymbol{r})$. In the following formulation, the indices of spin σ will be omitted for simplicity. The derivative for each spin can be obtained through the chain rule,

$$\begin{aligned}\frac{\delta E^e_{xc}[n^e]}{\delta\varphi_i^*(\boldsymbol{r})} &= \frac{\delta n(\boldsymbol{r})}{\delta\varphi_i^*(\boldsymbol{r})}\frac{\delta n^e(\epsilon)}{\delta n(\boldsymbol{r})}\frac{\delta E^e_{xc}[n^e]}{\delta n^e(\epsilon)} \\ &= \varphi_i(\boldsymbol{r})\delta(\epsilon - \upsilon_{\text{def}}(\boldsymbol{r}))\upsilon^e_{xc}\left[n^e\right](\epsilon) \\ &= \upsilon^e_{xc}\left[n^e\right](\epsilon)\big|_{\epsilon=\upsilon_{\text{def}}(\boldsymbol{r})}\varphi_i(\boldsymbol{r})\end{aligned} \tag{3.58}$$

where the relations of Eqs. (1.113) and (3.2) are used. Using the exchange-correlation potential $\upsilon^e_{xc}[n^e](\epsilon)$ dependent on the energy coordinate ϵ, we get

$$\left(-\frac{1}{2}\nabla^2 + \upsilon_{\text{H}}[n](\boldsymbol{r}) + \upsilon^e_{xc}\left[n^e\right](\epsilon) + \upsilon_{\text{ext}}(\boldsymbol{r})\right)\varphi_i(\boldsymbol{r}) = \epsilon_i\varphi_i(\boldsymbol{r}) \tag{3.59}$$

Thus, the exchange-correlation potential $\upsilon^e_{xc}[n^e]$ can be reasonably defined on the energy coordinate ϵ, and hence, the Kohn-Sham equation for the functional $E^e_{xc}[n^e]$ can also be formulated in parallel to the original one. In the next Subsection we provide explicit functionals $E^e_{xc}[n^e]$ which can be adopted in Eq. (3.59).

It is useful to reconsider the constraint minimization of Eq. (3.53). This minimization is the counterpart of Eq. (1.112) in the conventional KS-DFT. Equation (3.53) can be rewritten as

$$\begin{aligned}G^e\left[n^e\right] &= \min_{n\to n^e}\left\{\min_{\Psi_{\text{SD}}\to n}\left\langle\Psi_{\text{SD}}\left|\hat{T}\right|\Psi_{\text{SD}}\right\rangle + E_{\text{H}}\left[n\right]\right\} \\ &= \min_{\Psi_{\text{SD}}\to n^e}\left\{\left\langle\Psi_{\text{SD}}\left|\hat{T}\right|\Psi_{\text{SD}}\right\rangle + E_{\text{H}}\left[n\right]\right\}\end{aligned} \tag{3.60}$$

The constraints can be explicitly expressed as

$$\begin{cases} n^e(\epsilon) - \int d\boldsymbol{r}\, n(\boldsymbol{r})\,\delta\left(\epsilon - \upsilon_{\text{def}}(\boldsymbol{r})\right) = 0 \\ \langle\varphi_i|\,\varphi_j\rangle - \delta_{ij} = 0 \end{cases} \tag{3.61}$$

The first relation imposes that the projection of the density $n(\boldsymbol{r})$ on to the energy coordinate ϵ yields the energy electron density $n^e(\epsilon)$ given as an argument. The second one represents the orthonormalization condition of the wave functions. With these constraints, the Lagrangian $L[\{\varphi_i\}]$ is given by

$$
\begin{aligned}
L\left[\{\varphi_i\}\right] = 2\sum_{i}^{N/2}\left\langle\varphi_i\left|-\frac{1}{2}\nabla^2\right|\varphi_i\right\rangle + E_{\mathrm{H}}[n] \\
+ \int d\epsilon\, \lambda(\epsilon)\left(n^e(\epsilon) - \int d\boldsymbol{r}\, n(\boldsymbol{r})\,\delta(\epsilon - \upsilon(\boldsymbol{r}))\right) \\
- \sum_{i,j}^{N/2}\epsilon_{ij}\left(\langle\varphi_i|\,\varphi_j\rangle - \delta_{ij}\right)
\end{aligned} \tag{3.62}
$$

$\lambda(\epsilon)$ and ϵ_{ij} are Lagrange's multipliers corresponding to the first and the second constraints, respectively, in Eq. (3.61). In Eq. (3.62), it is supposed that N electrons occupies $N/2$ orbitals to form a closed shell structure. Then the minimization of Eq. (3.60) can be achieved when the following relations are fulfilled,

$$
\frac{\delta L\left[\{\varphi_i\}\right]}{\delta\varphi_i(\boldsymbol{r})} = 0 \quad (i = 1, 2, \cdots, N/2) \tag{3.63}
$$

which reads

$$
\begin{aligned}
\frac{\delta L\left[\{\varphi_i\}\right]}{\delta\varphi_i^*(\boldsymbol{r})} = \left(-\frac{1}{2}\nabla^2 + \upsilon_{\mathrm{H}}[n](\boldsymbol{r})\right)\varphi_i(\boldsymbol{r}) \\
+ \lambda(\epsilon)\,\delta(\epsilon - \upsilon_{\mathrm{def}}(\boldsymbol{r}))\,\varphi_i(\boldsymbol{r}) \\
- \sum_{j}^{N/2}\epsilon_{ij}\varphi_j(\boldsymbol{r}) = 0
\end{aligned} \tag{3.64}
$$

Note that the potential $\upsilon_{\mathrm{H}}[n]$ is the Hartree potential expressed by

$$
\upsilon_{\mathrm{H}}[n](\boldsymbol{r}) = \int d\boldsymbol{r}'\frac{n(\boldsymbol{r}')}{|\boldsymbol{r} - \boldsymbol{r}'|} \tag{3.65}
$$

Then we get

$$
\left(-\frac{1}{2}\nabla^2 + \upsilon_{\mathrm{H}}(\boldsymbol{r}) + \lambda(\epsilon)|_{\epsilon=\upsilon_{\mathrm{def}}(\boldsymbol{r})}\right)\varphi_i(\boldsymbol{r}) = \sum_{j}^{N/2}\epsilon_{ij}\varphi_j(\boldsymbol{r}) \tag{3.66}
$$

It is quite important to note that $\lambda(\epsilon)$ is the local potential on the energy coordinate ϵ defined by the potential $\upsilon_{\mathrm{def}}(\boldsymbol{r})$. Here we assume that $\lambda(\epsilon)$ is composed of two terms,

$$
\lambda(\epsilon) = \upsilon_{xc}\left[n^e\right](\epsilon) + \overline{\upsilon}^e_{\mathrm{ext}}(\epsilon) \tag{3.67}
$$

where $\upsilon_{xc}[n^e]$ is the exchange-correlation potential introduced in Eq. (3.58) and $\overline{\upsilon}^e_{\mathrm{ext}}$ is the external potential that uniquely corresponds to the non-interacting υ-rep.

density n^e given as an argument. $\overline{\upsilon}^e_{\text{ext}}$ can be reversed to the potential $\overline{\upsilon}_{\text{ext}}(\boldsymbol{r})$ on the real space using the relation

$$\overline{\upsilon}_{\text{ext}}(\boldsymbol{r}) = \overline{\upsilon}^e_{\text{ext}}(\epsilon)\Big|_{\epsilon=\upsilon_{\text{def}}(\boldsymbol{r})} \tag{3.68}$$

With this assumption, it is shown that the constraint minimization in the functional G^e in Eq. (3.60) is equivalent to solving the Kohn-Sham equation of Eq. (3.59). Thus $\lambda(\epsilon)$ is the effective potential represented on the energy coordinate excluding the Hartree potential $\upsilon_{\text{H}}[n](\boldsymbol{r})$. Furthermore, when the argument n^e for the functional G^e in Eq. (3.60) coincides with the ground state energy electron density n^e of the system of interest, the corresponding external potential $\overline{\upsilon}^e_{\text{ext}}$ also agrees with υ^e_{ext} of the system.

3.6 Development of Exchange-Correlation Functional Based on Energy Electron Density

In the previous Subsections, we discussed the theoretical framework of the novel DFT that utilizes energy electron density $n^e(\epsilon)$ as a fundamental variable. It was demonstrated that the foundation of the DFT is almost parallel to the conventional one. However, for the success of the DFT, the development of the effective functional $E^e_{xc}[n^e]$ for the exchange-correlation energy is of critical importance since the functional constitutes the energy due to the effect of the quantum correlations among the electrons. This Section will be devoted to construct explicit forms of the functionals. Actually, the exchange functional $E^{e,\text{LDA}}_x[n^e]$ based on the local density approximation (LDA) was already provided in Sect. 3.1.

In the first Subsection below, we formulate an exchange functional $E^{e,\text{B88}}_x[n^e]$ using the Becke88 exchange functional [5] as a template. The Becke88 functional is developed by Becke in 1988 and known as a representable GGA (generalized gradient approximation) [6] functional. The transformation of the GGA functional in the conventional DFT to that in the novel DFT using the energy electron density can be performed in a straightforward manner. Further, the extension of the Becke-Roussel(BR) exchange functional [7] to that defined on the energy coordinate ϵ is also performed to construct $E^{e,\text{BR}}_x[n^e]$.

In the second Subsection, we first examine the accuracy of the BR exchange functional $E^{e,\text{BR}}_x[n^e]$ combined with the LYP correlation functional [8] by computing the potential energy curve of a hydrogen molecule. We also compare the LDA-based exchange functional for the energy electron density $n^e(\epsilon)$ with the conventional LDA functional in describing the potential energy curves of chemical bonds in several small molecules.

3.6.1 Generalized Gradient Approximations for DFT with Energy Electron Density

The functional $E_x^{e,\text{LDA}}[n^e]$ based on the local-density approximation (LDA) was already provided in Eq. (3.7) in Sect. 3.1 for which the variable $\widetilde{n}^e(\epsilon)$(Eq. (3.6)) had been introduced as the average electron density on the energy coordinate ϵ. It should be reminded that $E_x^{e,\text{LDA}}[n^e]$ becomes equivalent to the conventional LDA functional $E_x^{\text{LDA}}[n]$ when the electron density $n(\boldsymbol{r})$ is always constant on the surface with the same energy coordinate ϵ. In a mathematical description, it can be written as follows. When the condition

$$n(\boldsymbol{r}) = \text{const.} : \boldsymbol{r} \in \mathcal{R}_\epsilon^I \text{ for all } \epsilon \tag{3.69}$$

holds for a given electron density $n(\boldsymbol{r})$, the energy $E_x^{e,\text{LDA}}[n^e]$ is exactly equal to $E_x^{\text{LDA}}[n]$. In Eq. (3.69), the spatial region $\mathcal{R}_\epsilon^I$ is explicitly defined in Eq. (3.16). One finds such a situation in the electron density of an atomic system with spherical symmetry. It is also possible to apply the generalized-gradient approximation (GGA) to improve the functional $E_x^{\text{LDA}}[n^e]$. To do this, we consider the dimensionless parameter $q_\sigma(\boldsymbol{r})$ for each spin σ

$$q_\sigma(\boldsymbol{r}) = \frac{|\nabla n_\sigma(\boldsymbol{r})|}{n_\sigma(\boldsymbol{r})^{\frac{4}{3}}} \tag{3.70}$$

Actually the parameter $q(\boldsymbol{r})$ was already introduced in Eq. (1.150) in Sect. 1.3.6. However, we dare to show it for the sake of convenience. Remind that $q(\boldsymbol{r})$ measures the inhomogeneity of the density at $\boldsymbol{r}$ and serves as a fundamental variable in the GGA functionals. Usually, the enhancement factor, defined as the ratio of the GGA exchange energy to the pure LDA energy, is a monotonically increasing function of the variable $q_\sigma(\boldsymbol{r})$. The inhomogeneity $q_\sigma^e(\epsilon)$ on the energy coordinate ϵ is similarly defined as

$$q_\sigma^e(\epsilon) = \frac{g_1^e[n_\sigma](\epsilon)}{\widetilde{n}_\sigma^e(\epsilon)^{\frac{4}{3}}} \tag{3.71}$$

where

$$g_1^e[n_\sigma](\epsilon) = \Omega(\epsilon)^{-1} \int d\boldsymbol{r}\delta(\epsilon - v_{\text{def}}(\boldsymbol{r}))\,|\nabla n_\sigma(\boldsymbol{r})| \tag{3.72}$$

gives the average of the magnitude of gradient of the density over the isosurface of the energy coordinate ϵ. It should be noticed that $g_1^e[n](\epsilon)$ is not a functional of n^e, but a functional of $n(\boldsymbol{r})$. With the variable $q^e(\epsilon)$, a conventional GGA exchange functional can be transformed to the functional in terms of the energy distribution $n^e(\epsilon)$. As was introduced in Eq. (1.153) in Sect. 1.3.6, the Becke88 exchange functional $E_x^{\text{B88}}[n]$ for each spin is explicitly expressed as

$$E_{x\sigma}^{\text{B88}}[n_\sigma] = E_{x\sigma}^{\text{LDA}}[n_\sigma] - \beta \int d\boldsymbol{r}\; n_\sigma(\boldsymbol{r})^{\frac{4}{3}} \frac{q_\sigma(\boldsymbol{r})^2}{1 + 6\beta q_\sigma(\boldsymbol{r}) \sinh^{-1} q_\sigma(\boldsymbol{r})} \tag{3.73}$$

in its original form [5]. The empirical parameter β was optimized specifically for the Becke88 functional [5]. This can be used directly in the corresponding functional for the energy electron density, thus,

$$E_{x\sigma}^{e,\mathrm{B88}}[\widetilde{n}_\sigma^e] = E_x^{e,\mathrm{LDA}}[\widetilde{n}_\sigma^e] - \beta \int d\epsilon\ \Omega(\epsilon)\widetilde{n}_\sigma^e(\epsilon)^{\frac{4}{3}} \frac{q_\sigma^e(\epsilon)^2}{1 + 6\beta q_\sigma^e(\epsilon)\sinh^{-1} q_\sigma^e(\epsilon)} \tag{3.74}$$

where $\Omega(\epsilon)$ is the volume of the spatial region with the energy coordinate ϵ and defined in Eq. (3.4). It can be readily recognized that the functional $E_{x\sigma}^{e,\mathrm{B88}}[\widetilde{n}_\sigma^e]$ gives the same result as $E_{x\sigma}^{\mathrm{B88}}[n_\sigma]$ when they are applied to the atomic systems with a spherical symmetry since the inhomogeneity parameter $q(\boldsymbol{r})$ is also constant over the isosurface with the energy coordinate ϵ, that is, $q(\boldsymbol{r}) = q^e(\epsilon)|_{\epsilon=\upsilon_{\mathrm{def}}(\boldsymbol{r})}$.

More sophisticated exchange functional of $n(\boldsymbol{r})$ can also be adapted to construct the functional of n^e. Here, we implement the exchange functional of n^e on the basis of the Becke-Roussel (BR) approach [7] introduced in Sect. 2.3.2. In the BR method, the quadratic term $Q(\boldsymbol{r})$ given by Eq. (2.52) plays a central role in determining the width of the exchange hole and the distance between the center of the hole and the reference point $\boldsymbol{r}$. As shown in Eq. (2.52), $Q(\boldsymbol{r})$ includes the second derivative $\nabla^2 n(\boldsymbol{r})$ of the density and the kinetic energy density $\tau(\boldsymbol{r})$ besides the gradient $\nabla n(\boldsymbol{r})$ of the density. Hence, the level of theory of the BR method is comparable to that of the meta-GGA functional [9]. To adapt the BR method to the functional with the energy electron density n^e, we consider the average of the quadratic term $Q(\boldsymbol{r})$ over the isosurface with the energy coordinate ϵ,

$$Q_\sigma^e[n_\sigma](\epsilon) = \Omega(\epsilon)^{-1} \int d\boldsymbol{r}\ \delta(\epsilon - \upsilon_{\mathrm{def}}(\boldsymbol{r}))Q_\sigma[n_\sigma](\boldsymbol{r}) \tag{3.75}$$

The quantities $Q_\sigma^e[n_\sigma](\epsilon)$ and $\widetilde{n}_\sigma^e(\epsilon)$ serve as inputs to the BR machinery that provides the non empirical parameters (ζ, r) of the model exchange hole in the form of the Slater function (see Eq. (2.50)). The computational procedure can be concisely summarized as follows.

1. A discrete energy coordinate ϵ is introduced for the system under consideration.
2. The energy electron density $n_\sigma^e(\epsilon)$ and $\widetilde{n}_\sigma^e(\epsilon)$ in Eq. (3.6) are constructed for the given electron density $n_\sigma(\boldsymbol{r})$.
3. The quadratic term $Q_\sigma^e[n_\sigma](\epsilon)$ in Eq. (3.75) is computed for the density $n_\sigma(\boldsymbol{r})$.
4. For a given grid point $\boldsymbol{r}$ in the real-space cell, the corresponding energy coordinate ϵ is specified. And the quantities $\widetilde{n}_\sigma^e(\epsilon)$ and $Q_\sigma^e[n_\sigma](\epsilon)$ are obtained accordingly.
5. For the point $\boldsymbol{r}$ in the Step 3, the BR parameters (ζ, r) in Eq. (2.50) are determined by the method described in Sect. 1.3.6.
6. For the point $\boldsymbol{r}$ in the Step 3, the exchange energy density $U_{x\sigma}^{\mathrm{BR}}(\boldsymbol{r})$ in Eq. (2.56) is computed.
7. Steps from 4–6 are applied to all the grid points in the real-space cell and $U_{x\sigma}^{\mathrm{BR}}(\boldsymbol{r})$ are accumulated to obtain the BR exchange energy $E_{x\sigma}^{\mathrm{BR}}$ in Eq. (2.57).

Thus the computational procedure is almost parallel to that used in the original BR approach described in Sect. 1.3.6. Again, it is important to note that the BR method with the density n^e yields the same result as the calculation with the original BR functional when it is applied to a spherically symmetric system since the quantity $Q(\boldsymbol{r})$ as well as $n_\sigma(\boldsymbol{r})$ is constant over the spatial region with the same energy coordinate. One might think that the implementation of the BR functional for the energy electron density is nothing more than degrading the calculation of the original BR method by averaging the electron density and its derivatives over the surface with the same energy coordinate. Actually, this is true. However, the major advantage of the DFT based on the energy electron density is that the novel DFT has the potential to overcome the difficulty of the approximate functionals of conventional DFT. As described in Sect. 2.3.1, limitation of the approximate functionals based on the LDA is that they cannot take into consideration the nonlocal effect intrinsic to the formal exchange-correlation functional. The flaw of the approximate functionals gives rise to the static correlation error in describing the dissociations of chemical bonds. The new fundamental variable $n^e(\epsilon)$ provided by the projection of $n(\boldsymbol{r})$ using Eq. (3.2) is local in the energy coordinate. However, it includes the information content that is nonlocal in the real space. Actually, in Chap. 4, the exchange functional that realizes the static correlation will be developed on the basis of the novel DFT. Furthermore, the DFT is also applied to the development of the nonlocal part of the kinetic energy functional in Chap. 5. Before these projects are to be undertaken, it will be useful to examine the effect of the projection of $n(\boldsymbol{r})$ to $n^e(\epsilon)$ on the descriptions of the chemical bonds around their equilibrium distances.

3.6.2 Application of the Functional to Chemical Bonds

Here we examine the accuracies of the functionals in terms of the energy electron density n^e in describing the potential energy curves of chemical bonds. Explicitly, we implement the BR exchange functional $E_x^{e,\mathrm{BR}}[n^e]$ as well as the LDA-based functional $E_x^{e,\mathrm{LDA}}[n^e]$ and applied them to the calculations of the potential energy curves of chemical bonds in simple molecules. We first applied the BR method [7] to the simplest chemical bond, i.e. the σ bond in an H_2 molecule. The results are presented in Fig. 3.11, where the potential energy curves of the H_2 molecule are plotted as functions of the bond length around its equilibrium distance. The curve with the legend 'BR-LYP' is the result given by the Becke-Roussel functional $E_x^{\mathrm{BR}}[n]$ using our house code 'Vmol' which utilizes the real-space grids (RSG) [10–14] to represent the one-electron wave functions in KS-DFT. We refer the readers to Appendix F and the literature [12–14] for the details of the code. In the original BR procedure the parameters (ζ, r) in Eq. (2.50) are numerically obtained for each real-space grid through the Newton-Raphson algorithm. Therefore, the functional derivative of $E_x^{\mathrm{BR}}[n]$ with respect to $n(\boldsymbol{r})$ is not feasible in general. However, in the calculations with LCAO (linear combination of atomic orbitals) basis sets, the derivative can be obtained through the method proposed in Ref. [15]. Furthermore, in our develop-

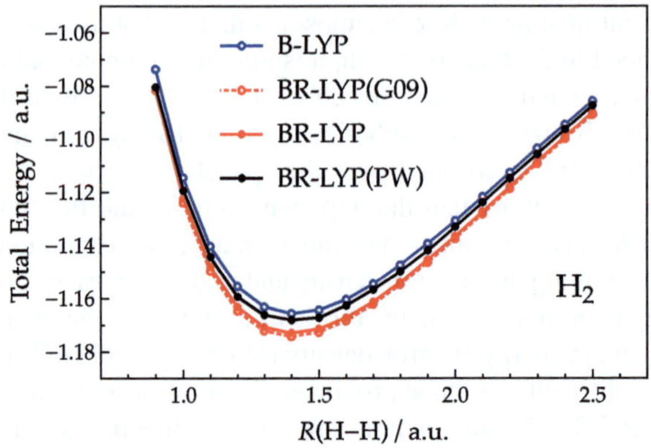

Fig. 3.11 Potential energy curves of H_2 molecule computed as functions of the distance $R(\mathrm{H{-}H})$. The LYP functional [8] based on $n(\boldsymbol{r})$ was utilized as the correlation functional in all of the calculations. The self-consistent electron density $n(\boldsymbol{r})$ given by the B-LYP functional is used in the BR-LYP calculations except for the calculation by the Gaussian 09 (G09) package. The numerical details for the production of the graphs are presented in Ref. [1]

ment [16] it was made possible to perform the SCF calculations for the modified Becke-Roussel model in the KS-DFT calculations with the RSG method. Anyway, in the present RSG calculations, we carry out the SCF calculation with the B-LYP functional beforehand, and the resultant wave functions and the electron density $n(\boldsymbol{r})$ or $n^e(\epsilon)$ serve to construct the Becke-Roussel exchange energy $E_x^{\mathrm{BR}}[n]$ or $E_x^{e,\mathrm{BR}}[n^e]$. The correlation energy $E_c[n]$ is evaluated utilizing the functional developed by Lee, Yang and Parr (LYP) [8] throughout the present calculations. The correlation energy is computed using solely the electron density $n(\boldsymbol{r})$. As a reference for the BR calculation, we also conduct the Gaussian 09 (G09) program suite, [17] where the SCF calculation for the electron density is performed with the BR exchange and the LYP correlation functionals. In the LCAO calculation with the G09 package, the basis set 'aug-cc-pVTZ' is employed. It is shown in Fig. 3.11 that the potential energy curve given by the BR-LYP calculation with the Vmol package is very close to that given by G09 regardless of the difference in the methods representing the wave functions. It can also be recognized that the use of the electron density optimized by the B-LYP functional is adequate as an input to the BR-LYP calculation. The B-LYP functional with the RSG method provides the slightly higher energies than the BR-LYP functional within the range of the H−H distance considered in the present calculation. It is of major importance to note that the present work (PW), which utilizes the BR exchange functional based on the energy electron density $n^e(\epsilon)$, provides the sound potential energy curve of H_2. Actually, the curve lies between those given by BR-LYP and B-LYP functionals which use $n(\boldsymbol{r})$ as the fundamental variable. This result clearly shows that the energy electron density n^e as a fundamental variable in the KS-DFT is sufficient to realize the potential energy curve around the equilibrium bond length of the hydrogen molecule.

Next we examine the performance of the exchange functional $E_x^e[n^e]$ in terms of n^e in describing the electronic energies of double and triple bonds. Since we are only interested in the fundamental property of the functional, we apply the LDA-based functional $E_x^{e,\mathrm{LDA}}[n^e]$ for this calculation. To this end, the potential energy curves for ethylene (C_2H_4) and acetylene (C_2H_2) are computed using $E_x^{e,\mathrm{LDA}}[n^e]$. As mentioned in the previous paragraph, we utilized the RSG to represent the wave functions, which necessitated the use of the pseudopotentials. As a consequence only the valence electrons are considered in the calculations. The core electrons are thus excluded in yielding the electron density $n(\boldsymbol{r})$, which gives rise to a minor problem in constructing the energy electron density $n^e(\epsilon)$ according to Eq. (3.2). The integration of the density $n(\boldsymbol{r})$ over the surface with an energy coordinate ϵ_{core} corresponding to the spatial region around the hydrogen atomic cores also involves the integration over the surface with the same ϵ_{core} around the carbon atomic cores.

Since the electron population on the C cores is completely lost due to the use of the pseudopotential, this gives rise to the artificial decrease in the energy electron density $n^e(\epsilon_{\mathrm{core}})$. As a consequence, the corresponding exchange energy density $U_x^{e,\mathrm{LDA}}[n^e](\epsilon_{\mathrm{core}})$ is also lowered. However, the effect of the artifact can be considered to be minor since the density of core electrons of C is rigid and its population hardly changes for the variation of the bond distance. Therefore, the deficit of the core electrons might cause only the constant shift of the potential energy curve with respect to the variation of the bond distance. We refer readers to Ref. [1] for more detailed discussions. The results for ethylene and acetylene are presented in Fig. 3.12a and b, respectively. In the computation of each molecule, only the C–C distance is varied with other geometrical parameters fixed at those of the equilibrium molecular structure. The figures show that the multiple bonds can also be correctly described

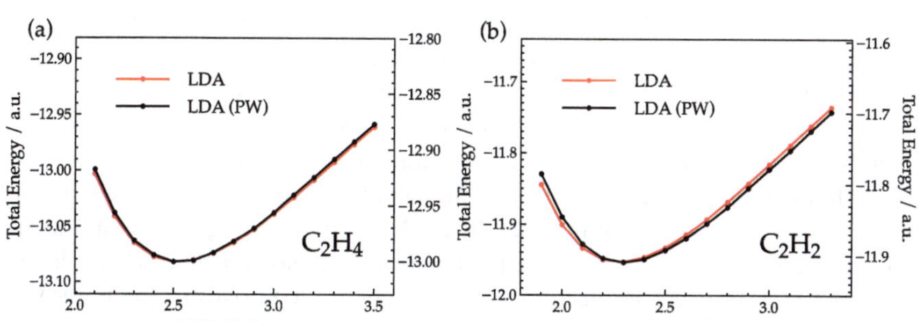

Fig. 3.12 Potential energy curves of **a** ethylene (C_2H_4) and **b** acetylene (C_2H_2) computed as functions of the distance R(C–C). The conventional LDA exchange functional given by Eq. (3.5) is used to yield the curve of 'LDA'. On the other hand, the 'LDA (PW)' is provided by Eq. (3.7). The vertical axis on the left is for 'LDA', while that on the right is for 'LDA (PW)'. The energy range of the axis on the left is strictly adjusted to that on the right. The self-consistent electron density $n(\boldsymbol{r})$ given by the Kohn-Sham DFT calculation using the LDA functional (Eq. (3.5)) is adopted for the calculations of these curves. The numerical details for the production of the graphs are presented in Ref. [1]

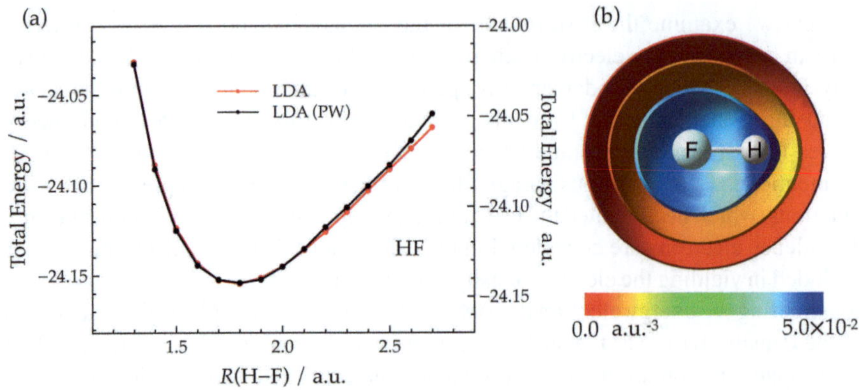

Fig. 3.13 **a** Potential energy curves of HF molecule computed as functions of the distance R(H−F). The other captions for the legends and the axes are the same as in Fig. 3.12. **b** The electron density mapped onto the isosurfaces of the bare Coulomb potential (Eq. (3.1)) of HF with R(H−F) = 1.8 a.u. The depicted isosurfaces are for the energies $\epsilon = 3.0, 3.8, and 5.4$ a.u. from outer to inner region of the molecule. The electron densities on the isosurfaces are identified with the color scale shown on the horizontal bar

with the functional $E_x^{e,\mathrm{LDA}}[n^e]$. Actually, the curves yielded by $E_x^{e,\mathrm{LDA}}[n^e]$ are in good agreements with those given by conventional LDA. Note, however, that both graphs for $E_x^{e,\mathrm{LDA}}[n^e]$ are shifted downward so that their bottoms coincide with those of the conventional LDA energies to compensate the error due to the loss of electrons on the carbon cores.

We also consider the chemical bond in a hetero diatomic molecule HF. The σ bond of HF is characterized by the charge polarization due to the large electronegativity of fluoride. In Fig. 3.13a the potential energy curve for HF computed using the conventional LDA functional is compared with that provided by the LDA functional formulated in the present work (PW). As shown in the figure the potential energy curve of 'LDA (PW)' shifted upward in energy shows good agreement with the curve given by the conventional LDA although some undulation appears in the curve of 'LDA (PW)' probably due to the lack of the sampling points for the construction of $n^e(\epsilon)$. Anyway, it is demonstrated that the LDA functional using $n^e(\epsilon)$ can describe properly the potential energy of the chemical bond with significant charge polarization. Figure 3.13b shows the electron density mapped onto the isosurfaces of the bare Coulomb potential of HF. It should be noticed that the color does not varies largely on each isosurface, which suggests that the electron density $n(\boldsymbol{r})$ of HF is almost constant on the isosurface of the potential energy. This fact, of course, underlies the success of the LDA-based functional using the energy electron density n^e.

Another issue of great interest is to examine a system where the LDA-based functional $E_x^{e,\mathrm{LDA}}[n^e]$ fails seriously. It is readily recognized that the conventional LDA functional $E_x^{\mathrm{LDA}}[n]$ can be equivalently transformed to the functional $E_x^{e,\mathrm{LDA}}[n^e]$ in the form of Eq. (3.7) when the isosurfaces of the electron density $n(\boldsymbol{r})$ are strictly parallel to those of the external potential. As shown by R. G. Parr and A. Berk in

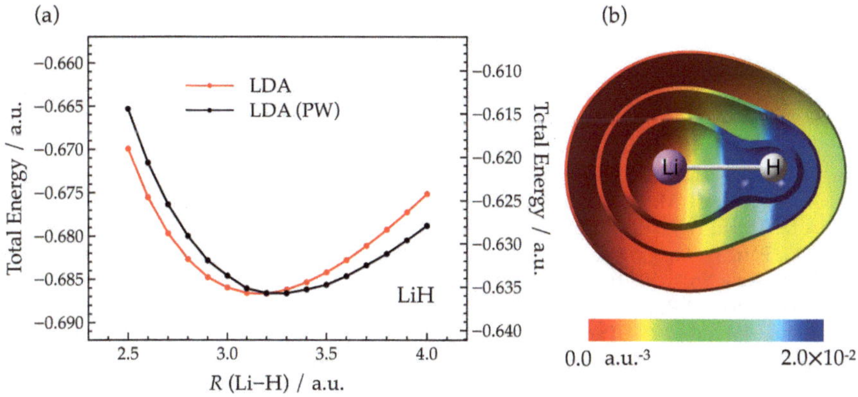

Fig. 3.14 **a** Potential energy curves of LiH molecule computed as functions of the distance R(Li−H). The other captions for the legends and the axes are the same as in Fig. 3.12. **b** The electron density mapped onto the isosurfaces of the bare Coulomb potential (Eq. (3.1)) of LiH with R(Li−H) = 3.0 a.u. The depicted isosurfaces are for the energies $\epsilon = 1.1, 1.5, and 2.1$ a.u. from outer to inner region of the molecule. The electron densities on the isosurfaces are identified with the color scale shown on the horizontal bar

Figs. 3 and 4 in Ref. [2], the contours of the bare Coulomb potential of LiH are exceptionally *not* parallel to those of the electron density given by the HF method. It is thus anticipated that the application of the functional $E_x^{e,\mathrm{LDA}}[n^e]$ in the form of Eq. (3.7) fails to describe the chemical bond of LiH. Actually, it is shown in Fig. 3.14a that the equilibrium bond distance of the molecule is elongated by the calculation with $E_x^{e,\mathrm{LDA}}[n^e]$ by $\sim$0.1 a.u. as compared to the conventional LDA functional. It is observed in Fig. 3.14b that the electron density on the isosurfaces of the bare Coulomb potential varies seriously, which shows a clear contrast to the density of HF molecule (Fig. 3.13b). This result suggests the limitation of the functional $E_x^{e,\mathrm{LDA}}[n^e]$ given by Eq. (3.7). It seems that the development of the exchange energy $E_x^e[n^e]$ merely by using the conventional LDA functional as a template is too crude to describe the system such as LiH which has the exceptional property. Note, however, that this does not mean the limitation of the novel DFT that utilizes the energy electron density n^e as a fundamental variable. To overcome the flaw in the LDA-based exchange functional $E_x^{e,\mathrm{LDA}}[n^e]$ will be an important issue in the future developments. At this stage, it has not yet been examined whether either the GGA or the BR approach improves the potential energy curve of LiH. However, as noted above, our major motivation of resorting to the novel DFT is to solve the problems arising from the static correlation errors. Next chapter will be devoted to describe the development of the static correlation functional $E_{\mathrm{sc}}^e[n^e]$.

References

1. H. Takahashi, J. Phys. B: Atomic, Molecular and Optical Physics **51**(5), 055102(11) (2018)
2. R.G. Parr, A. Berk, in *Chemical Applications of Atomic and Molecular Electrostatic Potentials*, ed. by P. Politzer, D.G. Truhlar (Plenum Press, 1981), pp. 51–62
3. N. Matubayasi, M. Nakahara, J. Chem. Phys. **113**(15), 6070 (2000)
4. N. Matubayasi, M. Nakahara, J. Chem. Phys. **117**(8), 3605 (2002)
5. A.D. Becke, Phys. Rev. A **38**(6), 3098 (1988)
6. R.G. Parr, W. Yang, *Density-Functional Theory of Atoms and Molecules* (Oxford University Press, New York, 1989)
7. A.D. Becke, M.R. Roussel, Phys. Rev. A **39**(8), 3761 (1989)
8. C. Lee, W. Yang, R.G. Parr, Phys. Rev. B **37**(2), 785 (1988)
9. J.P. Perdew, S. Kurth, Phys. Rev. Lett. **82**(12), 2544 (1999)
10. J.R. Chelikowsky, N. Troullier, Y. Saad, Phys. Rev. Lett. **72**(8), 1240 (1994)
11. J.R. Chelikowsky, N. Troullier, K. Wu, Y. Saad, Phys. Rev. B **50**(16), 11355 (1994)
12. H. Takahashi, T. Hori, T. Wakabayashi, T. Nitta, Chem. Lett. **3**, 222 (2000)
13. H. Takahashi, T. Hori, T. Wakabayashi, T. Nitta, J. Phys. Chem. A **105**(17), 4351 (2001)
14. H. Takahashi, T. Hori, H. Hashimoto, T. Nitta, J. Comp. Chem. **22**(12), 1252 (2001)
15. R. Neumann, R.H. Nobes, N.C. Handy, Mol. Phys. **87**(1), 1 (1996)
16. H. Takahashi, R. Kishi, M. Nakano, J. Chem. Theory Comput. **6**, 647 (2010)
17. M.J.F. et al., *GAUSSIAN 09, Revision C.01, Gaussian Inc.* (Wallingford, CT, 2010)

Chapter 4
Development of Static Correlation Functional

4.1 Property of the Energy Electron Density at the Dissociation Limit

The development of the density-functional theory based on the energy electron density $n^e(\epsilon)$ on the energy coordinate ϵ is motivated by a property intrinsic to the energy electron density n^e itself. An interesting feature of $n^e_\sigma(\epsilon)$ with spin σ appears clearly at the dissociation limit of a chemical bond, which makes a distinct contrast to the conventional electron density $n_\sigma(\boldsymbol{r})$. For simplicity, we consider the electron distributions $n_\sigma(\boldsymbol{r})$ and $n^e_\sigma(\epsilon)$ derived from the density of the simplest diatomic molecule, that is, an H_2 molecule at its dissociation limit. To illustrate the property of $n^e_\sigma(\epsilon)$, we introduce the reference system that consists of two isolated (infinitely separated) H atoms. It is supposed that each hydrogen atom has the ground state electron density with the opposite spins (spin polarized). On the other hand, the interacting (entangled) system is defined as an H_2 molecule with $R(\mathrm{H{-}H}) = \infty$, where the spin symmetry is supposed to be strictly adapted (spin depolarized). Note that the two electrons are fully entangled although there is no interaction between the two H atoms. The spatial electron densities of the reference and the interacting systems are indicated by n_0 and n_1, respectively. It is supposed that the total electron density $n_1(\boldsymbol{r})$ given by the sum of the contributions from both spins is identical to the density $n_0(\boldsymbol{r})$ for the reference system. To be precisely, the following relation holds,

$$
\begin{aligned}
n_1(\boldsymbol{r}) &= n_{1\alpha}(\boldsymbol{r}) + n_{1\beta}(\boldsymbol{r}) \\
&= n_{0\alpha}(\boldsymbol{r}) + n_{0\beta}(\boldsymbol{r}) \\
&= n_0(\boldsymbol{r})
\end{aligned}
\tag{4.1}
$$

The corresponding energy electron densities obtained by the projection of $n_0(\boldsymbol{r})$ and $n_1(\boldsymbol{r})$ onto the energy coordinate ϵ using Eq. (3.2) are denoted as $n^e_0(\epsilon)$ and $n^e_1(\epsilon)$, respectively.

H. Takahashi, *Density-Functional Theory for Electrons*,
https://doi.org/10.1007/978-981-95-1229-4_4

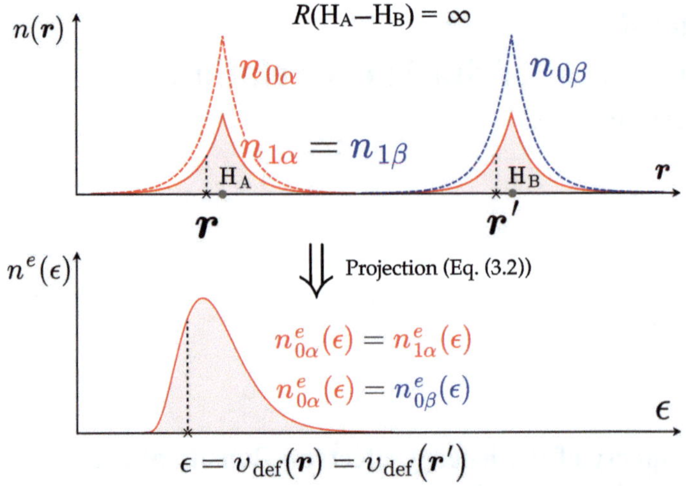

Fig. 4.1 Electron densities of an H_2 with the infinitely stretched chemical bond. Upper: Spatial electron densities $n_{0\sigma}(\boldsymbol{r})$ and $n_{1\sigma}(\boldsymbol{r})$, respectively, for the reference and the entangled states of an H_2 molecule with the infinitely stretched bond. Lower: Energy electron densities $n^e(\epsilon)$ obtained by the projection of $n(\boldsymbol{r})$ onto the energy coordinate ϵ. The objects with red color are for alpha spins, while those with blue are for beta spins. The broken curves are for the densities of the reference (symmetry broken) system. The red shaded regions show the electron populations with spin σ for the interacting system. The spatial coordinates $\boldsymbol{r}$ and $\boldsymbol{r}'$ are placed on the different atomic sites H_A and H_B, respectively, while it is assumed they have the same energy coordinate ϵ. Namely, the distance of $\boldsymbol{r}$ from the nucleus H_A is the same as that of $\boldsymbol{r}'$ from the nucleus H_B. Thus both the electron populations on $\boldsymbol{r}$ and $\boldsymbol{r}'$ are added in the density n^e at the same energy coordinate ϵ through the integration of Eq. (3.2)

In Fig. 4.1 the schematic illustrations are presented for the electron densities of the reference and the interacting systems of the infinitely stretched H_2 molecule. In the construction of the figure it is assumed that the total electron density n_0 is exactly equal to n_1, that is, $n_{0\alpha}(\boldsymbol{r}) + n_{0\beta}(\boldsymbol{r}) = n_{1\alpha}(\boldsymbol{r}) + n_{1\beta}(\boldsymbol{r})$. In the reference system the electron with a spin σ is localized on a single atomic site, while it completely delocalizes over the infinitely separated atomic sites in the interacting (entangled) system. It is supposed that the energy of the reference system with the density n_0 can be properly described with the conventional DFT. It is desirable for the exchange-correlation functional E_{xc} to evaluate correctly the energy difference ΔE_{xc} between the interacting and the reference systems. Explicitly, ΔE_{xc} is expressed as

$$\begin{aligned}\Delta E_{xc} &= E_{xc}[n_{1\alpha}] + E_{xc}[n_{1\beta}] - \left(E_{xc}[n_{0\alpha}] + E_{xc}[n_{0\beta}]\right) \\ &= 2E_{xc}[n_{1\alpha}] - 2E_{xc}[n_{0\alpha}]\end{aligned} \tag{4.2}$$

It is readily recognized that $\Delta E_{xc} = 0$ should be satisfied at the dissociation limit, namely, $E_{xc}[n_{1\alpha}] = E_{xc}[n_{0\alpha}]$ should hold. In other words, the exchange-correlation functional that incorporates the correct static correlation realizes the equation

$\Delta E_{xc} = 0$. However, it is quite difficult to achieve the relation within the conventional DFT because the density $n_{0\sigma}$ and $n_{1\sigma}$ are completely different in their spatial distributions as depicted in Fig. 4.1. In the level of the local density approximation (LDA) given in Eq. (3.5), $\Delta E_x^{\mathrm{LDA}}$ can be readily evaluated, thus,

$$
\begin{aligned}
\Delta E_x^{\mathrm{LDA}} &= E_x^{\mathrm{LDA}}[n_{1\alpha}] + E_x^{\mathrm{LDA}}[n_{1\beta}] - \left(E_x^{\mathrm{LDA}}[n_{0\alpha}] + E_x^{\mathrm{LDA}}[n_{0\beta}]\right) \\
&= 2\overline{C}_x \left(\int_{\Omega_{\mathrm{A}}} d\boldsymbol{r}\; n_{1\alpha}(\boldsymbol{r})^{\frac{4}{3}} + \int_{\Omega_{\mathrm{A}}} d\boldsymbol{r}\; n_{1\beta}(\boldsymbol{r})^{\frac{4}{3}}\right) \\
&\qquad - 2\overline{C}_x \int_{\Omega_{\mathrm{A}}} d\boldsymbol{r}\; n_{0\alpha}(\boldsymbol{r})^{\frac{4}{3}} \\
&= 2\overline{C}_x \int_{\Omega_{\mathrm{A}}} d\boldsymbol{r}\; 2 \times \left(\frac{1}{2} n_{0\alpha}(\boldsymbol{r})\right)^{\frac{4}{3}} - 2\overline{C}_x \int_{\Omega_{\mathrm{A}}} d\boldsymbol{r}\; n_{0\alpha}(\boldsymbol{r})^{\frac{4}{3}} \\
&= \left(\left(\frac{1}{2}\right)^{\frac{1}{3}} - 1\right) \times 2\overline{C}_x \int_{\Omega_{\mathrm{A}}} d\boldsymbol{r}\; n_{0\alpha}(\boldsymbol{r})^{\frac{4}{3}} \\
&= -0.20629\cdots \times 2\overline{C}_x \int_{\Omega_{\mathrm{A}}} d\boldsymbol{r}\; n_{0\alpha}(\boldsymbol{r})^{\frac{4}{3}}
\end{aligned}
\tag{4.3}
$$

where Ω_{A} represents the spatial region which encloses the whole electron density of the atom H_{A} and $\overline{C}_x$ is defined as $\overline{C}_x = 2^{1/3} C_x^{\mathrm{LDA}} = \frac{3}{4}\left(\frac{6}{\pi}\right)^{1/3}$ (see Appendix A). The multiplication of 2 in the right hand side of the second equality is to incorporate the contribution from the atom H_{B}. The first term of the right hand side of the third equality is obtained from the relation $n_{1\alpha}(\boldsymbol{r}) = \frac{1}{2} n_{0\alpha}(\boldsymbol{r})$. The last line of Eq. (4.3) shows that the difference $\Delta E_x^{\mathrm{LDA}}$ amounts to more than 20 % of the exchange energy $E_x^{\mathrm{LDA}}[n_0]$ itself of the whole system although $\Delta E_x^{\mathrm{LDA}} = 0$ must be satisfied. Actually, $\Delta E_x^{\mathrm{LDA}}$ is explicitly evaluated as −69.4 kcal/mol by adopting the exact ground state wavefunction of the hydrogen atom to $n_{0\alpha}(\boldsymbol{r})$ in Eq. (4.3). Seemingly, such a huge error cannot be overcome by merely applying any GGA correction to the LDA energy. Furthermore, it is also useful to make a remark on the intermediately coupled system at $R(\mathsf{H}-\mathsf{H}) = \infty$. The electron density $n_{\lambda\sigma}$ of the intermediate system can be represented e.g. by the linear combination of the densities of the reference and the fully interacting (entangled) systems, thus,

$$
n_{\lambda\sigma}(\boldsymbol{r}) = (1-\lambda) n_{0\sigma}(\boldsymbol{r}) + \lambda n_{1\sigma}(\boldsymbol{r}) \qquad (0 \le \lambda \le 1) \tag{4.4}
$$

Accordingly, the projection of $n_{\lambda\sigma}(\boldsymbol{r})$ onto the energy coordinate ϵ is given by

$$
n_{\lambda\sigma}^{e}(\epsilon) = (1-\lambda) n_{0\sigma}^{e}(\epsilon) + \lambda n_{1\sigma}^{e}(\epsilon) \qquad (0 \le \lambda \le 1) \tag{4.5}
$$

It is readily recognized that the *perfect* exchange-correlation functional offers the following relation irrespective of the value of λ,

$$
\Delta E_{xc}^{\lambda} = 2E_{xc}[n_{\lambda\alpha}] - 2E_{xc}[n_{0\alpha}] = 0, \tag{4.6}
$$

which imposes a stringent condition on the development of the actual exchange-correlation functional. Of course, there exists no perfect wave function or density-functional theory that is capable of realizing Eq. (4.6).

Remind that the failure of the conventional DFT functional can be understood in terms of the exchange hole distribution as discussed in Sect. 2.3.1. The on-top depth of the exchange hole by the LDA is estimated to be half of the true exchange hole although the hole is normalized to 1.0 and localized at the reference point $\boldsymbol{r}$. To compensate the deficit of the hole, e.g. A. D. Becke devised a method [1] in 2003 to deepen the hole with a spin σ by mixing the fraction of the exchange hole of the electron with the opposite spin at $\boldsymbol{r}$, for which the outline was illustrated in Sect. 2.3.2. We note, however, that any physical justification cannot be provided for the Becke03 functional [1] although it might be effective. The difficulty in describing the static correlation with the conventional DFT lies in the fact that the LDA-based approximate functionals refer only to the local or semi-local properties, i.e. the density n or its gradient $|\nabla n|$ at the reference point $\boldsymbol{r}$. Obviously, such quantities are not sufficient in principle to yield the correct exchange energy E_{xc} for the wave functions delocalized over the two infinitely separated atomic sites. In the next paragraph, we show that the DFT in terms of the energy electron density $n^e(\epsilon)$ has a potential to conquer the problems arising from the nonlocal property of the static correlation.

The lower figure in Fig. 4.1 shows the projections of the densities $n_{0\sigma}(\boldsymbol{r})$ and $n_{1\sigma}(\boldsymbol{r})$ onto the energy coordinate ϵ. It is of major importance to notice that both densities $n_{0\sigma}$ and $n_{1\sigma}$ have the same distributions on the energy coordinate, thus, $n^e_{0\sigma}(\epsilon) = n^e_{1\sigma}(\epsilon)$. The relation can be readily proved from the assumption that the two atoms are infinitely separated. It reads,

$$
\begin{aligned}
n^e_{1\alpha}(\epsilon) &= \int d\boldsymbol{r}\; n_{1\alpha}(\boldsymbol{r})\delta\left(\epsilon - \upsilon_{\mathrm{def}}(\boldsymbol{r})\right) \\
&= \int_{\Omega_{\mathrm{A}}} d\boldsymbol{r}\; n_{1\alpha}(\boldsymbol{r})\delta\left(\epsilon - \upsilon_{\mathrm{def}}(\boldsymbol{r})\right) + \int_{\Omega_{\mathrm{B}}} d\boldsymbol{r}\; n_{1\alpha}(\boldsymbol{r})\delta\left(\epsilon - \upsilon_{\mathrm{def}}(\boldsymbol{r})\right) \\
&= 2\int_{\Omega_{\mathrm{A}}} d\boldsymbol{r}\; n_{1\alpha}(\boldsymbol{r})\delta\left(\epsilon - \upsilon_{\mathrm{def}}(\boldsymbol{r})\right) \\
&= \int d\boldsymbol{r}\; n_{0\alpha}(\boldsymbol{r})\delta\left(\epsilon - \upsilon_{\mathrm{def}}(\boldsymbol{r})\right) \\
&= n^e_{0\alpha}(\epsilon)
\end{aligned}
\tag{4.7}
$$

where Ω_{A} stands for the spatial subspace that encloses only the wave function of $\mathrm{H_A}$. The same also applies to the electrons with β spin. It can also be readily understood that $n^e_{0\alpha}(\epsilon) = n^e_{0\beta}(\epsilon)$. The contents described in the above equation can be understood more easily by an illustration depicted in Fig. 4.2. The red circle placed on each contour graph for the spin-depolarized or polarized electron density n_σ represents the spatial region with the same energy coordinate ϵ defined by the potential $\upsilon_{\mathrm{def}}(\boldsymbol{r})$. When the projection of Eq. (3.2) is applied to the depolarized density $n_{1\sigma}$, the two electron populations on the red circles placed on the two atomic sites are projected onto a single population on the same energy coordinate ϵ. The population $n^e_{1\sigma}(\epsilon)$ is

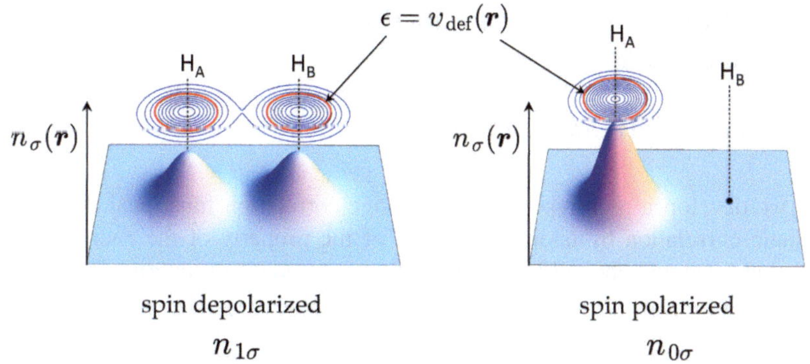

Fig. 4.2 Electron density $n_\sigma(\boldsymbol{r})$ for spin $\sigma = \alpha$ or β of an H_2 molecule at the dissociation limit. Left: $n_{1\sigma}$ in the system with the spin-depolarized (spin-adapted) density. Right: $n_{0\sigma}$ for the spin-polarized (symmetry broken) density. On the left, the density with the opposite spin has the same population as $n_{1\sigma}$. On the right, it is supposed that the density with the opposite spin is located at the atomic site of H_B. Red circles indicate the spatial regions with the energy coordinate ϵ

exactly the same as $n^e_{0\sigma}(\epsilon)$ for the spin-polarized density $n_{0\sigma}$. Thus, $n_{0\sigma}(\boldsymbol{r})$ in the spin-polarized density is different from $n_{1\sigma}(\boldsymbol{r})$ in the depolarized density, but they are identical in the distribution projected onto the energy coordinate.

Here, we revisit Eq. (4.2) for the exchange-correlation energy E^e_{xc} as a functional of the energy electron density n^e, thus,

$$\Delta E^e_{xc} = 2E^e_{xc}[n^e_{1\alpha}] - 2E^e_{xc}[n^e_{0\alpha}] = 0 \tag{4.8}$$

By virtue of the equality of Eq. (4.7), the relation of Eq. (4.8), that is desired at the dissociation limit, is guaranteed for any choice of the approximate functional E^e_{xc}. Furthermore, the projected density $n^e_{\lambda\sigma}$ of the intermediate density $n_{\lambda\sigma}$ also satisfies the relation

$$n^e_{\lambda\sigma} = n^e_{0\sigma} \tag{4.9}$$

regardless of the coupling parameter λ. Hence, the equation parallel to Eq. (4.6) is automatically realized, thus,

$$\Delta E^{e,\lambda}_{xc} = 2E^e_{xc}[n^e_{\lambda\alpha}] - 2E^e_{xc}[n^e_{0\alpha}] = 0. \tag{4.10}$$

where $n^e_{\lambda\alpha}$ is defined in Eq. (4.5). This property constitutes the major advantage in the construction of the exchange-correlation functional based on the energy electron density since the desired condition at the dissociation limit is intrinsically fulfilled. In the next Section, we develop a novel functional E^e_{xc} by taking advantage of the property of n^e.

4.2 Exchange-Correlation Functional Realizing Static Correlation

4.2.1 Static Correlation Functional

Here we develop a prototype of the exchange-correlation functional that realizes the static correlation by taking advantage of the property of the energy electron density $n^e(\epsilon)$ noted above [2]. To this end, we introduce an intermediate electron density $n_0(\boldsymbol{r})$ which serves as the density of a reference system. The density $n_0(\boldsymbol{r})$ is constructed merely by the overlap of the densities of the isolated fragments of the whole system of interest. Suppose that $n_0^p(\boldsymbol{r})$ is the density of the fragment p in the system. Then, $n_0(\boldsymbol{r})$ is given by

$$n_0(\boldsymbol{r}) = \sum_p^{N_{\text{frag}}} n_0^p(\boldsymbol{r}) \tag{4.11}$$

where N_{frag} is the number of the fragments comprising the system. The smallest unit fragment is obviously the constituent atoms in the system. However, as we are interested in describing a bond dissociation, the minimum fragmentation can be identified by $N_{\text{frag}} = 2$. Constructing the electron density of the whole system from the those of the constituent fragments is motivated by the method of Harris functional [3], the frozen density functional approach [4] and the partition DFT method [5]. Apart from the issue of the static correlation, it seems quite excellent to build the electron density of the whole system from a reference density n_0 given by its fragments (Eq. (4.11)) since we are only interested in the energy changes associated with some physical or chemical events occurring in the system and not in the total electronic energy itself.

To construct the exchange-correlation functional E_{xc} that realizes the static correlation, we exploit the reference density n_0 of Eq. (4.11) as an argument. It is assumed that the energies of the fragments as well as the densities are obtained beforehand by the conventional DFT or some wave function theory. To facilitate the explanation of the functional, we again consider a hydrogen molecule in the electronic ground state as the simplest model system (see Fig. 4.3). It is assumed that two hydrogen atoms $\mathsf{H_A}$ and $\mathsf{H_B}$ are separated with a distance R. Then the total electron density $n_0(\boldsymbol{r})$ of the reference system is given by

$$n_0(\boldsymbol{r}) = n_0^{\text{A}}(\boldsymbol{r}) + n_0^{\text{B}}(\boldsymbol{r}) \tag{4.12}$$

where n_0^{A} and n_0^{B} are the electron densities of the atoms $\mathsf{H_A}$ and $\mathsf{H_B}$ at their isolations, respectively. It is important to note that these two electrons have opposite spins so that the sum of s_z of the fragments matches S_z of the whole system, namely, $S_z = s_z^{\text{A}} + s_z^{\text{B}}$. Therefore, the densities $n_{0\alpha}$ and $n_{0\beta}$ are, respectively, given by n_0^{A} and n_0^{B}. Thus the reference system is *not* spin-adapted obviously. Suppose that these two isolated

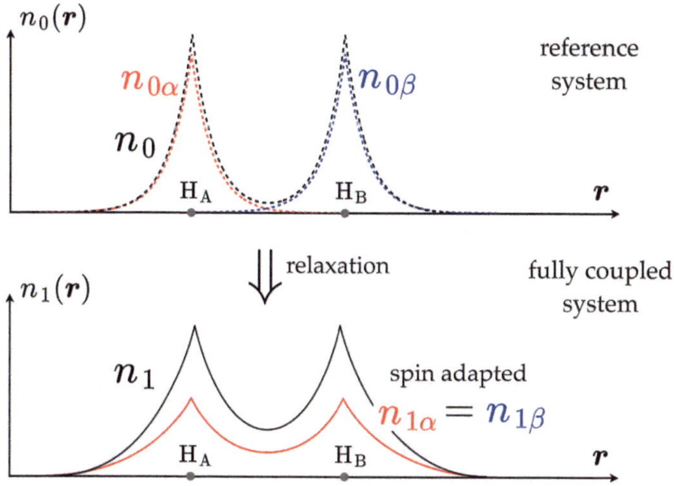

Fig. 4.3 Upper: Illustration of the construction of the reference electron density $n_0(\boldsymbol{r})$ for an H_2 molecule from its fragment densities $n_0^{\mathrm{A}}(\boldsymbol{r}) = n_{0\alpha}(\boldsymbol{r})$ and $n_0^{\mathrm{B}}(\boldsymbol{r}) = n_{0\beta}(\boldsymbol{r})$, respectively, for the constituent atoms H_A and H_B. n_0 is merely given by the overlap of the fragment densities. (see Eq. (4.12)) The reference density n_0 is symmetry broken. Lower: Two atoms H_A and H_B are fully coupled to form a σ bond with a spin-adapted electron density $n_1(\boldsymbol{r})$. When the distance between the two atoms are infinitely elongated n_0 coincides with n_1 in principle

systems identified by A and B are fully coupled to form the spin-adapted clcctron density $n_1(\boldsymbol{r})$. Now we formulate an exchange-correlation functional $E_{xc}[n_1, n_0]$ that realizes the static correlation thus

$$E_{xc}\left[n_1, n_0\right] = E_x\left[n_0\right] + \Delta E_x^{e,\mathrm{relax}}\left[n_1^e, n_0^e\right] \tag{4.13}$$

where only the static correlation is considered and the dynamical correlation is excluded for simplicity. It is assumed that the exchange energy $E_x\left[n_0\right]$ of the reference system can be computed adequately with a certain approximate functional of the conventional DFT. It is important to notice that the relaxation term $\Delta E_x^{e,\mathrm{relax}}\left[n_1^e, n_0^e\right]$ in Eq. (4.13) is described in terms of the energy electron densities $n_1^e(\epsilon)$ and $n_0^e(\epsilon)$. The difference $\Delta E_x^{e,\mathrm{relax}}$ is written as

$$\Delta E_x^{e,\mathrm{relax}}\left[n_1^e, n_0^e\right] = E_x^e\left[n_1^e\right] - E_x^e\left[n_0^e\right] \tag{4.14}$$

where the exchange functional E_x^e can be described with the LDA functional $E_x^{e,\mathrm{LDA}}$ in Eq. (3.7) or the GGA functional $E_x^{e,\mathrm{B88}}$ in Eq. (3.74). Note that the relaxation term includes the energy change associated with the change of the electron density from the broken symmetry to the spin-adapted distribution. It is valuable to consider the behavior of $\Delta E_x^{e,\mathrm{relax}}$ at the dissociation limit of H_2. Since the energy of the two isolated hydrogen atoms is in principle equal to the energy of the two infinitely separated but entangled hydrogen atoms, $\Delta E_x^{e,\mathrm{relax}}$ should vanish completely. As shown in Eq. (4.8), the relation of $\Delta E_x^{e,\mathrm{relax}} = 0$ is always fulfilled regardless of the choice of the functional E_x^e. By substituting this expression to Eq. (4.13) we obtain,

$$\begin{aligned} E_{xc}[n_1, n_0] &= E_x[n_0] + \left(E_x^e\left[n_1^e\right] - E_x^e\left[n_0^e\right]\right) \\ &= E_x^e\left[n_1^e\right] + \left(E_x[n_0] - E_x^e\left[n_0^e\right]\right) \end{aligned} \tag{4.15}$$

where the static-correlation functional E_{sc} can be defined as

$$E_{sc}[n_0] = E_x[n_0] - E_x^e\left[n_0^e\right] \tag{4.16}$$

Thus, in the present development, the E_{sc} functional is solely evaluated from the reference electron density n_0 and its projection onto the energy coordinate. The remaining contribution to Eq. (4.15) is the exchange energy $E_x^e\left[n_1^e\right]$ as a functional of the energy electron density. As demonstrated in the previous Subsection, the functional $E_x^e\left[n_1^e\right]$ can describe adequately the potential energy curves of chemical bonds around their equilibrium bond distances. We also make a remark on a numerical implementation of Eq. (4.15). It can be readily performed e.g. by adopting the local-density approximation (LDA) for the calculations of E_x and E_x^e. In this case, Eq. (4.15) can be expressed as

$$E_{xc}^{\mathrm{LDA}}[n_1, n_0] = E_x^{e,\mathrm{LDA}}\left[n_1^e\right] + \left(E_x^{\mathrm{LDA}}[n_0] - E_x^{e,\mathrm{LDA}}\left[n_0^e\right]\right) \tag{4.17}$$

where $E_x^{e,\mathrm{LDA}}$ is provided in Eq. (3.7) in Sect. 3.1. Of course, GGA (generalized gradient approximation) functional can also be applied to Eq. (4.15) in parallel to Eq. (4.17).

A notable feature of the functional of Eq. (4.15) is that the functional $E_{sc}[n_0]$ is just added to $E_x^e\left[n_1^e\right]$ and the density n_1 itself is to be optimized without the influence of the static correlation. Thus the density n_1 under the static correlation must be prepared by performing another calculation beforehand. Obviously this is a major drawback of Eq. (4.15). In the next Subsection, we provide another approach to construct an exchange-correlation functional by means of the coupling parameter integration. This allows one to incorporate the static correlation into the SCF variational calculation for the density n_1 of the correlated system.

4.2.2 Coupling Parameter Integration

This Subsection will be devoted to describe approximately the difference $\Delta E_x^{e,\mathrm{relax}}$ in Eq. (4.14) with the coupling-parameter integration scheme. Explicitly, $\Delta E_{xc}^{e,\mathrm{relax}}$ is formally expressed as

$$\Delta E_x^{e,\mathrm{relax}} = \int_0^1 d\lambda \frac{\partial E_x^e\left[n_\lambda^e\right]}{\partial \lambda} \tag{4.18}$$

Note that we omit the spin indexes in these equations for the sake of simplicity and n^e_λ is defined in Eq. (4.5). The major purpose of introducing the coupling-parameter integration is to obtain an approach to optimize self-consistently the density n_1 using n^e_0 as the reference density. Since, at the dissociation limit, the density n^e_0 serves as the distribution of the fully correlated electrons, it is reasonable to adopt n^e_0 as the reference distribution. The exchange-correlation energy E_{xc} in Eq. (4.13) can be rewritten as,

$$E_{xc}[n_1, n_0] = E_x[n_0] + \int_0^1 d\lambda \frac{\partial E^e_x\left[n^e_\lambda\right]}{\partial \lambda} \tag{4.19}$$

The right hand side of Eq. (4.18) can be transformed to

$$\begin{aligned}\int_0^1 d\lambda \frac{\partial E^e_x\left[n^e_\lambda\right]}{\partial \lambda} &= \int_0^1 d\lambda \int d\epsilon \frac{\partial \eta^e_\lambda(\epsilon)}{\partial \lambda} \cdot \frac{\delta E^e_x\left[n^e_\lambda\right]}{\delta n^e_\lambda(\epsilon)} \\ &= \int d\epsilon \left(n^e_1(\epsilon) - n^e_0(\epsilon)\right) \int_0^1 d\lambda\, v^e_x\left[n^e_\lambda\right](\epsilon)\end{aligned} \tag{4.20}$$

where v^e_x is the exchange potential defined by $v^e_x[n^e](\epsilon) = \frac{\delta E^e_x[n^e]}{\delta n^e(\epsilon)}$. In deriving the second equality in Eq. (4.20) it is assumed that the density n^e_λ varies linearly with respect to the coupling parameter λ as shown in Eq. (4.5). Further, we apply the linear-response approximation to the potential v^e_x, thus,

$$v^e_x\left[n^e_\lambda\right](\epsilon) = v^e_x\left[n^e_0\right](\epsilon) + \int d\epsilon' \left.\frac{\delta v^e_x\left[\eta^e_\lambda\right](\epsilon)}{\delta n^e_\lambda(\epsilon')}\right|_{\lambda=0} \left(n^e_\lambda(\epsilon') - n^e_0(\epsilon')\right) \tag{4.21}$$

where the potential $v^e_x\left[n^e_0\right](\epsilon)$ is given by

$$v^e_x\left[n^e_0\right](\epsilon) = \frac{\delta E^e_x[n^e_0]}{\delta n^e_0(\epsilon)} \tag{4.22}$$

With this approximation, it is possible to evaluate the integral in Eq. (4.20) thus,

$$\begin{aligned}E_{xc}\left[n_1, n_0\right] &= E_x\left[n_0\right] \\ &\quad + \int d\epsilon \left(n^e_1(\epsilon) - n^e_0(\epsilon)\right) \frac{1}{2}\left(v^e_{xc}\left[n^e_1\right](\epsilon) + v^e_x\left[n^e_0\right](\epsilon)\right)\end{aligned} \tag{4.23}$$

In this equation we note that the potential at $\lambda = 1$ is denoted as $v^e_{xc}\left[n^e_1\right](\epsilon)$ specifically since the potential is supposed to include the static-correlation potential. By the functional derivative of $E_{xc}\left[n_1, n_0\right]$ with respect to the density n^e_1, we obtain the exchange-correlation potential

$$
\begin{aligned}
v_{xc}^{e}\left[n_1^e\right](\epsilon) &= \frac{\delta E_{xc}\left[n_1, n_0\right]}{\delta n_1^e(\epsilon)} \\
&= \frac{1}{2}\left(v_{xc}^{e}\left[n_1^e\right](\epsilon) + v_x^e\left[n_0^e\right](\epsilon)\right) \\
&\quad + \frac{1}{2}\int d\epsilon' \left.\frac{\delta v_x^e\left[n^e\right](\epsilon)}{\delta n^e(\epsilon')}\right|_{n^e=n_1^e}\left(n_1^e(\epsilon') - n_0^e(\epsilon')\right)
\end{aligned}
\tag{4.24}
$$

After a simple algebraic manipulation, one obtains

$$
v_{xc}^{e}\left[n_1^e\right](\epsilon) = v_x^e\left[n_0^e\right](\epsilon) + \int d\epsilon' \left.\frac{\delta v_x^e\left[n^e\right](\epsilon)}{\delta n^e(\epsilon')}\right|_{n^e=n_1^e}\left(n_1^e(\epsilon') - n_0^e(\epsilon')\right) \tag{4.25}
$$

Since the linearity of the exchange potential v_x^e is assumed, the equation above can be equivalently transformed to

$$
v_{xc}^{e}\left[n_1^e\right](\epsilon) = v_x^e\left[n_0^e\right](\epsilon) + \int d\epsilon' \left.\frac{\delta v_x^e\left[n^e\right](\epsilon)}{\delta n^e(\epsilon')}\right|_{n^e=n_0^e}\left(n_1^e(\epsilon') - n_0^e(\epsilon')\right) \tag{4.26}
$$

It is valuable to notice the coincidence of Eq. (4.26) with Eq. (4.21) at $\lambda = 1$. Since the difference $\delta n^e = n_1^e - n_0^e$ in the energy electron density is anticipated to be sufficiently small even at the dissociation limit, it is quite natural to introduce the approximation that the potential v_{xc}^e varies linearly with respect to δn^e. Of course, the approximation does not work when one uses the conventional density $n(\boldsymbol{r})$ as the variable because the spatial distribution $n_1(\boldsymbol{r})$ differs from $n_0(\boldsymbol{r})$ significantly at the dissociation limit.

Now let us construct an explicit functional based on Eqs. (4.23) and (4.26) for numerical applications. The most preliminary approximation, namely the local-density approximation (LDA) of Eq. (3.7), is utilized to construct a prototype of the functional $E_{xc}\left[n_1, n_0\right]$. Note that our major concern will be placed on the computation of the second term of Eq. (4.23) described with the energy electron density. On the other hand, the exchange energy $E_x[n_0]$ for the reference system is to be computed using a conventional DFT functional. Anyway, the exchange potential $v_x^e[n^e](\epsilon)$ for each spin is obtained as

$$
\begin{aligned}
v_x^{e,\mathrm{LDA}}\left[\widetilde{n}^e\right](\epsilon) = \frac{\delta \widetilde{n}^e(\epsilon)}{\delta n^e(\epsilon)}\frac{\delta E_x^{e,\mathrm{LDA}}\left[\widetilde{n}^e\right]}{\delta \widetilde{n}^e(\epsilon)} &= \frac{1}{\Omega(\epsilon)} \times \left(-\frac{4}{3}\right) 2^{\frac{1}{3}} C_x^{\mathrm{LDA}} \widetilde{n}^e(\epsilon)^{\frac{1}{3}}\, \Omega(\epsilon) \\
&= -\frac{4}{3} 2^{\frac{1}{3}} C_x^{\mathrm{LDA}} \widetilde{n}^e(\epsilon)^{\frac{1}{3}}
\end{aligned}
\tag{4.27}
$$

where the relation of Eq. (3.6) is used. The factor C_x^{LDA} is that defined in Appendix A. Then the functional derivative of the potential v_x with respect to n^e becomes

$$\left.\frac{\delta v_x^{e,\mathrm{LDA}}\left[\widetilde{n}^e\right](\epsilon)}{\delta n^e(\epsilon')}\right|_{n^e=n_0^e} = \frac{\delta \widetilde{n}_0^e(\epsilon')}{\delta n_0^e(\epsilon')}\frac{\delta v_x^{e,\mathrm{LDA}}\left[\widetilde{n}_0^e\right](\epsilon)}{\delta \widetilde{n}_0^e(\epsilon')}\delta(\epsilon-\epsilon')$$
$$= \frac{1}{\Omega(\epsilon)}\left(-\frac{4}{9}\right)2^{\frac{1}{3}}C_x^{\mathrm{LDA}}\widetilde{n}_0^e(\epsilon)^{-\frac{2}{3}} \quad (4.28)$$

By substituting Eqs. (4.27) and (4.28) to Eq. (4.23) we get $E_{xc}^{\mathrm{LDA}}[n_1, n_0]$ thus

$$E_{xc}^{\mathrm{LDA}}[n_1, n_0] = E_x^{\mathrm{LDA}}[n_0]$$
$$-\frac{4}{3}2^{\frac{1}{3}}C_x^{\mathrm{LDA}}\int d\epsilon\ \delta n^e(\epsilon)\left(\widetilde{n}_0^e(\epsilon)^{\frac{1}{3}} + \frac{1}{6}\widetilde{n}_0^e(\epsilon)^{\frac{1}{3}}\frac{\delta n^e(\epsilon)}{n_0^e(\epsilon)}\right) \quad (4.29)$$

where $\delta n^e(\epsilon)$ is defined as $\delta n^e = n_1^e - n_0^e$. The density n_1 is thus optimized under the influence of the reference electron density n_0.

Before we provide the results of the test calculations for the static-correlation functionals, we make some remarks on the E_{xc} functional of Eq. (4.29). First, although the construction of Eq. (4.23) assumes the linearity of the exchange potential v_x^e with respect to the density n_λ^e, the LDA potential $v_x^{e,\mathrm{LDA}}$ is obviously not linear. However, it is reasonably anticipated that the coupling parameter integration assuming the linearity works adequately because the density difference δn^e itself is considered to be sufficiently small. Second, the functionals in Eqs. (4.17) and (4.29) are not designed to guarantee that n_1^e coincides with n_0^e at the dissociation limit of the chemical bond under consideration, which is considered a major flaw in the functional.

In Fig. 4.4 the potential energy curves (PECs) for the dissociation of the σ bond in H_2 are presented. For the computation of the PECs, we conduct our house code 'Vmol' that utilizes the real-space grids (RSGs) [7–11] to represent the wave functions and densities. We refer the readers to Appendix F and the literature [9–11] for the details of the code. The energy electron densities $n^e(\epsilon)$ are also constructed on a discrete energy coordinate. We refer readers to Ref. [2] for the computational details. As shown in the figure the dissociation energy $D_e = 145.4$ kcal/mol by the conventional LDA functional E_x^{LDA} seriously overshoots the experimental value 109 kcal/mol with the zero-point vibrational energy removed. The PEC of the LDA functional $E_x^{e,\mathrm{LDA}}$ in Eq. (3.7) shows a faithful agreement with that given by the functional E_x^{LDA}. However, it is demonstrated that the PEC provided by the functional $E_{xc}^{e,\mathrm{LDA}}$ in Eq. (4.17), which is supposed to realize the static correlations, substantially alleviates the static-correlation error in the dissociation curves of the LDA functional. Actually, D_e is obtained as 109.6 kcal/mol, which shows a good agreement with the experimental value. However, it should be noticed that the dynamical correlation E_{dc} is not considered in the construction of Eq. (4.17). Since the correlation E_{dc} becomes significant in the bonding region, D_e will become larger when E_{dc} is somehow included in the potential energy. E_{dc} can be evaluated e.g. by subtracting the energy given by 'CAS(2,2)' from that by 'CAS(2,2)-MP2', which leads to $E_{dc} = -11.4$ kcal/mol at $R(\mathsf{H}-\mathsf{H}) = 1.4$ a.u. Anyway, it is shown that Eq. (4.17) improves the PEC of the conventional LDA (Eq. (3.5)) with the static-correlation error by ~36 kcal/mol. The other static-correlation functional in Eq. (4.29), derived from the coupling-parameter integration method, gives the PEC that is in good agreement with that given by Eq. (4.17). Using Eq. (4.29), D_e is obtained as 109.4 kcal/mol. Thus, it is shown that the use of Eq. (4.29) reduces the dissociation energy D_e of the LDA functional by 36 kcal/mol, which directly demonstrates the effectiveness of the equation as a static-correlation functional. It should also be stressed that the static-correlation functionals do *not* employ any adjustable parameters. However, as can be seen from the comparison in Fig. 4.4 between the PEC given by 'CAS(2,2)' and those by the static-correlation functionals $E_{xc}^{e,\text{LDA}}$ in Eqs. (4.17) and (4.29), $E_{xc}^{e,\text{LDA}}$ overestimate the depth of the potentials of the intermediate atomic distances, namely around 2.0 a.u. $\leq R(\text{H}-\text{H}) \leq$ 5.0 a.u. As a consequence the equilibrium bond distance of H_2 is elongated in the PECs of the functionals $E_{xc}^{e,\text{LDA}}$ by ~0.2 a.u. Obviously, the artificial lowering of the potential energies in the intermediately dissociated region is a flaw of the present static-correlation functionals. A development to improve the functional(s) is a major subject of the next subsection.

The functional in Eq. (4.29) is also applied to the bond dissociations of the π bond in ethene (C_2H_4). The dissociation curve for the double bond is presented in Fig. 4.5 where the PEC computed by the UCCSD(T) calculation is also provided to

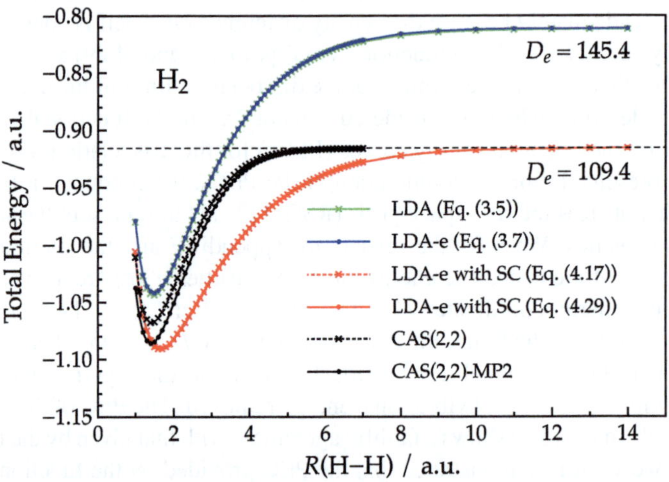

Fig. 4.4 Potential energy curves (PECs) for the dissociation of H_2 molecule computed by the conventional LDA functional E_x^{LDA} in Eq. (3.5), and the LDA functional with the energy electron density, namely $E_x^{e,\text{LDA}}$ in Eq. (3.7). The PECs computed with the static-correlation functionals $E_{xc}^{e,\text{LDA}}$ in Eqs. (4.17) and (4.29) are also presented [2]. The values D_e are the dissociation energies in the units of kcal/mol computed by Eqs. (3.5) and (4.29). The PEC with the legend 'CAS(2,2)' represents the result given by CASSCF calculation where the 2 electrons are correlated within the 2 orbitals (HOMO and LUMO) to realize the static correlation. 'CAS(2,2)-MP2' stands for the 'CAS(2,2)' calculation with the dynamical correlation computed by Møller-Plesset 2nd-order perturbation theory. The CASSCF calculations are performed by conducting the 'Gaussian 09' program package [6] with aug-cc-pVQZ basis set. The PECs by the CASSCF calculations are shifted upwards by 0.084 a.u. to match the energy of its dissociation limit to that given by the static-correlation functional based on the energy electron density

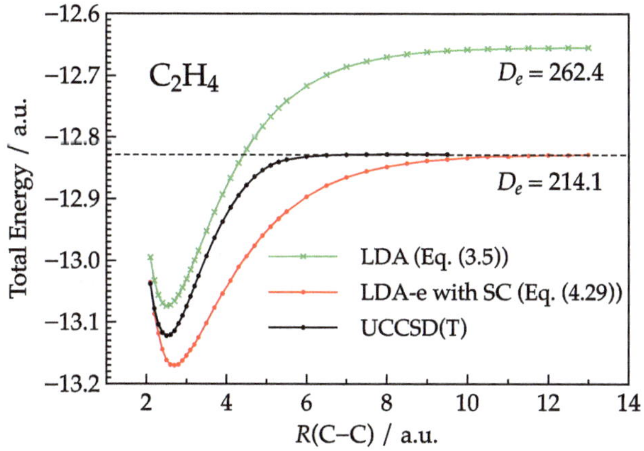

Fig. 4.5 Potential energy curves (PECs) for the dissociation of the double bond in ethene molecule (C_2H_4) computed by the conventional LDA functional $E_x^{\rm LDA}$ in Eq. (3.5) and the static correlation functionals $E_{xc}^{e,\rm LDA}$ in Eq. (4.29) [2]. The values D_e are the dissociation energies in the units of kcal/mol computed by Eqs. (3.5) and (4.29). The PEC computed by the UCCSD(T) calculation is also shown as a reference. The UCCSD(T) calculation is performed by conducting the 'Gaussian 09' program package [6] with cc-pVQZ basis set. The PECs by UCCSD(T) are shifted upwards by 65.3394 a.u. to match the energy at the dissociation limit to that of the static-correlation functional based on the energy electron density. In stretching the C-C bond, the internal geometries of the CH_2 groups are being fixed at those in the molecule(C_2H_4) at the equilibrium structure

make comparisons. The dissociation energy D_e is computed as 262.4 kcal/mol by the LDA functional, while it is reduced to 214.1 kcal/mol by considering the static correlation using the functional of Eq. (4.29). On the other hand, D_e was computed as 184.7 kcal/mol by the UCCSD(T) calculation with the cc-pVQZ basis set although the direct comparison cannot be made since the PEC by the UCCSD(T) also includes the dynamical correlation E_{dc}. It is anticipated that the inclusion of E_{dc} will deepen the PEC around the bonding region. Thus, the functional of Eq. (4.29) still suffers from the static-correlation error by ~30 kcal/mol at least. The error can be attributed to the overestimation of the static correlation from the bonding to the intermediately dissociated region.

In Fig. 4.6, the dissociation curves of a σ bond in methane (CH_4) are presented. In this heterolytic dissociation, as differed from H_2, a care must be taken for the construction of the exchange potential in Eq. (4.29). In the homolytic fission as in the H_2 molecule, the projected electron density n_0^e for each spin is automatically spin adapted. However, in the heterolytic fission, n_0^e cannot maintain the spin symmetry. Thus, the energy due to the spin symmetrization must be considered separately. Explicitly, the energy difference

$$E_x^{e,\rm LDA}[\widetilde{n}_0^{e,\rm SA}] - E_x^{e,\rm LDA}[\widetilde{n}_0^{e,\rm SB}] \qquad (4.30)$$

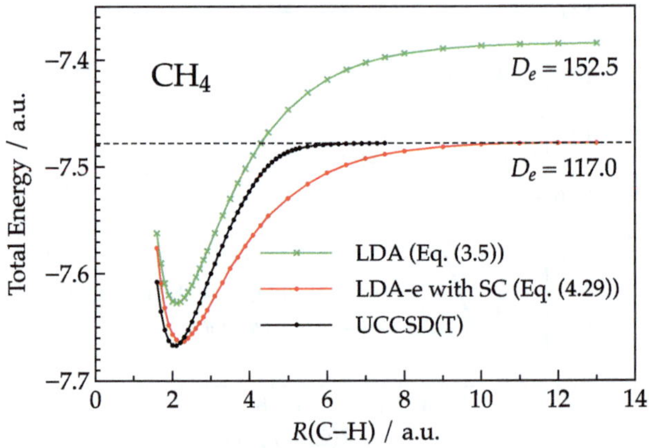

Fig. 4.6 Potential energy curves (PECs) for the dissociation of the σ bond in methane molecule (CH_4) computed by the conventional LDA functional E_x^{LDA} in Eq. (3.5) and the static-correlation functionals $E_{xc}^{e,\mathrm{LDA}}$ in Eq. (4.29) [2]. The values D_e are the dissociation energies in the units of kcal/mol. The PECs by UCCSD(T) are shifted upwards by 32.7835 a.u. to match the energy of the dissociation limit to that given by the static-correlation functional. In stretching the C-H bond, the internal geometry of the CH_3 groups is being fixed at that in the molecule (CH_4) at the equilibrium structure. The other notations are the same as in Fig. 4.5

must be added to Eq. (4.29), where $\widetilde{n}^{e,\mathrm{SA}}(\epsilon)$ is the spin-adapted energy electron density and $\widetilde{n}^{e,\mathrm{SB}}(\epsilon)$ is the symmetry-broken energy electron density. The spin-adapted densities for each spin (α, β) can be merely constructed by

$$\widetilde{n}_{0\alpha}^{e,\mathrm{SA}}(\epsilon) = \widetilde{n}_{0\beta}^{e,\mathrm{SA}}(\epsilon) = \frac{1}{2}\left(\widetilde{n}_{0\alpha}^{e,\mathrm{SB}}(\epsilon) + \widetilde{n}_{0\beta}^{e,\mathrm{SB}}(\epsilon)\right) \tag{4.31}$$

where the symmetry broken electron density $\widetilde{n}_{0\sigma}^{e,\mathrm{SB}}(\epsilon)$ is, of course, given by the projection of the electron density $n_{0\sigma}(\boldsymbol{r})$ of each spin σ onto the energy coordinate. The dissociation energy D_e is computed as 152.5 kcal/mol with the LDA functional (Eq. (3.5)), while it is decreased to $D_e = 117.0$ kcal/mol. It suggests that the SC energy in the dissociation is evaluated as 35.5 kcal/mol. On the other hand, D_e is obtained as 118.6 kcal/mol by the UCCSD(T) calculation with the cc-pVQZ basis set. It is thus shown that D_e of the $\mathrm{C-H}$ bond in methane can be computed correctly by the SC functional of Eq. (4.29). However, the potential energies in the middle range of the bond distance are much deeper than those given by the UCCSD(T) calculation, which is a common trend seen in the PECs of H_2 and the π bond in C_2H_4. Remember that the SC functional developed in this Section has been designed to correctly describe the dissociation limit of a chemical bond on the basis of the DFT with the energy electron density as the fundamental variable. Importantly, no new empirical parameter was introduced for the development. In the next Subsection, an effort will be made to improve the SC functional [12] by utilizing the degree of delocalization of the exchange hole to tune the strength of the SC included in the exchange energy.

4.2.3 Improvement of Static-Correlation Functional

As shown, the static-correlation (SC) functional developed in Sect. 4.2.2 overestimates the SC energies in the intermediate distances of chemical bonds. The objective of the present subsection is to improve the functional [12] by means of the degree of the delocalization of the exchange hole defined at a given coordinate of electrons. The delocalization factor given in the present development is derived from the generalized Becke-Roussel exchange-hole model discussed in Sect. 2.3.2. The method of the numerical implementation of the functional will be provided and the results of the applications to the dissociations of chemical bonds are also presented.

Delocalization Factor

It is somewhat surprising that the static-correlation (SC) functional developed in Sect. 4.2.2 shows good performances without using any empirical parameters. However, obviously the SC energies are overestimated in the middle ranges of the bond distances of the molecules considered. In the present Subsection, a method will be developed to improve the SC functional by means of a factor that measures the degree of delocalization of the exchange hole at an electron coordinate $\boldsymbol{r}$. The delocalization factor $\mathcal{D}(\boldsymbol{r})$ is directly derived from the normalization $N_x^{\mathrm{GBR}}(\boldsymbol{r})$ of the generalized Becke-Roussel (BR) exchange hole [1] of which center is placed at $\boldsymbol{r}$. The explicit definition of $N_x^{\mathrm{GBR}}(\boldsymbol{r})$ is given in Eq. (2.62). As described in Sect. 2.3.2, the normalization $N_x^{\mathrm{GBR}}(\boldsymbol{r})$ of an exchange hole for a spin goes to 0.5 regardless of $\boldsymbol{r}$ when the bond of H_2 is stretched infinitely as a consequence that the exchange hole is split into two fragments at the dissociation limit. $N_x^{\mathrm{GBR}}(\boldsymbol{r})$ is always less or equal to one. And it becomes much less than one when the exchange hole is seriously delocalized due to the fragmentations. For the method to construct the generalized BR exchange hole, [1] we refer readers to Sect. 2.3.2. The degree of the delocalization $\mathcal{D}(\boldsymbol{r})$ of the exchange hole at $\boldsymbol{r}$ can be naturally defined as

$$\mathcal{D}(\boldsymbol{r}) = N_s \left(1 - N_x^{\mathrm{GBR}}(\boldsymbol{r})\right) \tag{4.32}$$

where N_s is a real number representing the number of sites into which the exchange hole is fragmented. To see how Eq. (4.32) works, we consider again the fission of an H_2 molecule. At the infinite separation of the hydrogen atoms, number of the sites $N_s = 2$ and the normalization $N_x^{\mathrm{GBR}}(\boldsymbol{r}) = 0.5$, which leads to $\mathcal{D}(\boldsymbol{r}) = 1.0$ regardless of $\boldsymbol{r}$. It should be noticed that the relation $\mathcal{D}(\boldsymbol{r}) = 1.0$ holds even when the exchange hole is to be fully fragmented into three or more sites. The number of sites N_s can be evaluated by the inverse of the weighted average of $N_x^{\mathrm{GBR}}(\boldsymbol{r})$, thus,

$$N_s = N \left(\int d\boldsymbol{r}\; N_x^{\mathrm{GBR}}(\boldsymbol{r})\, n(\boldsymbol{r}) \right)^{-1} \tag{4.33}$$

where $n(\boldsymbol{r})$ is the electron density and N is the number of electrons with spin σ.

A possible strategy to improve the static-correlation (SC) functional developed in Sect. 4.2.2 is to attenuate the energy evaluated by the functional in the middle range of the distance of the chemical bond of interest. To do this, a certain functional $F[\mathcal{D}(\boldsymbol{r})]$ of the delocalization factor $\mathcal{D}(\boldsymbol{r})$ will be proposed later. To utilize the functional F in the DFT based on the energy electron density $n^e(\epsilon)$, the delocalization factor itself must be projected onto the energy coordinate ϵ. The projected factor $\overline{\mathcal{D}}(\epsilon)$ is given by a weighted average of $\mathcal{D}(\boldsymbol{r})$ over an energy coordinate,

$$\overline{\mathcal{D}}(\epsilon) = \frac{1}{n^e(\epsilon)} \int d\boldsymbol{r}\, \mathcal{D}(\boldsymbol{r})\, n(\boldsymbol{r})\, \delta(\epsilon - \upsilon_{\mathrm{def}}(\boldsymbol{r})) \tag{4.34}$$

where $\upsilon_{\mathrm{def}}(\boldsymbol{r})$ is the defining potential for the coordinate ϵ as introduced in Sect. 3.1. To see whether the projection of the delocalization factor onto the energy coordinate is reasonable, the factor $\mathcal{D}(\boldsymbol{r})$ is mapped onto the contour surfaces of the energy coordinate ϵ for an H_2 molecule at various bond distances R (Fig. 4.7). It is seen in the figure that the factor $\mathcal{D}(\boldsymbol{r})$ as well as the density $n_\sigma(\boldsymbol{r})$ are almost constant on the contour surfaces with the larger energy coordinates ϵ. However, on the contour surface with $\epsilon = 0.6$ a.u. at $R = 5.0$ a.u. these quantities vary significantly. As expected $\mathcal{D}(\boldsymbol{r})$ and $n_\sigma(\boldsymbol{r})$ are nearly constant on the contour surfaces at the larger bond length ($R = 10.0$ a.u.) regardless of the values of the contours. The factor $\mathcal{D}(\boldsymbol{r})$ is larger on the contour surfaces with the larger energy coordinates. It is also interesting to note that the factor becomes larger as R increases, due to the fact that the exchange hole is being split into two atomic sites. It can be concluded that the projection of Eq. (4.34) cannot be justified at least when the factor $\mathcal{D}(\boldsymbol{r})$ is small. However, as will be shown below, the SC energy is attenuated in the region where $\mathcal{D}(\boldsymbol{r})$ is small. Therefore, the problem associated with the projection is considered to be marginal.

Construction of Static-correlation Functional

The next issue is how to incorporate the projected delocalization factor $\overline{\mathcal{D}}(\epsilon)$ into the static-correlation (SC) functionals developed in the previous Subsection. In the present development, we utilize the SC functional $E_{sc}[n_0]$ in Eq. (4.15) as the basic form. To this end, we rewrite Eq. (4.15) in terms of the exchange energy density $U_x^e(\epsilon)$ defined on the energy coordinate ϵ,

$$\begin{aligned} E_{xc}[n_1, n_0] = & \int d\epsilon U_x^e\left[n_1^e\right](\epsilon)\, n_1^e(\epsilon) \\ & + \int d\epsilon \left(\overline{U}_x[n_0](\epsilon) - U_x^e\left[n_0^e\right](\epsilon)\right) n_0^e(\epsilon) \end{aligned} \tag{4.35}$$

Remind that n_1 stands for the spin adapted density of the fully interacting electrons while n_0 represents the overlap of the electron densities of the constituent atoms as shown in Eq. (4.12). $U^e[n^e](\epsilon)$ in Eq. (4.35) is expressed as

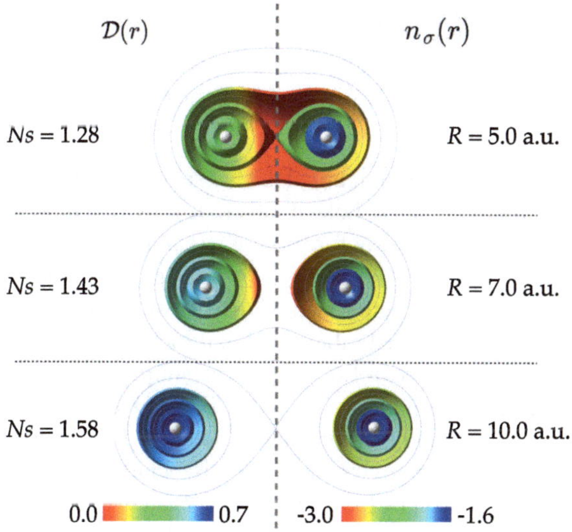

Fig. 4.7 (Left) Delocalization factor $\mathcal{D}(\boldsymbol{r})$ defined in Eq. (4.32) and (Right) logarithm of electron density $n_\sigma(\boldsymbol{r})$ of spin σ of an H_2 molecule with various bond lengths R. The values are mapped with colors onto the contour surfaces with $\epsilon = 0.6, 0.8$, and 1.2 a.u. of the absolute of the external potential $\epsilon = |\upsilon_{\text{ext}}(\boldsymbol{r})|$. The contours drawn with gray thin lines are for $\epsilon = 0.4, 0.5, 0.7, 1.0, 1.2, 2.0$, and 4.0 a.u. N_s stands for the number of sites defined in Eq. (4.33). The numerical details to produce these figures are shown in Ref. [12]. Reprinted with permission by IOP publishing for Fig. 3 in Ref. [12]

$$E_x^e\left[n^e\right] = \int d\epsilon\ U_x^e\left[n^e\right](\epsilon)\, n^e(\epsilon) \tag{4.36}$$

and $\overline{U}_x[n_0](\epsilon)$ is the weighted average of the exchange energy density $U_x[n](\boldsymbol{r})$ over the energy coordinate ϵ,

$$\overline{U}_x[n](\epsilon) = \frac{1}{n^e(\epsilon)} \int d\boldsymbol{r}\ U_x[n](\boldsymbol{r})\, n(\boldsymbol{r})\, \delta\left(\epsilon - \upsilon_{\text{def}}(\boldsymbol{r})\right) \tag{4.37}$$

It is easy to see that the relation

$$E_x[n] = \int d\epsilon\ \overline{U}_x[n](\epsilon)\, n^e(\epsilon) \tag{4.38}$$

holds for the electron density $n(\boldsymbol{r})$. Then, the improved SC functional $\widetilde{E}_{xc}[n_1, n_0]$ that utilizes the delocalization factor can be written as

$$\widetilde{E}_{xc}[n_1, n_0] = \int d\epsilon\ U_x^e\left[n_1^e\right](\epsilon)\, n_1^e(\epsilon) \\ + \int d\epsilon\ R\left(\overline{\mathcal{D}}(\epsilon)\right)\left(\overline{U}_x[n_0](\epsilon) - U_x^e\left[n_0^e\right](\epsilon)\right) n_0^e(\epsilon) \tag{4.39}$$

The function R is to regulate the amount of the SC energy depending on the degree of the delocalization of the exchange hole at the energy coordinate ϵ. Therefore, $R(x)$ must be a monotonically increasing function of a real number x and supposed to be within the range of $0 \le R(x) \le 1$ for $0 \le x \le 1$. $\mathcal{D}(\epsilon)$ should be constructed from the density n_1, that is, the symmetry-adapted density of the fully interacting system.

The remaining task to complete the functional of Eq. (4.39) is to determine the explicit form of the function R in Eq. (4.39). The simplest form of R is obviously to take R as the delocalization factor itself, that is $R\left(\overline{\mathcal{D}}(\epsilon)\right) = \overline{\mathcal{D}}(\epsilon)$. However, as will be shown later, it is found that this naive choice of R does not work favorably in describing the potential energy curve of the dissociation of H_2. More precisely, the contribution of E_{sc} becomes too small when R is chosen as the identity transformation. Hence, we decided to employ an upward-convex kth-order polynomial function $P_k(x)$ for R. Explicitly, P_k is given by

$$P_k\left(\overline{\mathcal{D}}\right) = 1 - \left(1 - \overline{\mathcal{D}}\right)^k \quad (k = 0, 1, 2, \cdots) \tag{4.40}$$

For later references, the graphs $P_k(x)$ for $k = 1, 2, 3$, and 5 are presented in Fig. 4.8. When $k = 0$, $P_0(x) = 0$ holds regardless of the value of x, and for $k \to \infty$, $P_\infty(x)$ asymptotically goes to 1.0 in the range of $0 < x \le 1$. That is, the functional form of Eq. (3.7) is realized by taking $k = 0$ in Eq. (4.39) using Eq. (4.40), and Eq. (4.17) is achieved when $k \to \infty$. Thus, it is found that the series of the polynomials P_k in Eq. (4.40) covers the full range of the SC energy. As k increases, $P_k(x)$ increases for all x in $0 \le x \le 1$. Thus, by changing the value of k, the amount of the SC energy included in the total energy can be regulated through Eq. (4.39). Of course, k can

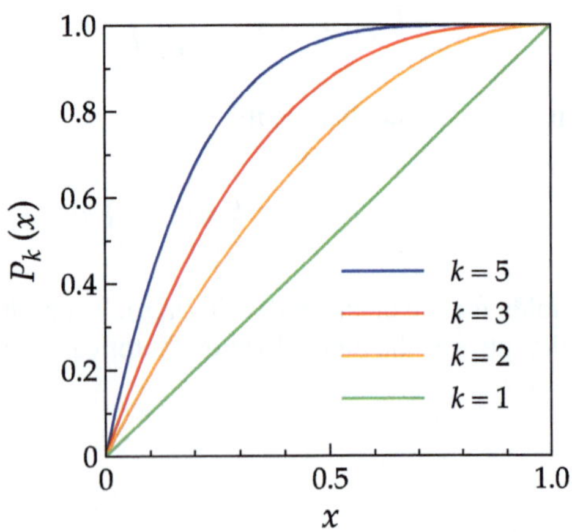

Fig. 4.8 The polynomial functions $P_k(x) = 1 - (1 - x)^k$ for $k = 1, 2, 3$, and 5. For $k \to \infty$, $P_\infty(x) = 0$ $(0 < x \le 1)$

be any real number. However, the major interest of the present development is not focused on a sophistication of the SC functional, but on the assessment of Eq. (4.39). Thus, we consider only the integer numbers for k in Eq. (4.40). Further, we also note that Eq. (4.40) is merely one of the possible forms of the function to attenuate the SC energy. Another simple form, such as the sigmoid function, can be a possible candidate as the function. The tests for these possibilities remain to be the subjects in the future works.

Before describing the numerical implementation of Eq. (4.39), we also introduce an effort to reduce the effect of the dynamical correlations between electrons on the dissociation energies D_e of chemical bonds. Explicitly, the LYP functional, [13] known as a representative GGA correlation functional, is employed to evaluate the dynamical correlation and the Becke88 (B88) exchange functional [14] introduced in Sect. 1.3.6.2 is adapted for the construction of SC energy in Eq. (4.39). To be precise, the functional $\widetilde{E}_{xc}^{\mathrm{BLYP}}$ based on the BLYP functional is written as

$$\begin{aligned}\widetilde{E}_{xc}^{\mathrm{BLYP}}[n_1, n_0] = &\int d\epsilon\ U_{xc}^{e,\mathrm{BLYP}}\left[n_1^e\right](\epsilon)\, n_1^e(\epsilon) \\ &+ \int d\epsilon\ P_k\left(\overline{\mathcal{D}}(\epsilon)\right)\left(\overline{U}_{xc}^{\mathrm{BLYP}}[n_0](\epsilon) - U_{xc}^{e,\mathrm{BLYP}}\left[n_0^e\right](\epsilon)\right) n_0^e(\epsilon)\end{aligned} \tag{4.41}$$

Some explanation will be needed to implement the components in the right hand side. $U_{xc}^{e,\mathrm{BLYP}}[n^e](\epsilon)$ is an exchange-correlation energy density for the energy electron density n^e defined on the energy coordinate ϵ. Thus, the integration of the first term amounts to the BLYP exchange-correlation energy $E_{xc}^{e,\mathrm{BLYP}}[n^e]$ in terms of the energy electron density n_σ^e,

$$E_{xc}^{e,\mathrm{BLYP}}\left[n_1^e\right] = \int d\epsilon\ U_{xc}^{e,\mathrm{BLYP}}\left[n_1^e\right](\epsilon)\, n_1^e(\epsilon) \tag{4.42}$$

For simplicity, $U_{xc}^{e,\mathrm{BLYP}}[n^e](\epsilon)$ is divided into the contributions from the exchange and the correlation energies,

$$U_{xc}^{e,\mathrm{BLYP}}\left[n^e\right](\epsilon) = U_x^{e,\mathrm{B88}}\left[n^e\right](\epsilon) + U_c^{e,\mathrm{LYP}}\left[n^e\right](\epsilon) \tag{4.43}$$

The quantity $U_x^{e,\mathrm{B88}}[n^e](\epsilon)$ is, of course, derived from the original Becke88 exchange functional $E_x^{\mathrm{B88}}[n]$. As apparent from Eq. (1.149), the exchange energy density $U_x^{\mathrm{GGA}}[n_\sigma](\boldsymbol{r})$ of a conventional GGA functional for spin σ can be expressed as

$$U_x^{\mathrm{GGA}}[n_\sigma](\boldsymbol{r}) = -2^{\frac{1}{3}} C_x^{\mathrm{LDA}} n_\sigma(\boldsymbol{r})^{\frac{1}{3}} - n_\sigma(\boldsymbol{r})^{\frac{1}{3}} F^{\mathrm{GGA}}(q_\sigma(\boldsymbol{r})) \tag{4.44}$$

where $q_\sigma(\boldsymbol{r})$ represents the inhomogeneity of the electron density and defined in Eq. (1.150) and F^{GGA} is the enhancement factor F for a GGA functional as introduced in Eq. (1.149). Specifically, the factor $F^{\mathrm{B88}}(q_\sigma)$ for the Becke88 functional is explicitly given in Eq. (1.153). Anyway, once the value q_σ is given, the exchange energy density for a GGA functional can be obtained in general. In fact, the inhomogeneity parameter $q^e_\sigma(\epsilon)$ on the energy coordinate is provided in Eq. (3.71) in Sect. 3.6.1. Thus, the exchange energy density on ϵ is given by

$$U_x^{e,\mathrm{B88}}\left[n^e\right](\epsilon) = -2^{\frac{1}{3}} C_x^{\mathrm{LDA}} \widetilde{n}_\sigma^e(\epsilon)^{\frac{1}{3}} - \widetilde{n}_\sigma^e(\epsilon)^{\frac{1}{3}} F^{\mathrm{B88}}\left(q_\sigma^e(\epsilon)\right) \tag{4.45}$$

For the evaluation of the correlation term $U_c^{e,\mathrm{LYP}}[n^e](\epsilon)$ in Eq. (4.43), the average of the second derivative $\nabla^2 n_\sigma(\boldsymbol{r})$ over the energy coordinate ϵ is needed besides that of the first derivative,

$$g_2^e[n_\sigma](\epsilon) = \Omega(\epsilon)^{-1} \int d\boldsymbol{r}\, \delta(\epsilon - \upsilon(\boldsymbol{r}))\nabla^2 n_\sigma(\boldsymbol{r}) \tag{4.46}$$

where $\Omega(\epsilon)$ is the volume of the spatial region with the coordinate ϵ and defined in Eq. (3.4). Thus, the terms $U_{xc}^{e,\mathrm{BLYP}}\left[n_1^e\right](\epsilon)$ and $U_{xc}^{e,\mathrm{BLYP}}\left[n_0^e\right](\epsilon)$ in Eq. (4.41) can be evaluated. The remaining term $\overline{U}_{xc}^{\mathrm{BLYP}}[n_0](\epsilon)$ is merely obtained by using Eq. (4.37), thus,

$$\overline{U}_{xc}^{\mathrm{BLYP}}[n_0](\epsilon) = \frac{1}{n_0^e(\epsilon)} \int d\boldsymbol{r}\; U_{xc}^{\mathrm{BLYP}}[n_0](\boldsymbol{r})\, n_0(\boldsymbol{r})\, \delta\left(\epsilon - \upsilon_{\mathrm{def}}(\boldsymbol{r})\right) \tag{4.47}$$

In the LYP functional [13], different forms of the correlation functionals are provided for the spin-polarized and depolarized densities. The LYP functional for spin-depolarized density is applied to evaluate the functional $U_{xc}^{e,\mathrm{BLYP}}$, while that for spin-polarized density is employed in the computation of $\overline{U}_{xc}^{\mathrm{BLYP}}$ in Eq. (4.41). It should be noted that the newly introduced parameter in the functional of Eq. (4.41) is only the order k of the polynomial function P_k which regulates the strength of the SC energy according to the delocalization of the exchange hole at a given energy coordinate.

Applications to Dissociations of Chemical Bonds

The static-correlation (SC) functional in Eq. (4.41) is implemented on our house code 'Vmol' which utilizes the real-space grids (RSG) method[7–11]. The numerical details of the real-space cell, grid size, and the construction of the energy electron densities $n^e(\epsilon)$ are provided in the Supplementary material for Ref. [12]. We note that the spin-adapted electron density $n(\boldsymbol{r})$ used as an input to Eq. (4.41) is constructed from the SCF calculation using the conventional BLYP functional throughout the applications noted below. On the other hand, the reference electron density n_0 is built from the isolated fragments according to Eq. (4.12). Thus, the electron density as an argument is provided from the outset and is not self-consistently optimized under the influence of the SC functional. This issue will be discussed at the end of this paragraph.

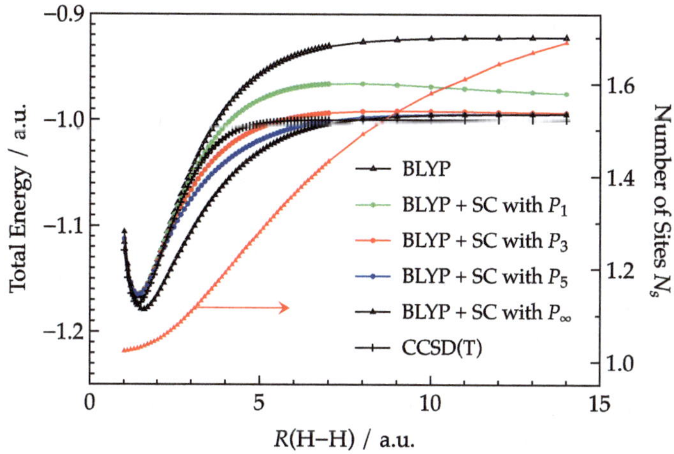

Fig. 4.9 The legends 'BLYP+SC with $P_k(k = 1, 3, 5)$' represent the results obtained using Eq. (4.41). P_k are the polynomial functions in Eq. (4.40) (see also Fig. 4.8). The notation P_∞ indicates that $P_k = 1$ identically. 'BLYP' is obtained from the pure conventional BLYP calculation. The result given by the CCSD(T) calculation using 'Gaussian 09' package [6] with the aug-cc-pVQZ basis set is also provided as a reference. The number of sites N_s defined in Eq. (4.33) is also shown

We first consider the fission of a single chemical bond. To this end, the potential energy curves (PECs) for the dissociation of H_2 molecule are presented in Fig. 4.9 where the number of sites N_s (Eq. (4.33)) onto which the exchange hole is being split is also shown. As expected, the pure BLYP functional overshoots the energy at the dissociation limit. D_e was evaluated as 153.7 kcal/mol by the BLYP functional, while the CCSD(T) calculation gives 109.2 kcal/mol. Thus, the spin-adapted (spin-restricted) BLYP calculation overestimates D_e by 44.5 kcal/mol. When the 1st-order polynomial P_1 is used in Eq. (4.41), D_e is decreased by 33.0 kcal/mol at $R(\mathrm{H-H}) = 14.0$ a.u. It is obvious, however, that the amount of the SC energy included in the total energy is too small. It is found that the choice of $k = 3$ further improves the dissociation energy D_e. In fact, D_e is obtained as 107.8 kcal/mol, that is very close to the CCSD(T) value. More importantly, the middle-range ($2.0 \leq R(\mathrm{H-H}) \leq 5.0$ a.u.) behavior of the PEC for $k = 3$ is also close to that of the CCSD(T) calculation. However, it is found that the use of polynomial function P_5 overestimate the SC energies in the middle range although the dissociation limit is described correctly. It is, thus, shown that the polynomial P_3 among the series of P_k in Eq. (4.40) will be appropriate to describe the dissociation of a hydrogen molecule. The PEC that fully includes the SC energy ('BLYP + SC with P_∞') is also shown in the figure, which is obtained simply by setting the factor $P_k = 1$ regardless of the energy coordinate ϵ in Eq. (4.41). In the legend, it is indicated by P_∞ since $P_k = 1$ is realized when $k \to \infty$ in Eq. (4.40). The PEC shows the same trend as that observed in Fig. 4.4. It is clearly demonstrated that the multiplication of the polynomial to the SC energy effectively suppresses the excess lowering of the energy in the middle range of the bond length. It is also interesting to discuss the behavior of the quantity N_s defined in Eq. (4.33). As mentioned above, N_s delivers the number of the atomic sites over which the exchange hole is delocalized. When the bond length is shorter than that at the equilibrium structure, N_s is nearly 1.0 and it increases monotonically as the increase in the bond distance. In the present case, it is anticipated that N_s becomes 2.0 at the dissociation limit. However, as shown in the graph, N_s increases up to $\sim$ 1.7 and does not converge even at $R(\mathsf{H}-\mathsf{H}) = 14.0$ a.u.

The SC functional in Eq. (4.41) is also applied to the dissociation of the double bond in an ethene molecule(C_2H_4). For the variation of the $\mathsf{C}=\mathsf{C}$ bond, the structure of the CH_2 groups is fixed at that in ethene with the equilibrium structure. In the construction of the PEC for the ethene with long bond lengths, it sometimes occurs that the nearly degenerated anti-bonding π^* orbital comes in the set of the occupied orbitals for the valence electrons in place of the π orbital, which leads to numerical instabilities in the SCF calculation for KS-DFT. To avoid the instabilities, the set of wave functions optimized at a bond length is used as an initial guess for the next structure created by a small increment of the original bond length. The results are shown in Fig. 4.10. First, the spin-restricted BLYP calculation gives the dissociation energy $D_e = 257.5$ kcal/mol although it seems that the BLYP energy does not con-

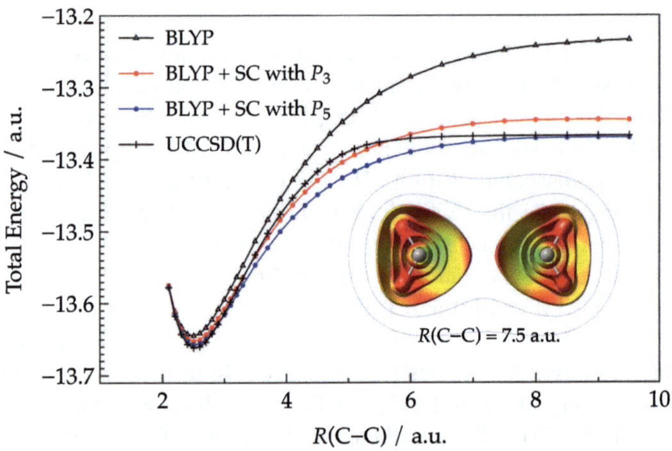

Fig. 4.10 The potential energy curves for the dissociation of the $\mathsf{C}=\mathsf{C}$ bond in ethene molecule [12]. The legends 'BLYP+SC with $P_k (k = 3$ and $5)$' represent the results obtained using Eq. (4.41). P_k are the polynomial functions in Eq. (4.40) (see also Fig. 4.8). 'BLYP' is obtained from the pure conventional BLYP calculation. The result given by the 'UCCSD(T)' calculation using 'Gaussian 09' package [6] with the cc-pVQZ basis set is also provided as a reference. Contour surfaces of ϵ, defined as the absolute of the external potential, i.e. $\epsilon = |\upsilon_{\text{ext}}(\boldsymbol{r})|$, are also shown in the inset ($\epsilon = 3.0, 4.0, 5.0$, and 7.0 a.u.). The contour lines are for $\epsilon = 2.0$ and 2.5 a.u. on the molecular plane. The values of the delocalization factor $\mathcal{D}(\boldsymbol{r})$ in Eq. (4.32) at $R(\mathsf{C}-\mathsf{C}) = 7.5$ a.u. are mapped onto the contour surfaces. The color assignment for the value $\mathcal{D}(\boldsymbol{r})$ is the same as in Fig. 4.7. For comparisons all the graphs are shifted so that the left ends of the graphs match that of 'BLYP + SC with P_5'. The numerical details to produce these figures are shown in Ref. [12]. Reprinted with permission by IOP publishing for Fig. 5 in Ref. [12]

verge at $R(\mathrm{C}-\mathrm{C}) = 9.5$ a.u. On the other hand, the UCCSD(T) calculation gives D_e as 184.7 kcal/mol. Thus, it is found that the BLYP calculation overestimates D_e by 72.8 kcal/mol. The figure shows that using polynomial P_3 in the SC functional of Eq. (4.40) gives the PEC that reasonably agrees with that given by the UCCSD(T) calculation although the dissociation energy D_e obtained with the SC functional is larger than the UCCSD(T) calculation by 13.7 kcal/mol. On the other hand, D_e is evaluated as 180.4 kcal/mol by the functional with the polynomial P_5, which shows a good agreement with the value 184.7 kcal/mol by the UCCSD(T) calculation. However, the functional with P_5 underestimates the total energy in the intermediate region of $R(\mathrm{C}-\mathrm{C})$. These trends are common to those observed in Fig. 4.9 for the dissociation of a hydrogen molecule. Therefore, the origin of the trends is worth describing. First, it is possible that the series of polynomials $P_k(x)$ in Eq. (4.40) is not suitable to accurately describe the SC energy as a function of the degree of the delocalization of the exchange hole. In fact, there will be a lot of other possible analytical functions to regulate the amount of the SC energy included in the total energy. Second, the delocalization factor $\mathcal{D}(\mathrm{r})$ defined in Eq. (4.32), based on the normalization N_x^{GBR} in the generalized Becke-Roussel (GBR) exchange-hole model, possibly does not perform in the way as expected. As suggested in Ref. [15], the hole population is tends to be overestimated by the GBR model, which leads to the underestimation of the delocalization factor $\mathcal{D}(\boldsymbol{r})$. As shown in the inset of Fig. 4.10, the values of $\mathcal{D}(\boldsymbol{r})$ remain mostly within 0.3 even at $R(\mathrm{C}-\mathrm{C}) = 7.5$ a.u. Consequently, the number of sites N_s will be underestimated. In fact, N_s is evaluated as only 1.12 even at $R(\mathrm{C}-\mathrm{C}) = 9.5$ a.u. in contrast to the case of the H_2 fission. Thus, the SC energy at the dissociation of ethene will possibly be underestimated. It is possible that another choice of the method to evaluate the population of the exchange hole will remedy the drawback of the SC functional and improve the behavior of the dissociation curve. Anyway, it is demonstrated that the functional in Eq. (4.41) is capable of reasonably describing the PEC of a dissociating double bond.

Finally, the SC functional in Eq. (4.41) is applied to the dissociation of a triple bond in N_2 molecule. The results are presented in Fig. 4.11. The dissociation energy D_e is evaluated as 385.9 kcal/mol by the spin-restricted BLYP functional, while the UCCSD(T) calculation gives $D_e = 222.4$ kcal/mol. Thus, the BLYP overestimates D_e by 163.5 kcal/mol as compared to the UCCSD(T) calculation. Due to the fact that the bond stretch involves the fission of a triple bond, the SC error is found to be larger than those of the single and the double bonds. The SC functional with the polynomial P_3 gives D_e as 272.5 kcal/mol. It is thus found that the inclusion of the SC energy improves the value D_e by more than 110 kcal/mol. However, the functional still overestimates the D_e by ~50 kcal/mol. By using the SC functional with P_5, D_e is found to be reduced to 250.7 kcal/mol. Thus, the SC error associated with the functional with P_5 is estimated to be about 10 % of the dissociation energy. As expected, describing a triple bond is found to be the most difficult test for the functional among the three calculations. The delocalization factors $\mathcal{D}(\boldsymbol{r})$ projected onto the contour surfaces of the external potential are also shown in the inset for the system with $R(\mathrm{N}-\mathrm{N}) = 7.0$ a.u. Similar to the case of the dissociation of the ethene, $\mathcal{D}(\boldsymbol{r})$ remains rather small even in the region of the bond length for the nearly

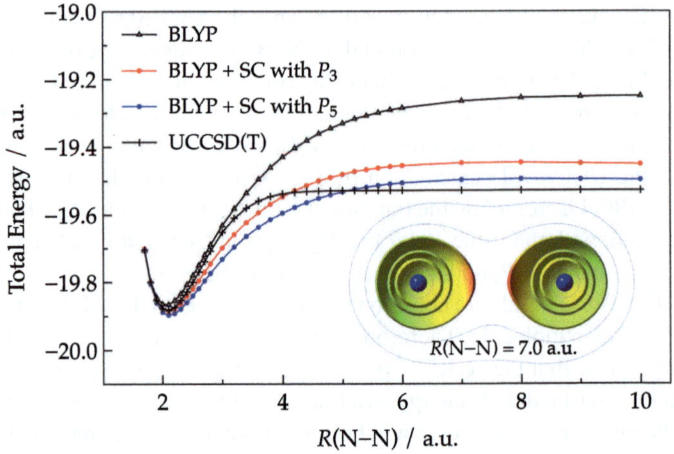

Fig. 4.11 The potential energy curves for the dissociation of the bond in N_2 molecule [12]. The notations for the legends 'BLYP+SC with $P_k (k = 3$ and $5)$', 'BLYP', and 'UCCSD(T)' are the same as in Fig. 4.10. Contour surfaces of ϵ, defined as the absolute of the external potential, i.e. $\epsilon = |v_{\text{ext}}(\boldsymbol{r})|$, are also shown in the inset ($\epsilon = 3.0, 4.0$, and 5.0 a.u.). The contour lines are for $\epsilon = 2.0, 2.5$, and 8.0 a.u. The values of the delocalization factor $\mathcal{D}(\boldsymbol{r})$ in Eq. (4.32) at $R(\mathrm{N}-\mathrm{N}) = 7.0$ a.u. are mapped onto the contour surfaces. The color assignment for the value $\mathcal{D}(\boldsymbol{r})$ is the same as in Fig. 4.7. For comparisons all the graphs are shifted so that the left ends of the graphs match that of 'BLYP + SC with P_5'. The numerical details to produce these figures are shown in Ref. [12]. Reprinted with permission by IOP publishing for Fig. 6 in Ref. [12]

dissociated molecule. The number of sites N_s in Eq. (4.33) is evaluated as only 1.27 at $R(\mathrm{N}-\mathrm{N}) = 10.0$ a.u. As mentioned above, the underestimation of the SC energy can also be attributed to the overestimation of the population of the exchange hole described with the generalized Becke-Roussel model [1]. It should be noted again that the SC functional developed in this Subsection contains only a single adjustable number k, that designates the order of the polynomial P_k. Therefore, the functional can possibly be improved by a certain sophistication. This issue will be a major subject in the future works.

In the above three applications to the molecules H_2, C_2H_4, and N_2, the electron density for each configuration of each system is provided by the conventional BLYP functional and is not self-consistently optimized using the SC functional as noted at the top of this paragraph. The extension of the functional to the SCF calculation is rather straightforward. For instance, we shall utilize the SC functional of Eq. (4.29). To this end, we extract the LDA exchange energy from the conventional BLYP functional,[13, 14] thus,

$$E_{xc}^{\text{BLYP}}[n] = \int d\epsilon\, \overline{U}_x^{\text{LDA}}[n](\epsilon)\, n^e(\epsilon) - \int d\boldsymbol{r}\, n(\boldsymbol{r})^{\frac{4}{3}}\, F^{\text{B88}}[n](\boldsymbol{r}) + E_c^{\text{LYP}}[n] \tag{4.48}$$

where the enhancement factor F^{B88} for spin σ is defined in Eq. (1.153) in Chap. 1. $E_c^{\mathrm{LYP}}[n]$ is the correlation functional by Lee et al. [13]. The quantity $\overline{U}_x^{\mathrm{LDA}}[n](\epsilon)$ is the weighted average of the LDA exchange energy density over the energy coordinate ϵ. The definition of the weighted average is explicitly shown in Eq. (4.37). It important to note that the first term on the right of Eq. (4.48) amounts to the exchange energy given by the conventional LDA functional $E_x^{\mathrm{LDA}}[n]$. In the construction of the SC functional for the variational calculation, the second and third terms on the right hand side in Eq. (4.48) are fully included in the total energy regardless of the energy coordinate. On the other hand, only the first term contributes to the SC energy depending on the degree of the delocalization of the exchange hole. The form of the second SC energy functional $\widetilde{E}_{xc}^{\mathrm{BLYP2}}[n_1, n_0]$ based on a GGA functional BLYP is expressed as,

$$\begin{aligned}\widetilde{E}_{xc}^{\mathrm{BLYP2}}[n_1, n_0] &= \int d\epsilon \left(1 - P_k\left(\overline{\mathcal{D}}(\epsilon)\right)\right) \overline{U}_x^{\mathrm{LDA}}[n_1](\epsilon)\, n_1^e(\epsilon) \\ &+ \int d\epsilon\, P_k\left(\overline{\mathcal{D}}(\epsilon)\right) \left\{\overline{U}_x^{\mathrm{LDA}}[n_0](\epsilon)\, n_0^e(\epsilon) - \zeta^e(\epsilon)\right\} \\ &- \int d\boldsymbol{r}\, n_1(\boldsymbol{r})^{\frac{4}{3}} F^{\mathrm{B88}}[n_1](\boldsymbol{r}) + E_c^{\mathrm{LYP}}[n_1] \end{aligned} \tag{4.49}$$

where $\zeta^e(\epsilon)$ stands for the integrand of the second term on the right hand side in Eq. (4.29) including the pre-factors. It should be noted that the correlation energy $E_c^{\mathrm{LYP}}[n_1]$ in Eq. (4.49) unfortunately does not vanish even at the dissociation limit of H_2, since n_1 is the spin-adapted density. On the contrary, the correlation energy in Eq. (4.49) vanishes at the dissociation due to its construction. To perform the SCF calculation with Eq. (4.49), the functional derivative of the equation with respect to the density $n_1^e(\epsilon)$ is needed. A difficulty will arise in the differentiation of the projected delocalization factor $\overline{\mathcal{D}}(\epsilon)$ defined in Eq. (4.34). A simple and efficient way to circumvent the bothering work to obtain the derivative is to freeze the factor in differentiating Eq. (4.49). Then, the factor $\overline{\mathcal{D}}(\epsilon)$ shall be renewed after the energy electron density n_1^e is updated. Thus, the evaluation of $\overline{\mathcal{D}}$ and the update of n_1^e must be repeated self-consistently to achieve the convergence in energy. Unfortunately, the calculations with the SC functional of Eq. (4.49) have not yet been performed.

Spin Polarization and Depolarization in Atom

In Ref. [15], Becke proposed a strongly correlated system as a stringent test for his SC functional combined with the dynamical-correlation functional. Explicitly, a carbon atom with the polarized and depolarized spin states served to assess the performance of the functional Becke13. The test was motivated by the work by Yang et al. [16] where it is proved that the exact energy functional $E_{\mathrm{exct}}[n]$ satisfies the following equation

$$E_{\mathrm{exct}}[n_{\mathrm{mix}}] = E_0 \tag{4.50}$$

E_0 is the energy of the ground state of a system of N electrons subject to an external potential υ. And n_{mix} is a linear combination of the densities of the g-degenerated ground state wave functions for the system. Explicitly, n_{mix} is expressed as

$$n_{\mathrm{mix}}(\boldsymbol{r}) = \sum_{i}^{g} p_i n_0^i(\boldsymbol{r}) \tag{4.51}$$

where n_0^i $(i = 1, 2, \cdots, g)$ stands for the electron density of the ith wave function that belongs to the eigenvalue E_0. The coefficients p_i are the positive real numbers and satisfy the normalization condition

$$\sum_{i}^{g} p_i = 1 \tag{4.52}$$

And each density also satisfies

$$E_{\mathrm{exct}}[n_0^i] = E_0 \quad (i = 1, 2, \cdots, g) \tag{4.53}$$

As described in Ref. [15], a surprising conclusion can be directly derived from Eq. (4.50) for example for the ground state of a carbon atom. The KS ground state wave function of an isolated carbon atom can be identified with the occupancies of the $2p$ orbitals: $2p_x(\alpha), 2p_y(\alpha), 2p_z(0)$. Of course, the occupancies of the $2s$ orbital is presumed. Another wave function with the configuration: $2p_x(\beta), 2p_y(\beta), 2p_z(0)$ also has the same eigenvalue. Then, the half-and-half mixture of the electron densities corresponding to these wave functions gives E_0 by virtue of Eq. (4.50). To be more precise, the mixed density n_{HandH},

$$\begin{aligned} n_{\mathrm{HandH}} = &\frac{1}{2} n_{\mathrm{px}}(\alpha) + \frac{1}{2} n_{\mathrm{px}}(\beta) \\ &+ \frac{1}{2} n_{\mathrm{py}}(\alpha) + \frac{1}{2} n_{\mathrm{py}}(\beta) \end{aligned} \tag{4.54}$$

fulfills the relation $E_{\mathrm{exct}}[n_{\mathrm{HandH}}] = E_0$. In Eq. (4.54) the contribution from the $2s$ orbital to the density is included as an offset. Somewhat surprisingly, the spin-depolarized electron density also provides the ground state energy E_0 through the exact functional. In the following, we assess the performance of the SC functional of Eq. (4.41) in computing the difference between the energies of the spin-polarized and depolarized densities [12]. The key feature of the functional, which plays a central role in computing the energy difference, can be reduced to the simple relation satisfied by the energy electron densities of the $2p$ orbitals. The energy electron density $n_{px}^e(\epsilon)$ for the $2p_x$ orbital is given by

$$n_{\mathrm{px}}^e(\epsilon) = \int d\boldsymbol{r}\, n_{\mathrm{px}}(\boldsymbol{r})\, \delta(\epsilon - \upsilon(\boldsymbol{r})) \tag{4.55}$$

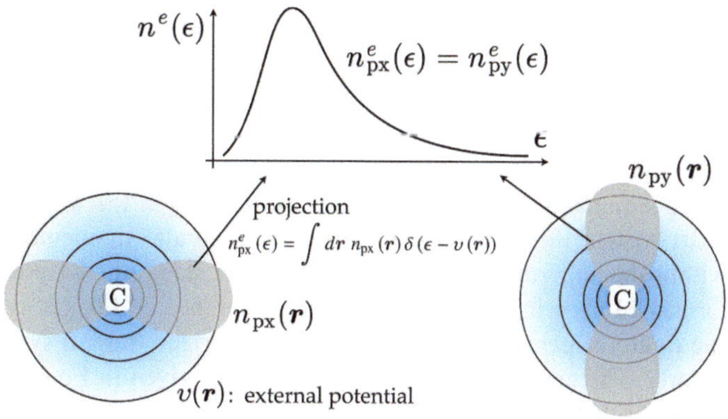

Fig. 4.12 Electron densities n_{px} and n_{py} for the $2p_x$ and $2p_y$ orbitals in a carbon atom. The projections of these densities onto the energy coordinate ϵ are $n^e_{px}(\epsilon)$ and $n^e_{py}(\epsilon)$. Since the external potential υ of the carbon atom is spherically symmetric, the relation: $n^e_{px}(\epsilon) = n^e_{py}(\epsilon)$ holds in principle. The contour lines of the potential υ are shown as concentric circles with blue shading

where $\upsilon(\boldsymbol{r})$ represents the external potential of the atom of interest. Due to the spherical symmetry of the external potential around the nucleus, the following relations hold exactly for the energy electron densities for the $2p$ orbitals,

$$n^e_{px}(\epsilon) = n^e_{py}(\epsilon) = n^e_{pz}(\epsilon) \tag{4.56}$$

We provide in Fig. 4.12 an illustration which shows the projections of the densities for $2p_x$ and $2p_y$ orbitals in a carbon atom onto the energy coordinate ϵ. Therefore, the half-and-half mixing of the energy electron densities $n^e_{px}(\epsilon)$ and $n^e_{py}(\epsilon)$ for the $2p_x$ and $2p_y$ orbitals is exactly the same as that for the $2p_x$ or $2p_y$ orbital, namely,

$$n^e_{px}(\epsilon) = \frac{1}{2}n^e_{px}(\epsilon) + \frac{1}{2}n^e_{py}(\epsilon) \tag{4.57}$$

Of course, it can be extended to a general form,

$$n^e_{px}(\epsilon) = n^e_{\lambda}(\epsilon) = \lambda n^e_{px}(\epsilon) + (1-\lambda)\, n^e_{py}(\epsilon) \tag{4.58}$$

where λ is the mixing parameter satisfying $0 \le \lambda \le 1$. This equation reminds us of the relation in Eq. (4.9) that was derived from the discussion on the electron density of the dissociating H_2. The similar equation can also be applied to the correlated electrons on an atomic site.

However, a problem arises when Eq. (4.41) is applied to a single atom with the strongly correlated electrons. Remind that, in the case of calculating the SC energy associated with the bond dissociation, the reference density is taken as the overlap of those of the isolated fragments as described above. Then, what density should be taken as the reference for the calculation of a *single* atom? Furthermore, the number of α-spin electrons in the spin-depolarized density n_{HandH} differs from that of the spin-polarized density $n_{\mathrm{px}}(\alpha) + n_{\mathrm{py}}(\alpha)$. A simple solution to deal with such a single atom is to place the identical atom at an infinite distance from the original atom. Then, the entire system, namely, the 'supermolecule' is regarded as the reference system which serves the reference density n_0. All the spin orientations in the original wave function are reversed in the construction of the wave function of its counterpart so that the z component of the spin angular momentum are fully canceled within the supermolecule. For reader's understanding, we provide illustrations in Fig. 4.13 showing the densities of the reference and target systems. As shown in the figure, supermolecule for the spin-depolarized (target) system is also created similarly. The energy $E[n_{\mathrm{HandH}}]$ for the density given in Eq. (4.54) is then obtained as half of that for the supermolecule. Further, since the α and β spin populations in the supermolecule are identical, the exchange energy in $E[n_{\mathrm{HandH}}]$ can be evaluated as that for the α-spin electrons of the supermolecule. Importantly, the number of α-spin electrons in the supermolecule for the target system is the same as that for the reference system. Thus, the number of each spin is maintained for the change of the supermolecule from the reference to the target system. Then, it is possible to apply Eq. (4.41) to the isolated carbon atoms by using the density n_0 in Fig. 4.13 as the reference. The energy difference ΔE between the carbon atom with

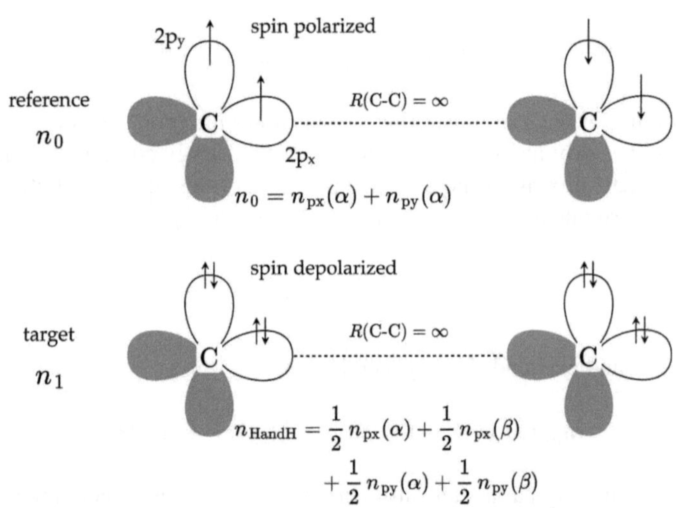

Fig. 4.13 Upper: spin-polarized electron density n_0 in a supermolecule that consists of infinitely separated carbon atoms. The α spin density of the system is taken as the reference density n_0 which serves to construct the energy in Eq. (4.41). Lower: spin-depolarized electron density n_{HandH} of the supermolecule. The density n_{HandH} is obtained by the overlap of the equally weighted electron densities identified with the configurations $2p_x(\alpha), 2p_y(\alpha), 2p_z(0)$ and $2p_x(\beta), 2p_y(\beta), 2p_z(0)$. n_{HandH} constitutes the target density n_1 in Eq. (4.41). In these densities, the contribution from the $2s$ orbital to the density is included as an offset

the spin-depolarized density $n_{\mathrm{HandH}}(\boldsymbol{r})$ and that with the spin-polarized density n_0 is computed using Eqs. (4.41) and (4.42). Explicitly, ΔE is given by

$$\Delta E = \widetilde{E}_{xc}^{\mathrm{BLYP}}\left[n_1, n_0\right] - E_{xc}^{e,\mathrm{BLYP}}\left[n_0^e\right] \tag{4.59}$$

where n_0 is spin-polarized reference density and n_1 is the target density, namely, $n_1 = n_{\mathrm{HandH}}$. Note that the terms, such as the kinetic energy of the electrons, the classical Coulomb energy, and the interaction between the electrons and the external potential are completely canceled between the two systems. As shown in the second term in the right hand side of Eq. (4.59), the exchange-correlation energy of the spin-polarized system is also computed with the DFT based on the energy electron density. This is to exclude from ΔE the effect of projecting the density onto the energy coordinate. In fact, the energy difference $E_{xc}^{\mathrm{BLYP}}\left[n_0\right] - E_{xc}^{e,\mathrm{BLYP}}\left[n_0^e\right]$ is evaluated as -12.4 kcal/mol. The α-spin density in $n_{\mathrm{HandH}}(\boldsymbol{r})$ is obtained as the half of the total electron density computed by the UBLYP functional for the isolated carbon atom. The results are summarized in Table 4.1. ΔE in Eq. (4.59) is computed as 6.7 kcal/mol when the polynomial P_3 is used in the functional $\widetilde{E}_{xc}^{\mathrm{BLYP}}\left[n_1, n_0\right]$. However, it is reduced to -2.5 kcal/mol by using the polynomial P_5. This is due to the fact that $P_5(\overline{\mathcal{D}})$ is always larger than $P_3(\overline{\mathcal{D}})$ for the delocalization factor $\overline{\mathcal{D}}$ as shown in Fig. 4.8. It is thus demonstrated that the SC functional of Eq. (4.41) almost realizes Eq. (4.50) for the mixed density given by the overlap of the two electron densities with the same weight. In principle the same result can be obtained when the overlap density n_1 is created from the densities with different weights since the energy electron density n_1^e for each spin in the supermolecule does not change with respect to the variation of the weights as long as the normalization condition of Eq. (4.52) is fulfilled. We also evaluate ΔE when the SC energy is fully included in Eq. (4.41). This can be realized by setting $P_k(\overline{\mathcal{D}}) = 1$ for all $\overline{\mathcal{D}}$. In Table 4.1, the column is indicated by P_∞ since $P_k(x) = 1$ for $0 < x \leq 1$ when $k \to \infty$. ΔE is obtained as -17.8 kcal/mol for P_∞. As shown in the dissociations of the hydrogen molecule (see Fig. 4.9), the full inclusions of the SC energy also underestimates the potential energies of the molecule at intermediate bond distances. The similar trends observed in the two types of calculations suggest the universality of Eq. (4.41) in the calculation of strongly correlated electron systems. In other words, it is expected that a development to

Table 4.1 Energy difference ΔE between the systems with the spin-depolarized and polarized densities of carbon atom. Explicitly, ΔE is defined as $\Delta E = \widetilde{E}_{xc}^{\mathrm{BLYP}}\left[n_1, n_0\right] - E_{xc}^{e,\mathrm{BLYP}}\left[n_0^e\right]$ where the densities n_1 and n_0 are illustrated in Fig. 4.13. The functional $\widetilde{E}_{xc}^{\mathrm{BLYP}}\left[n_1, n_0\right]$ and $E_{xc}^{e,\mathrm{BLYP}}\left[n_0^e\right]$ are defined in Eqs. (4.41) and (4.42), respectively. In the computation of $\widetilde{E}_{xc}^{\mathrm{BLYP}}\left[n_1, n_0\right]$, the polynomials P_3 and P_5 in Eq. (4.40) are employed. For the column P_∞, see the main text. The result in the column 'BLYP' is given by $\Delta E = E_{xc}^{\mathrm{RBLYP}}\left[n_1\right] - E_{xc}^{\mathrm{UBLYP}}\left[n_0\right]$ using the conventional BLYP functional. The energies are in the unit of kcal/mol. See also Fig. 7 in Ref. [12]

Method	P_3	P_5	P_∞	BLYP
ΔE	6.7	−2.5	−17.8	37.4

improve the SC functional for a set of limited systems can also be applied to other systems successfully. The energy difference $E_{xc}^{\mathrm{RBLYP}}[n_1] - E_{xc}^{\mathrm{UBLYP}}[n_0]$ using the GGA functional with the conventional electron density is computed as 37.4 kcal/mol. It is thus shown that the energy difference between the spin-depolarized and polarized systems of an atom can be reasonably computed by the SC functional in contrast to the conventional DFT functional. Another interesting test is to consider the electron density where an α spin electron is symmetrically placed on each $2p$ orbital with a population of $1/3$. The detailed discussion for such a system is provided in Ref. [12].

Comparison of the Present Approach with Becke03 functional

In closing this chapter, we make comparisons of the Becke03 functional [1] introduced in Sect. 2.3.2 with the present approach based on the energy electron density. For the sake of simplicity, we consider an H_2 molecule at the dissociation limit. In the following, we use notation n_1 for the spin-depolarized (spin-adapted) and n_0 for the spin-polarized (symmetry-broken) density. In addition, it is assumed that $n_{0\alpha}$ is the electron density of the ground state of a hydrogen atom and $n_{0\beta}$ is that of the other. As described in Sect. 2.3.2, in the construction of the exchange-correlation hole function $n_{xc\alpha}^{\mathrm{static}}$ for α spin with the Becke03 functional, the fraction of the exchange hole function $n_{x\beta}$ is added to $n_{x\alpha}$, which leads to the relation of Eq. (2.65),

$$U_{x\alpha}^{\mathrm{static}}(\boldsymbol{r})n_{1\alpha}(\boldsymbol{r}) = U_{x\alpha}^{\mathrm{exact}}(\boldsymbol{r})n_{1\alpha}(\boldsymbol{r}) + f U_{x\beta}^{\mathrm{exact}}(\boldsymbol{r})n_{1\alpha}(\boldsymbol{r}) \tag{4.60}$$

Note that $U_{x\sigma}(\boldsymbol{r})$ stands for the exchange energy density $U_x[n_{1\sigma}](\boldsymbol{r})$ for spin σ at $\boldsymbol{r}$ and the superscript 'exact' represents the generalized Becke-Roussel (GBR) method. As was described in Sect. 2.3.2, remind that $U_x^{\mathrm{GBR}} = U_x^{\mathrm{exact}}$ holds where U_x^{exact} stands for the exchange energy density of the Hartree-Fock wave function. The factor f $(0 \leq f \leq 1)$ measures the strength of the static correlation and, hence $f = 1$ at the dissociation limit. Then, we get

$$U_{x\alpha}^{\mathrm{static}}(\boldsymbol{r})n_{1\alpha}(\boldsymbol{r}) = \left(U_{x\alpha}^{\mathrm{exact}}(\boldsymbol{r}) + U_{x\beta}^{\mathrm{exact}}(\boldsymbol{r})\right) n_{1\alpha}(\boldsymbol{r}) = 2U_{x\alpha}^{\mathrm{exact}}(\boldsymbol{r})n_{1\alpha}(\boldsymbol{r}) \tag{4.61}$$

because of the symmetry of the molecule.

Next we consider the integrand of the right hand side of Eq. (4.35) which is based on the energy electron density. Explicitly, it is given by

$$U_x^e\left[n_{1\sigma}^e\right](\epsilon)\, n_{1\sigma}^e(\epsilon) + \left(\overline{U}_x\left[n_{0\sigma}\right](\epsilon) - U_x^e\left[n_{0\sigma}^e\right](\epsilon)\right) n_{0\sigma}^e(\epsilon) \tag{4.62}$$

The notations $n_{0\sigma}^e$ and $n_{1\sigma}^e$ stand for the energy electron density with spin σ for the spin polarized and depolarized densities, respectively. Remind that the relations $n_{1\alpha}^e = n_{1\beta}^e$, $n_{0\alpha}^e = n_{0\beta}^e$ and $n_{1\alpha}^e = n_{0\alpha}^e$ hold in principle for H_2 with the infinitely stretched bond as shown in Eq. (4.7). Therefore, Eq. (4.62) becomes

$$U_x^e\left[n_{0\sigma}^e\right](\epsilon)\,n_{0\sigma}^e(\epsilon)+\left(\overline{U}_x\left[n_{0\sigma}\right](\epsilon)-U_x^e\left[n_{0\sigma}^e\right](\epsilon)\right)n_{0\sigma}^e(\epsilon)$$
$$=\overline{U}_x\left[n_{0\sigma}\right](\epsilon)\,n_{0\sigma}^e(\epsilon) \qquad (4.63)$$

It is quite interesting to note the similarity of Eq. (4.60) with Eq. (4.62). Actually, the first term of each equation represents the exchange energy density for a spin without static correlation. Regardless of the fact that the former utilizes the energy electron density and the latter uses the conventional electron density in the GBR approach, they are the same as energy contents. Actually, it is possible to prove that the terms are equivalent when the GBR method is employed to evaluate the energy $U_x^e[n_{0\sigma}^e](\epsilon)$. On the other hand, the second term in each equation represents the static correlation (SC) associated with the bond dissociation. In the Becke03 functional, the SC energy for a spin σ is created by incorporating the fraction of the exchange hole with the opposite spin, while in the present work, it is obtained as the difference between the exchange energy computed with the densities n_0 and n_0^e. It is useful to prove that Eq. (4.63) coincides with Eq. (4.61) when the GBR method is applied to evaluate $\overline{U}_x$ in Eq. (4.63). By inserting the definition of $\overline{U}_x$ given in Eq. (4.37), the right hand side of the second equality in Eq. (4.63) becomes,

$$\begin{aligned}
&\int d\epsilon\,\overline{U}_x^{\text{exact}}\left[n_{0\alpha}\right](\epsilon)\,n_{0\alpha}^e(\epsilon)\\
&=\int d\epsilon\,\frac{1}{n_{0\alpha}^e(\epsilon)}\left\{\int d\boldsymbol{r}\,U_x^{\text{exact}}\left[n_{0\alpha}\right](\boldsymbol{r})\,n_{0\alpha}(\boldsymbol{r})\,\delta\left(\epsilon-v_{\text{def}}(\boldsymbol{r})\right)\right\}n_{0\alpha}^e(\epsilon)\\
&=\int d\boldsymbol{r}\,U_x^{\text{exact}}\left[n_{0\alpha}\right](\boldsymbol{r})\,n_{0\alpha}(\boldsymbol{r})\\
&=\int_{\Omega_A} d\boldsymbol{r}\,U_x^{\text{exact}}\left[n_{0\alpha}\right](\boldsymbol{r})\times 2n_{1\alpha}(\boldsymbol{r})\\
&=\int_{\Omega_A} d\boldsymbol{r}\,U_x^{\text{exact}}\left[n_{0\alpha}\right](\boldsymbol{r})\,n_{1\alpha}(\boldsymbol{r})+\int_{\Omega_B} d\boldsymbol{r}\,U_x^{\text{exact}}\left[n_{0\beta}\right](\boldsymbol{r})\,n_{1\alpha}(\boldsymbol{r})\\
&=\int d\boldsymbol{r}\,2U_x^{\text{exact}}\left[n_{1\alpha}\right](\boldsymbol{r})\,n_{1\alpha}(\boldsymbol{r})
\end{aligned} \qquad (4.64)$$

where Ω_A is the spatial region that encloses entire density of $n_{0\alpha}$ and the region Ω_B is that for $n_{0\beta}$. It should also be reminded that $U_x^{\text{exact}}[n_{1\alpha}]$ is identical to $U_{x\alpha}^{\text{exact}}$ by definition. Thus, the equality of Eqs. (4.63) to (4.61) is proved. It can be concluded that the two different approaches, that is, the present work and the Becke03 functional, are numerically equivalent for the computation of the dissociated H_2 molecule.

In Fig. 4.14 we present the potential energy curves (PECs) of the dissociating H_2 molecule, where the comparisons are made between the energy given by the Becke03 functional (Eq. (4.60)) and that by the present work (Eq. (4.62)). For references, the results of the spin-restricted HF(RHF), CAS(2,2) and CAS(2,2)-MP2 calculations are also presented. The graphs for the CAS(2,2) and CAS(2,2)-MP2 calculations are the same as those shown in Fig. 4.4 and the computational details are provided in the corresponding figure caption. The RHF calculation is performed using the

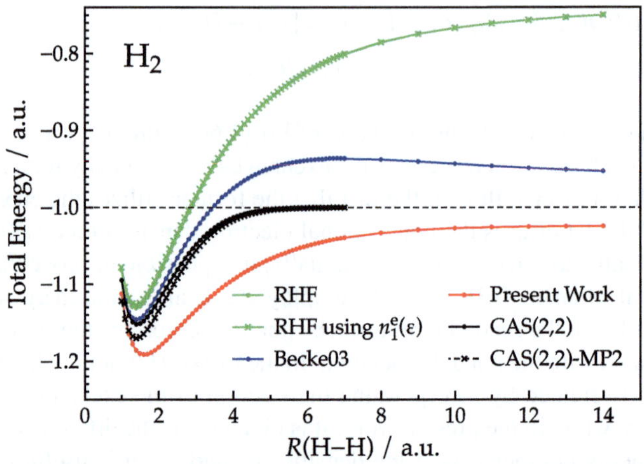

Fig. 4.14 The potential energy curves for the bond dissociation in H_2 molecule. The RHF calculations are performed using the 'Vmol' program. In the RHF calculation with $n_1^e(\epsilon)$, the exchange energy density of Eq. (4.65) is applied. The notations for the legends 'CAS(2,2)'and 'CAS(2,2)-MP2' are the same as in Fig. 4.4. Equation (4.60) is used in the 'Becke03' calculation and Eq. (4.62) is employed for the calculation of 'Present Work'. The RHF wave functions are employed for these calculations

'Vmol' program based on the real-space grids[9–11, 17]. The exchange energy density $U_x^e[n_{1\sigma}^e](\epsilon)$ in Eq. (4.62) is described in terms of the HF exchange energy density $U_{x\sigma}^{\mathrm{HF}}(\boldsymbol{r})$, thus,

$$U_{x\sigma}^{e,\mathrm{HF}}(\epsilon) = \frac{1}{\Omega(\epsilon)} \int d\boldsymbol{r}\, U_{x\sigma}^{\mathrm{HF}}(\boldsymbol{r})\,\delta(\epsilon - \upsilon(\boldsymbol{r})) \tag{4.65}$$

It shows that $U_{x\sigma}^{e,\mathrm{HF}}(\epsilon)$ is obtained merely by averaging $U_{x\sigma}^{\mathrm{HF}}(\boldsymbol{r})$ over the spatial region with the energy coordinate ϵ. $U_{x\sigma}^{\mathrm{HF}}(\boldsymbol{r})$ is constructed from the spin-adapted HF wave function, namely, the RHF wave function. The energy density $U_x^e[n_{0\sigma}^e](\epsilon)$ in Eq. (4.62) for the reference system is computed similarly using the HF wave function for an electron with spin σ in an isolated hydrogen atom. Thus, in the construction of the graphs in Fig. 4.14, the relationship $n_1(\boldsymbol{r}) = n_0(\boldsymbol{r})$ does not hold for the total electron density since the density given by the RHF method does not match the UHF density. Therefore, it is *not* guaranteed that Eq. (4.62) provides the correct dissociation limit. The same RHF wave function is employed for the evaluation of the Becke03 functional of Eq. (4.60). It is seen in the figure that the PEC given by the Becke03 functional is larger in energy than that by the CAS(2,2) calculation, suggesting that the absolute of the SC energy is underestimated in the Becke03 functional. In Ref. [1], it is advocated that the SC energy should be half of that given in Eq. (4.60)) when one considers the kinetic correlation energy and the virial theorem. Although the claim might be correct, it is not applied in our implementation. The calculation with the legend of 'Present Work' utilizes the functional of Eq. (4.62). As a common trend seen in Fig. 4.4, it is shown that the SC energy is overestimated in the middle range of the bond distance. At $R(\mathsf{H}-\mathsf{H}) = 14.0$ a.u., the energy is lower than that given by the CAS(2,2) or CAS(2,2)-MP2 calculation. This can be attributed to the fact that the kinetic energy and the Hartree potential of the electrons are lowered in the RHF wave function as compared with those of the wave functions for the isolated atoms. It can be concluded that the present approach, based on the energy electron density, gives the PEC for H_2 that is significantly different from that given by the Becke03 functional, at least when their original forms are adopted. The present work overestimates the SC energy in the intermediate region of the bond length, while the Becke03 functional underestimates it in the region of the longer bond distance. This is somewhat surprising because both functionals are designed so that they can correctly describe the dissociation limit of the molecule.

References

1. A.D. Becke, J. Chem. Phys. **119**(6), 2972 (2003)
2. H. Takahashi, J. Phys. B: Atomic, Mol. Optic. Phys. **51**(5), 055102(11) (2018)
3. J. Harris, Phys. Rev. B **31**(4), 1770 (1985)
4. T.A. Wesolowsk, A. Warshel, J. Phys. Chem. **97**, 8050 (1993)
5. P. Elliott, M.H. Cohen, A. Wasserman, K. Burke, J. Chem. Theory Comput. **5**(4), 827 (2009)
6. M.J.F. et al., *GAUSSIAN 09, Revision C.01, Gaussian Inc.* (Wallingford, CT, 2010)
7. J.R. Chelikowsky, N. Troullier, Y. Saad, Phys. Rev. Lett. **72**(8), 1240 (1994)
8. J.R. Chelikowsky, N. Troullier, K. Wu, Y. Saad, Phys. Rev. B **50**(16), 11355 (1994)
9. H. Takahashi, T. Hori, T. Wakabayashi, T. Nitta, Chem. Lett. **3**, 222 (2000)
10. H. Takahashi, T. Hori, T. Wakabayashi, T. Nitta, J. Phys. Chem. A **105**(17), 4351 (2001)
11. H. Takahashi, T. Hori, H. Hashimoto, T. Nitta, J. Comp. Chem. **22**(12), 1252 (2001)
12. H. Takahashi, J. Phys. B: At. Mol. Opt. Phys. **53**, 245101(9) (2020)
13. C. Lee, W. Yang, R.G. Parr, Phys. Rev. B **37**(2), 785 (1988)
14. A.D. Becke, Phys. Rev. A **38**(6), 3098 (1988)
15. A.D. Becke, J. Chem. Phys. **138**(7), 074109(10) (2013)
16. W. Yang, Y. Zhang, P.W. Ayers, Phys. Rev. Lett. **84**(22), 5172 (2000)
17. H. Takahashi, S. Sakuraba, A. Morita, J. Chem. Inf. Model. **60**, 1376 (2020)

Chapter 5
Development of Kinetic Energy Functional

5.1 Overview of the Developments of Kinetic-Energy Density Functional

The development of the kinetic-energy functional [3, 4] is obviously one of the subjects with great significance in the density-functional theory (DFT). As discussed in Sect. 1.3.5, in the Kohn-Sham (KS) DFT [5], the kinetic energy (KE) of the N-electrons system under consideration is described in terms of the set of one-electron wave functions $\{\varphi_i\}(i = 1, 2, \cdots, N/2)$ which identify the non-interacting reference system. By introducing the wave functions, most part of the KE of the interacting system can be obtained as the KE of its reference system. However, the use of the set $\{\varphi_i\}$ necessitates the additional computational procedure to impose the orthonormality condition, i.e. $\langle \varphi_i | \varphi_j \rangle = \delta_{ij}$, which scales as $O(N^2)$ besides the system size. Thus, the computational cost can be drastically reduced provided that the KE can be computed in terms of the electron density instead of the wave functions. Such an approach, often referred to as orbital-free (OF) DFT, could be a milestone in the history of DFT provided that it can be successfully applied to various systems in place of KS-DFT. Indeed, the development of the successful kinetic energy functional will extend the frontier of DFT to the materials with realistic system sizes [6, 7]. The classical force fields, widely utilized in the molecular dynamics simulations, will be made obsolete with the success of the OF-DFT.

5.1.1 Thomas-Fermi Model and Weizsäcker Correction

However, the development of the kinetic energy functional is known to be a quite difficult issue. Although there has been a lot of works to develop the approximate KE functionals and considerable progress has been made in this field, it seems that there exists no KE functional that produces the results with the accuracy and reli-

H. Takahashi, *Density-Functional Theory for Electrons*,
https://doi.org/10.1007/978-981-95-1229-4_5

ability comparable to the KS-DFT. In the following, we will clarify the source of the difficulty in developing the KE functional by briefly reviewing its history. The most primitive KE functional was that provided by Thomas [1] and Fermi [8] for the homogeneous electron gas (HEG). The explicit form of the Thomas-Fermi (TF) functional is given by

$$E_{\mathrm{TF}}[n] = C_{\mathrm{TF}} \int d\boldsymbol{r} n\left(\boldsymbol{r}\right)^{\frac{5}{3}} \tag{5.1}$$

where $n\left(\boldsymbol{r}\right)$ is the total electron density at the position $\boldsymbol{r}$ of the electrons and $C_{\mathrm{TF}} = \frac{3}{10}(3\pi^2)^{\frac{2}{3}}$. The concise derivation of Eq. (5.1) can be found in a standard textbook [9] for DFT. For readers' convenience, we provide the details of the derivation in Appendix D. Anyway, $C_{\mathrm{TF}}\ n\left(\boldsymbol{r}\right)^{\frac{5}{3}}\ d\boldsymbol{r}$ in Eq. (5.1) describes the kinetic energy at $\boldsymbol{r}$ of electrons with the density $n(\boldsymbol{r})$. Equation (5.1) works perfectly when it is applied to a region in a homogeneous non-interacting electron gas as in the case of the local-density approximation (LDA) for the exchange energy. A preliminary correction to incorporate the gradient of the density is first proposed by von Weizsäcker in 1935 [10]. The energy correction $E_{\mathrm{vW}}[n]$ is simply given by

$$E_{\mathrm{vW}}\left[n\right] = -\frac{1}{2} \int d\boldsymbol{r} n\left(\boldsymbol{r}\right)^{\frac{1}{2}} \nabla^2 n\left(\boldsymbol{r}\right)^{\frac{1}{2}} \tag{5.2}$$

It is important to note that the correction itself represents the exact kinetic energy of a one-electron system or a closed shell two-electrons system. The TF model corrected with the fraction of the energy $E_{\mathrm{vW}}[n]$ is given by

$$E_{\mathrm{TF}\lambda\mathrm{W}}\left[n\right] = E_{\mathrm{TF}}\left[n\right] + \lambda E_{\mathrm{vW}}\left[n\right] \tag{5.3}$$

where the prefactor λ regulates the amount of the correction $E_{\mathrm{vW}}[n]$ to be included in the kinetic energy. In fact, Eq. (5.3) makes a considerable contribution to reduce the error of the TF model. Unfortunately, however, it is known that the inclusion of the higher order terms in Eq. (5.3) gives only minor contribution to improve the accuracy of the KE. It is widely believed that the realization of the shell structures in the electron densities of the system is essential for the accurate KE functional. The shell structure in atoms is often discussed in the context of the Friedel oscillations in metals, which are observed as the density response to a perturbation. To get physical insight into the KE functional of Eq. (5.3), we consider below the linear response of the HEG on the basis of the analyses in Ref. [3].

For a system of interest, the linear response $\delta n\left(\boldsymbol{r}\right)$ of the density at $\boldsymbol{r}$ to the variation $\delta v(\boldsymbol{r}')$ of the potential at $\boldsymbol{r}'$ is described with the response function $\chi\left(\boldsymbol{r}, \boldsymbol{r}'\right)$, thus,

$$\delta n(\boldsymbol{r}) = \int d\boldsymbol{r}' \chi\left(\boldsymbol{r}, \boldsymbol{r}'\right) \delta v(\boldsymbol{r}') \tag{5.4}$$

This can be considered as the definition of the response function. Another form of the definition was provided in Eq. (1.183) in the context of the optimized effective potential (OEP) method in Chap. 1. For the HEG, Eq. (5.4) can be written as

$$\delta n(\boldsymbol{r}) = \int d\boldsymbol{r}' \chi(\boldsymbol{r}-\boldsymbol{r}')\delta \upsilon(\boldsymbol{r}') \tag{5.5}$$

By applying the Fourier transformation to the both sides, we get

$$\begin{aligned}\delta\widetilde{n}(\boldsymbol{k}) &= \int d\boldsymbol{r} \exp(i\boldsymbol{k}\cdot\boldsymbol{r})\,\delta n(\boldsymbol{r}) \\ &= \int d\boldsymbol{r} \exp(i\boldsymbol{k}\cdot\boldsymbol{r}) \int d\boldsymbol{r}'\chi(\boldsymbol{r}-\boldsymbol{r}')\delta\upsilon(\boldsymbol{r}') \\ &= \widetilde{\chi}(\boldsymbol{k})\,\delta\widetilde{\upsilon}(\boldsymbol{k})\end{aligned} \tag{5.6}$$

The response function for the HEG is known as the Lindhard function $\widetilde{\chi}_{\mathrm{Lind}}(\boldsymbol{k})$ [11] for which derivation is provided in Appendix E. Explicitly, $\widetilde{\chi}_{\mathrm{Lind}}(\boldsymbol{k})$ is written as

$$\widetilde{\chi}_{\mathrm{Lind}}(\boldsymbol{k}) = -\frac{k_F}{\pi^2}\left(\frac{1}{2} + \frac{1-\eta^2}{4\eta}\ln\left|\frac{1+\eta}{1-\eta}\right|\right) \tag{5.7}$$

with η being the dimensionless momentum $\eta = k/2k_F$. Our purpose here is to compare $\widetilde{\chi}_{\mathrm{Lind}}(\boldsymbol{k})$ with the response functions in the momentum space for the functionals $E_{\mathrm{TF}}[n]$ and $E_{\mathrm{vW}}[n]$. Thus, the response functions corresponding to these functionals should be obtained. To this end, we first derive the general relationship between the response function and the kinetic energy functional. The density functional $E_{\mathrm{tot}}[n]$ for the total energy of a system under consideration is decomposed into two contributions, thus,

$$E_{\mathrm{tot}}[n] = E_{\mathrm{kin}}[n] + E_{\mathrm{eff}}[n] \tag{5.8}$$

The first term in the right hand side is the kinetic energy functional and the second one stands for the residual energies. Explicitly, $E_{\mathrm{eff}}[n]$ is described as

$$E_{\mathrm{eff}}[n] = E_{\mathrm{H}}[n] + E_{xc}[n] + E_{\mathrm{en}}[n] \tag{5.9}$$

The three terms $E_{\mathrm{H}}[n]$, $E_{xc}[n]$, and $E_{\mathrm{en}}[n]$ in the right hand side are the Hartree energy, exchange-correlation energy, and the electron-nuclei interaction energy, respectively. At the equilibrium of the electronic states, the differentiation of the functional Eq. (5.8) gives the chemical potential μ of the electrons, thus,

$$\begin{aligned}\left.\frac{\delta E_{\mathrm{tot}}[n]}{\delta n(\boldsymbol{r})}\right|_{n_0} &= \left.\frac{\delta E_{\mathrm{kin}}[n]}{\delta n(\boldsymbol{r})}\right|_{n_0} + \upsilon_{\mathrm{eff}}[n_0](\boldsymbol{r}) \\ &= \upsilon_{\mathrm{kin}}[n_0](\boldsymbol{r}) + \upsilon_{\mathrm{eff}}[n_0](\boldsymbol{r}) = \mu\end{aligned} \tag{5.10}$$

where n_0 is the density of the equilibrium state (not necessarily the HEG at this stage). $v_{\mathrm{kin}}[n_0](\boldsymbol{r})$ is the kinetic potential of the system with the density n_0. The second derivative becomes

$$\left.\frac{\delta^2 E_{\mathrm{tot}}[n]}{\delta n(\boldsymbol{r}')\,\delta n(\boldsymbol{r})}\right|_{n_0} = \left.\frac{\delta^2 E_{\mathrm{kin}}[n]}{\delta n(\boldsymbol{r}')\,\delta n(\boldsymbol{r})}\right|_{n_0} + \left.\frac{\delta v_{\mathrm{eff}}[n](\boldsymbol{r})}{\delta n(\boldsymbol{r}')}\right|_{n_0} = 0 \tag{5.11}$$

which directly leads to

$$\left.\frac{\delta^2 E_{\mathrm{kin}}[n]}{\delta n(\boldsymbol{r}')\,\delta n(\boldsymbol{r})}\right|_{n_0} = -\left.\chi^{-1}(\boldsymbol{r}',\boldsymbol{r})\right|_{n_0} \tag{5.12}$$

By applying the Fourier transformation on the both sides, we have

$$\mathcal{F}\left(\left.\frac{\delta^2 E_{\mathrm{kin}}[n]}{\delta n(\boldsymbol{r}')\,\delta n(\boldsymbol{r})}\right|_{n_0}\right) = -\left.\frac{1}{\widetilde{\chi}(\boldsymbol{k})}\right|_{n_0} \tag{5.13}$$

In deriving this equation, we apply the procedure parallel to Eq. (5.6) to the inverse of χ. Specifically for the HEG we have

$$\mathcal{F}\left(\left.\frac{\delta^2 E_{\mathrm{kin}}[n]}{\delta n(\boldsymbol{r}')\,\delta n(\boldsymbol{r})}\right|_{n_0:\mathrm{HEG}}\right) = -\frac{1}{\widetilde{\chi}_{\mathrm{Lind}}(\boldsymbol{k})} \tag{5.14}$$

For the sake of notational simplicity, we define the dimensionless function C thus

$$C(\eta) = -\frac{k_F}{\pi^2}\frac{1}{\widetilde{\chi}(\boldsymbol{k})} \tag{5.15}$$

Then the function C_{Lind} corresponding to the Lindhard function is explicitly written as

$$C_{\mathrm{Lind}}(\eta) = \left(\frac{1}{2} + \frac{1-\eta^2}{4\eta}\ln\left|\frac{1+\eta}{1-\eta}\right|\right)^{-1} \tag{5.16}$$

Next we obtain the functions $C_{\mathrm{TF}}(\eta)$ and $C_{\mathrm{vW}}(\eta)$, respectively, for the TF model and the Weizsäcker correction by exploiting Eq. (5.13). For the TF model, it reads

$$\begin{aligned}\left.\frac{\delta^2 E_{\mathrm{TF}}[n]}{\delta n(\boldsymbol{r}')\,\delta n(\boldsymbol{r})}\right|_{n_0:\mathrm{HEG}} &= \left.\frac{\delta}{\delta n(\boldsymbol{r}')}\left(\frac{5}{3}C_{\mathrm{TF}} n_0(\boldsymbol{r})^{\frac{2}{3}}\right)\right|_{n_0:\mathrm{HEG}} \\ &= \left.\frac{10}{9}C_{\mathrm{TF}} n_0(\boldsymbol{r})^{-\frac{1}{3}}\,\delta\left(\boldsymbol{r}-\boldsymbol{r}'\right)\right|_{n_0:\mathrm{HEG}} \\ &= \frac{10}{9}\frac{3}{10}\left(3\pi^2\right)\frac{1}{k_F}\delta\left(\boldsymbol{r}-\boldsymbol{r}'\right) \\ &= \frac{\pi^2}{k_F}\delta\left(\boldsymbol{r}-\boldsymbol{r}'\right)\end{aligned} \tag{5.17}$$

where we use the relation Eq. (D.7), that is, $n_0 = k_F^3/(3\pi^2)$ for the HEG. By applying Eq. (5.13) and the definition for C given by Eq. (5.15), we have

$$\mathcal{F}\left(\left.\frac{\delta^2 E_{\mathrm{TF}}[n]}{\delta n(\boldsymbol{r}')\,\delta n(\boldsymbol{r})}\right|_{n_0:\mathrm{HEG}}\right) = -\frac{1}{\widetilde{\chi}_{\mathrm{TF}}(\boldsymbol{k})} = \frac{\pi^2}{k_F}C_{\mathrm{TF}}(\eta) = \frac{\pi^2}{k_F} \tag{5.18}$$

which leads to the relation $C_{\mathrm{TF}}(\eta) = 1$.

For the Weizsäcker term $E_{\mathrm{vW}}[n]$ we also consider its second derivative with respect to the density as in Eq. (5.17),

$$\begin{aligned}\left.\frac{\delta^2 E_{\mathrm{vW}}[n]}{\delta n(\boldsymbol{r}')\,\delta n(\boldsymbol{r})}\right|_{n_0:\mathrm{HEG}} &= \left.\frac{\delta}{\delta n(\boldsymbol{r}')}\frac{\delta}{\delta n(\boldsymbol{r})}\int d\boldsymbol{r}\left(-\frac{1}{2}\right) n(\boldsymbol{r})^{\frac{1}{2}}\,\nabla^2 n(\boldsymbol{r})^{\frac{1}{2}}\right|_{n_0:\mathrm{HEG}} \\ &= \frac{\delta}{\delta n(\boldsymbol{r}')}\left\{\left(-\frac{1}{4}\right) n(\boldsymbol{r})^{-\frac{1}{2}}\,\nabla^2 n(\boldsymbol{r})^{\frac{1}{2}}\right. \\ &\quad \left.\left.+\left(-\frac{1}{4}\right) n(\boldsymbol{r})^{-\frac{1}{2}}\,\nabla^2 n(\boldsymbol{r})^{\frac{1}{2}}\right\}\right|_{n_0:\mathrm{HEG}} \\ &= -\frac{1}{2}\frac{\delta}{\delta n(\boldsymbol{r}')}\left. n(\boldsymbol{r})^{-\frac{1}{2}}\,\nabla^2 n(\boldsymbol{r})^{\frac{1}{2}}\right|_{n_0:\mathrm{HEG}}\end{aligned} \tag{5.19}$$

It needs careful consideration to obtain the functional derivative of the last term in Eq. (5.19). The difficulty arises in differentiating $\nabla^2 n(\boldsymbol{r})^{\frac{1}{2}}$ with respect to $n(\boldsymbol{r}')$. We apply Taylor expansion to $n(\boldsymbol{r})^{1/2}$ around n_0, thus,

$$\left.\frac{\delta}{\delta n(\boldsymbol{r}')}\nabla^2 n(\boldsymbol{r})^{\frac{1}{2}}\right|_{n_0:\mathrm{HEG}} \tag{5.20}$$

$$= \left.\frac{\delta}{\delta n(\boldsymbol{r}')}\nabla^2\left(n_0(\boldsymbol{r})^{\frac{1}{2}} + \frac{1}{2}n_0(\boldsymbol{r})^{-\frac{1}{2}}\,\delta n(\boldsymbol{r}) + \cdots\right)\right|_{n_0:\mathrm{HEG}} \tag{5.21}$$

Since the reference density n_0 is assumed to be the HEG, we obtain,

$$\left.\frac{\delta}{\delta n(\boldsymbol{r}')}\nabla^2 n(\boldsymbol{r})^{\frac{1}{2}}\right|_{n_0:\mathrm{HEG}} = \delta\left(\boldsymbol{r}-\boldsymbol{r}'\right)\frac{1}{2}n_0(\boldsymbol{r})^{-\frac{1}{2}}\,\nabla^2 \tag{5.22}$$

Finally, the second derivative in Eq. (5.19) is obtained as

$$\left.\frac{\delta^2 E_{\mathrm{vW}}[n]}{\delta n(\boldsymbol{r}')\,\delta n(\boldsymbol{r})}\right|_{n_0:\mathrm{HEG}} = \delta\left(\boldsymbol{r}-\boldsymbol{r}'\right)\left(-\frac{1}{4}\right) n_0(\boldsymbol{r})^{-1}\,\nabla^2 \tag{5.23}$$

The Fourier transformation of the equation becomes

$$\begin{aligned}
\mathcal{F}\left(\left.\frac{\delta^2 E_{\mathrm{vW}}[n]}{\delta n(\boldsymbol{r}')\,\delta n(\boldsymbol{r})}\right|_{n_0:\mathrm{HEG}}\right) &= -\frac{1}{\widetilde{\chi}_{\mathrm{vW}}(\boldsymbol{k})} \\
&= \frac{1}{4}\frac{3\pi^2}{k_F^3}k^2 = 3\frac{\pi^2}{k_F}\eta^2 \\
&= \frac{\pi^2}{k_F}C_{\mathrm{vW}}(\eta)
\end{aligned} \tag{5.24}$$

Thus, it is found that $C_{\mathrm{vW}}(\eta)$ for the Weizsäcker term $E_{\mathrm{vW}}[n]$ is given by $C_{\mathrm{vW}}(\eta) = 3\eta^2$.

To make comparisons between the response functions of the HEG in the momentum space and of the TF model corrected with the Weizsäcker term, we consider the following Fourier transformation,

$$\begin{aligned}
-\frac{1}{\widetilde{\chi}_{\mathrm{TF}\lambda\mathrm{W}}(\boldsymbol{k})} &= \mathcal{F}\left(\left.\frac{\delta^2 E_{\mathrm{TF}\lambda\mathrm{W}}[n]}{\delta n(\boldsymbol{r}')\,\delta n(\boldsymbol{r})}\right|_{n_0:\mathrm{HEG}}\right) \\
&= \mathcal{F}\left(\left.\frac{\delta^2 E_{\mathrm{TF}}[n]}{\delta n(\boldsymbol{r}')\,\delta n(\boldsymbol{r})}\right|_{n_0:\mathrm{HEG}}\right) + \lambda\mathcal{F}\left(\left.\frac{\delta^2 E_{\mathrm{vW}}[n]}{\delta n(\boldsymbol{r}')\,\delta n(\boldsymbol{r})}\right|_{n_0:\mathrm{HEG}}\right) \\
&= -\frac{1}{\widetilde{\chi}_{\mathrm{TF}}(\boldsymbol{k})} - \lambda\frac{1}{\widetilde{\chi}_{\mathrm{vW}}(\boldsymbol{k})} \\
&= \frac{\pi^2}{k_F}\left(C_{\mathrm{TF}}(\eta) + \lambda C_{\mathrm{vW}}(\eta)\right)
\end{aligned} \tag{5.25}$$

Then, the response function $\widetilde{\chi}_{\mathrm{TF}\lambda\mathrm{W}}(\boldsymbol{k})$ in the momentum space for the kinetic energy functional in Eq. (5.3) can be given by

$$\begin{aligned}
-\widetilde{\chi}_{\mathrm{TF}\lambda\mathrm{W}}(\boldsymbol{k}) &= \frac{k_F}{\pi^2}\left(\frac{1}{C_{\mathrm{TF}}(\eta) + \lambda C_{\mathrm{vW}}(\eta)}\right) \\
&= \frac{k_F}{\pi^2}\frac{1}{C_{\mathrm{TF}\lambda\mathrm{W}}(\eta)}
\end{aligned} \tag{5.26}$$

In Fig. 5.1, plot of the inverse of the $C_{\mathrm{Lind}}(\eta)$ is presented and compared with those for the functionals E_{TF} and $E_{\mathrm{TF}\lambda\mathrm{W}}$ with $\lambda = 1, 1/9$. First of all, it should be noted that the response function of the TF model is constant regardless of the value of η and does not show the trend of the behavior of the Lindhard function. The TF

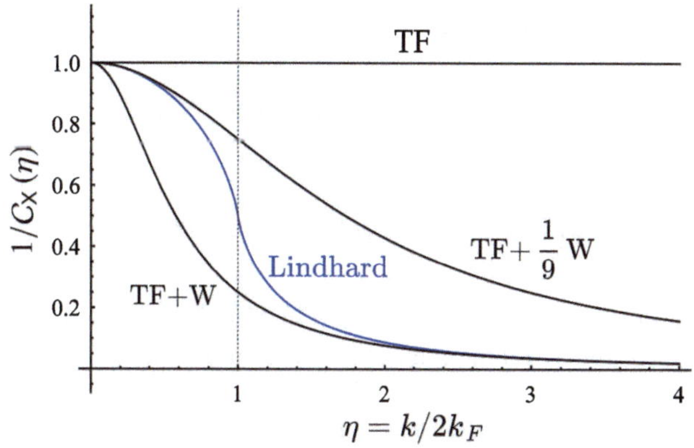

Fig. 5.1 The plots of the various response functions in the momentum space are shown. Explicitly, the inverses of the $C_{\mathsf{X}}(\eta)$ functions are presented for $\mathsf{X} =$ Lindhard, Thomas-Fermi (TF), and the TF corrected with the Weizsäcker term (TF+λW) with $\lambda = 1, 1/9$ (see Eq. (5.3)). The dashed line represents $\eta = 1$ where the slope of $1/C_{\text{Lind}}$ becomes divergent

model realizes exactly the kinetic energy of the non-interacting HEG. However, it does not reproduce the response property of the HEG. The response function of the 'TF+W' model ($\lambda = 1$ in Eq. (5.3)) does not reproduce the function $1/C_{\text{Lind}}(\eta)$ in the region of small η although it asymptotically realizes the function in the region of large η. On the other hand, 'TF+1/9W' model ($\lambda = 1/9$ in Eq. (5.3)) reproduces the behavior of thc $1/C_{\text{Lind}}(\eta)$ only for the small η. One might consider that the inclusion of the higher order gradient terms into the kinetic energy functional in addition to the Weizsäcker correction will improve the response function. As noted above, however, the higher order gradient correction gives only minor correction. A notable feature of the Lindhard function given in Eq. (5.7) is that it has a pole at $\eta = 1$ in the argument of the logarithmic function. As discussed in Ref. [3], the singularity gives rise to the Friedel oscillations, which occur in the density over a certain distance from a perturbation. It is quite unlikely that the inclusion of the higher gradient terms plays a decisive role in the realization of such a nonlocal effect. We are not sure that the Friedel oscillations occurring in the HEG is relevant to the formation of the shell structures in materials or atoms. However, it is possible to state that the shell structure itself can only be realized by a kinetic energy operator with some nonlocal properties. This constitutes the motivation for the development of the nonlocal functional by Wang and Teter [2]. The brief review of their work will be helpful to the readers to clarify the point of the developments provided in the next Subsection.

5.1.2 Wang-Teter Functional

The important feature of the KE functional $E_{\mathrm{WT}}[n]$ by Wang and Teter [2] lies in its functional form described as

$$E_{\mathrm{WT}}[n] = \int d\boldsymbol{r}d\boldsymbol{r}' n(\boldsymbol{r})^{\frac{5}{6}} f(\boldsymbol{r}, \boldsymbol{r}') n(\boldsymbol{r}')^{\frac{5}{6}} \tag{5.27}$$

where the function $f(\boldsymbol{r}, \boldsymbol{r}')$ carries the nonlocal effect of the functional. In the original version of the WT functional, the function f is described in terms of the linear response function $\widetilde{\chi}_{\mathrm{Lind}}$ of the HEG that is solely identified with the average density n_0 of the system. More precisely, $f(\boldsymbol{r}, \boldsymbol{r}')$ is approximated as $f(k_F; |\boldsymbol{r} - \boldsymbol{r}'|)$ where k_F is the Fermi momentum of the HEG with the density n_0. Specifically when Eq. 5.27 is applied to the HEG with density n_0, it is readily understood that the following relationship must be satisfied,

$$F_0 \equiv \int drf(r) = \frac{3}{10}(3\pi^2)^{\frac{2}{3}} \tag{5.28}$$

The major purpose of introducing the form of Eq. (5.27) is to extract the terms corresponding to the TF model and the Weizsäcker correction from the response function for the HEG. For this purpose, we derive the density response $\delta n(\boldsymbol{r})$ of the *noninteracting* system of which kinetic energy is described with Eq. (5.27). The following derivation is almost parallel to that in Ref. [2].

The Lagrangian L of the noninteracting system is expressed by

$$L = \int d\boldsymbol{r}d\boldsymbol{r}' n(\boldsymbol{r})^{\frac{5}{6}} f(\boldsymbol{r}, \boldsymbol{r}') n(\boldsymbol{r}')^{\frac{5}{6}} - \mu\left(\int d\boldsymbol{r}n(\boldsymbol{r}) - N\right) = 0 \tag{5.29}$$

where μ represents the chemical potential to maintain the number of the electron in the system at N. At the equilibrium, δL for the small variation δn of the density becomes zero, which reads

$$\begin{aligned}\delta L &= \frac{5}{3}n_0{}^{-\frac{1}{6}} \int d\boldsymbol{r}' f(\boldsymbol{r}, \boldsymbol{r}') n_0{}^{\frac{5}{6}} - \mu \\ &= \frac{5}{3}n_0^{\frac{2}{3}} F_0 - \mu = 0\end{aligned} \tag{5.30}$$

where n_0 is the equilibrium density. Suppose that a perturbation $\delta v(\boldsymbol{r})$ is applied to the unperturbed system. Then, $\delta L[n_0 + \delta n]$ up to the first order of the perturbation is obtained as

$$\delta L\left[n_0+\delta n\right]=\left(\frac{5}{3}n_0^{\frac{2}{3}}F_0-\mu\right)-\frac{5}{18}n_0^{-\frac{1}{3}}F_0\delta n\left(\boldsymbol{r}\right)$$
$$+\frac{25}{18}n_0^{-\frac{1}{3}}\int d\boldsymbol{r}'f\left(\left|\boldsymbol{r}-\boldsymbol{r}'\right|\right)\delta n\left(\boldsymbol{r}'\right)+\delta\upsilon\left(\boldsymbol{r}\right)=0 \quad (5.31)$$

As shown in Eq. (5.30), it is already confirmed that the zeroth order term in the equation is zero. Then, for the first order term, we have

$$-\frac{1}{12}\left(3\pi^2\right)^{\frac{2}{3}}n_0^{-\frac{1}{3}}\delta n\left(\boldsymbol{r}\right)+\frac{25}{18}n_0^{-\frac{1}{3}}\int d\boldsymbol{r}'f\left(\left|\boldsymbol{r}-\boldsymbol{r}'\right|\right)\delta n\left(\boldsymbol{r}'\right)+\delta\upsilon\left(\boldsymbol{r}\right)=0 \quad (5.32)$$

where Eq. (5.28) is used. The Fourier transformation of the equation becomes

$$-\frac{1}{12}\left(3\pi^2\right)^{\frac{2}{3}}n_0^{-\frac{1}{3}}\delta\widetilde{n}\left(\boldsymbol{k}\right)+\frac{25}{18}n_0^{-\frac{1}{3}}\widetilde{f}\left(\boldsymbol{k}\right)\delta\widetilde{n}\left(\boldsymbol{k}\right)+\delta\widetilde{\upsilon}\left(\boldsymbol{k}\right)=0 \quad (5.33)$$

This can be directly compared with Eqs. (5.6) and (5.7), which leads to

$$\left(-\frac{1}{12}\left(3\pi^2\right)^{\frac{2}{3}}n_0^{-\frac{1}{3}}+\frac{25}{18}n_0^{-\frac{1}{3}}\widetilde{f}\left(\boldsymbol{k}\right)\right)\widetilde{\chi}_{\mathrm{Lind}}\left(\boldsymbol{k}\right)+1=0 \quad (5.34)$$

Then, $\widetilde{f}\left(\boldsymbol{k}\right)$ is expressed as

$$\widetilde{f}\left(\boldsymbol{k}\right)=-\frac{18}{25}n_0^{\frac{1}{3}}\frac{1}{\widetilde{\chi}_{\mathrm{Lind}}\left(\boldsymbol{k}\right)}+\frac{3}{50}\left(3\pi^2\right)^{\frac{2}{3}} \quad (5.35)$$

Here, we introduce a function $\widetilde{w}(\boldsymbol{k})$ defined as

$$\widetilde{w}\left(\boldsymbol{k}\right)=-\frac{\pi^2}{k_F}\widetilde{\chi}_{\mathrm{Lind}}\left(\boldsymbol{k}\right)=\left(\frac{1}{2}+\frac{1-\eta^2}{4\eta}\ln\left|\frac{1+\eta}{1-\eta}\right|\right) \quad (5.36)$$

where η is defined as $\eta=k/(2k_F)$. Then, $\widetilde{f}\left(\boldsymbol{k}\right)$ can be rewritten as

$$\widetilde{f}\left(\boldsymbol{k}\right)=\frac{6}{25}\left(3\pi^2\right)^{\frac{2}{3}}\left(\frac{1}{\widetilde{w}\left(\boldsymbol{k}\right)}+\frac{1}{4}\right) \quad (5.37)$$

Since $\widetilde{w}\left(\boldsymbol{k}\right)^{-1}$ asymptotically behaves as $3\eta^2$ when η is large. This can be confirmed using the following series expansion,

$$\ln\left(\frac{1+x}{1-x}\right)=2\sum_{n=1}^{\infty}\frac{x^{2n-1}}{2n-1}=2\left(x+\frac{x^3}{3}+\frac{x^5}{5}+\cdots\right) \quad (5.38)$$

for $|x|<1$. We subtract $3\eta^2$ term and a constant from $\widetilde{w}\left(\boldsymbol{k}\right)^{-1}$ to define $\widetilde{w}_1(k)$, which reads,

$$\widetilde{w}_1(k) = \frac{1}{\widetilde{w}(k)} - 3\eta^2 + \frac{3}{5} \tag{5.39}$$

where we use k instead of $\boldsymbol{k}$ using the fact that $\widetilde{w}(\boldsymbol{k})$ is dependent only on the magnitude k. This subtraction is equivalent to the separation of the von Weizsäcker term from the linear response function for the HEG as will be shown later. In fact, we already know that the inverse of the response function of the von Weizsäcker functional is given by $3\eta^2$ as derived in Eq. (5.24). The coincidence of the Lindhard function with the von Weizsäcker functional is also exhibited in Fig. 5.1. We also note that the constant value $-3/5$ is also extracted from $\widetilde{w}(\boldsymbol{k})^{-1}$ in the WT approach. Later, it will turn out that the value corresponds to the term proportional to the TF model. With this definition, $\widetilde{f}(k)$ can be rewritten as

$$\widetilde{f}(k) = \frac{6}{25}\left(3\pi^2\right)^{\frac{2}{3}}\left(\widetilde{w}_1(k) + 3\eta^2 - \frac{7}{20}\right) \tag{5.40}$$

The function $\widetilde{f}(k)$ in Eq. (5.27) is thus decomposed into three terms in the k space. This equation will be transformed to the real-space through the backward Fourier transformation. Accordingly, Eq. (5.27) is rewritten as,

$$\begin{aligned} E_{\mathrm{WT}}[n] = \frac{6}{25}\left(3\pi^2\right)^{\frac{2}{3}} & \int\int d\boldsymbol{r}d\boldsymbol{r}' n(\boldsymbol{r})^{\frac{5}{6}}\, w_1\left(\boldsymbol{r}-\boldsymbol{r}'\right) n\left(\boldsymbol{r}'\right)^{\frac{5}{6}} \\ & - \frac{9}{50} n_0^{-\frac{2}{3}} \int d\boldsymbol{r}\; n(\boldsymbol{r})^{\frac{5}{6}}\, \nabla^2 n(\boldsymbol{r})^{\frac{5}{6}} \\ & - \frac{21}{250} \int d\boldsymbol{r}\; n(\boldsymbol{r})^{\frac{5}{3}} \end{aligned} \tag{5.41}$$

In deriving this equation, remind that the operator k^2 in the momentum space corresponds to ∇^2 in the real-space, and the constant number is transformed to the δ function. It should also be noted that the factor $5/8$ is not multiplied by $\widetilde{w}_1(k)$ in Eq. (5.39), which differs from the definition given in the original work by Wang and Teter [2]. Multiplying by the factor is merely to set the integral $\int d\boldsymbol{r}' w_1(\boldsymbol{r}-\boldsymbol{r}')$ to 1. Further, the second term in Eq. (5.41) can be simplified. To this end, it should be reminded that

$$\int d\boldsymbol{r} f(\boldsymbol{r})\nabla^2 f(\boldsymbol{r}) = -\int d\boldsymbol{r}\left(\nabla f(\boldsymbol{r})\right)^2 \tag{5.42}$$

as shown in Eq. (1.148). The square root of the density can be expanded in a Taylor series around the average density n_0, thus,

$$n(\boldsymbol{r})^{\frac{1}{2}} = n_0 + \frac{1}{2} n_0^{-\frac{1}{2}} \delta n(\boldsymbol{r}) + \cdots \tag{5.43}$$

and similarly for $n(\boldsymbol{r})^{5/6}$ we have

$$n(\boldsymbol{r})^{\frac{5}{6}} = n_0 + \frac{5}{6} n_0^{-\frac{1}{6}} \delta n(\boldsymbol{r}) + \cdots \tag{5.44}$$

Therefore, the gradient $\nabla n(\boldsymbol{r})^{5/6}$ can be expressed as

$$\nabla n(\boldsymbol{r})^{\frac{5}{6}} = \frac{5}{3} n_0^{\frac{1}{3}} \nabla n(\boldsymbol{r})^{\frac{1}{2}} \tag{5.45}$$

Using this expression, Eq. (5.41) can be equivalently replaced by

$$\begin{aligned} E_{\mathrm{WT}}[n] = & \frac{6}{25}\left(3\pi^2\right)^{\frac{2}{3}} \int\int d\boldsymbol{r} d\boldsymbol{r}' n(\boldsymbol{r})^{\frac{5}{6}} w_1\left(\boldsymbol{r}-\boldsymbol{r}'\right) n\left(\boldsymbol{r}'\right)^{\frac{5}{6}} \\ & - \frac{1}{2}\int d\boldsymbol{r} n(\boldsymbol{r})^{\frac{1}{2}} \nabla^2 n(\boldsymbol{r})^{\frac{1}{2}} \\ & - \frac{21}{250}\int d\boldsymbol{r} n(\boldsymbol{r})^{\frac{5}{3}} \end{aligned} \tag{5.46}$$

It turns out that both the von Weizsäcker correction and the term proportional to the TF model appear automatically in the WT functional. This is a direct consequence of the fact that $3\eta^2$ and the constant are subtracted from $\widetilde{w}(\boldsymbol{k})^{-1}$. It is important to notice that the response property of the HEG with the average electron density n_0 of the system is utilized in the evaluation of the nonlocal term of the functional, that is, w_1 in Eq. (5.46) is constructed from the response function of the HEG with the density n_0. On the other hand, the TF model and the Weizsäcker term refer to the variable density $n(\boldsymbol{r})$ depending on $\boldsymbol{r}$. This is made possible by the subtraction of the higher k term from the Lindhard function as done in Eq. (5.39).

The effect of introducing the nonlocal term in the WT functional is of great interest. In Ref. [2] Wang and Teter applied variations of their functional to Ne($1s^2 2s^2 2p^6$) and Cu$^+$($1s^2 2s^2 2p^6 3s^2 3p^6 3d^{10}$) atoms and the radial distribution functions (RDF): $rn(r)^{1/2}$ of electrons around the atoms were computed. The results were compared with that obtained by TF+1/5W functional. It was demonstrated in Figs. 8 and 9 in Ref. [2] that the density oscillations in the RDFs appear although very small along the radial distances r for these atoms. In contrast, no oscillation was formed in the RDFs provided by TF+1/5W functional. It was thus shown that the shell structures of the atoms can be realized by incorporating the nonlocal term to the TF+λW functional. For reader's reference, the RDFs for electrons in Ne atom [2] is presented in Fig. 5.2.

5.2 Nonlocal Kinetic-Energy Functional

Now let us consider the kinetic energy functional with a nonlocal term in a general framework without the reference to the homogeneous electron gas (HEG). Suppose that we have a reference system of which ground state electron density is given by $n_0(\boldsymbol{r})$. Then, the kinetic potential $v_{\mathrm{kin}}[n](\boldsymbol{r})$ for the perturbed system with the density

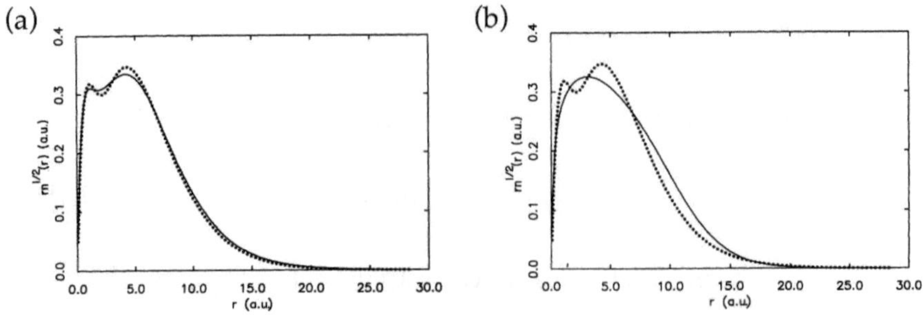

Fig. 5.2 Radial distribution functions $rn(r)^{1/2}$ for electrons in Ne atom with the configuration of $1s^2 2s^2 2p^6$. The dotted lines are those obtained from the exact density. For real lines, **a** a variation of the WT functional and **b** TF+1/5W functional. See Ref. [2] for more details. Reprinted with permission by APS publishing for Fig. 8 in Ref. [2]

n can be simply expressed as

$$\upsilon_{\text{kin}}[n](\boldsymbol{r}) = \upsilon_{\text{kin}}[n_0](\boldsymbol{r}) - \int d\boldsymbol{r}' \chi_0^{-1}(\boldsymbol{r}, \boldsymbol{r}')\left(n(\boldsymbol{r}') - n_0(\boldsymbol{r}')\right) \tag{5.47}$$

where χ_0 is the linear response function of the reference system and defined by

$$\chi_0(\boldsymbol{r}, \boldsymbol{r}') \equiv \left.\frac{\delta n(\boldsymbol{r})}{\delta \upsilon_{\text{eff}}[n](\boldsymbol{r}')}\right|_{n_0} \tag{5.48}$$

And the corresponding kinetic energy $E_{\text{kin}}[n]$ can be given by

$$\begin{aligned} E_{\text{kin}}[n] = E_{\text{kin}}[n_0] &+ \int d\boldsymbol{r} \upsilon_{\text{kin}}[n_0](\boldsymbol{r})\delta n(\boldsymbol{r}) \\ &- \frac{1}{2}\int d\boldsymbol{r} d\boldsymbol{r}' \delta n(\boldsymbol{r}) \chi_0^{-1}(\boldsymbol{r}, \boldsymbol{r}') \delta n(\boldsymbol{r}') \end{aligned} \tag{5.49}$$

where $\delta n(\boldsymbol{r})$ is defined as $\delta n(\boldsymbol{r}) = n(\boldsymbol{r}) - n_0(\boldsymbol{r})$. It is easy to see that Eq. (5.49) is a Taylor expansion of the kinetic energy up to the second order of δn, using the relationship given in Eq. (5.12). It is helpful to compare the form of the WT functional with that of Eq. (5.49). It is possible to consider that the first and second terms in Eq. (5.49) are described in the WT functional with the sum of the TF model and the von Weizsäcker correction, while the third term, which carries the nonlocal effects, is evaluated using the response function of the HEG. Thus, the WT functional can be regarded as a special case of the kinetic energy functional in the form of Eq. (5.49). In the following, we will construct a nonlocal functional of Eq. (5.49) introducing a reference system different from the HEG. It is of great interest to compare the kinetic energy of Eq. (5.49) with that of the Kohn-Sham DFT.

The introduction of a reference system with an *inhomogeneous* electron density, however, gives rise to a serious numerical problem in representing the response function $\chi_0(\boldsymbol{r}, \boldsymbol{r}')$ of the system. When the reference density is inhomogeneous, the response of the density at $\boldsymbol{r}'$ for the perturbation at $\boldsymbol{r}$ is not spherically symmetric around $\boldsymbol{r}$. Thus, $\chi_0(\boldsymbol{r}, \boldsymbol{r}')$ must be represented with a function with six variables in contrast to the response function for the HEG. Of course, the number of dimensions can be reduced by employing a certain appropriate set of basis functions. However, the basis functions as a whole must cover the entire spatial region of the system. Therefore, representing the response function with the basis set cannot be an essential solution to the problem. In the next Section, we will develop a method based on the new formalism of the DFT described in Chap. 3, where the response function of the inhomogeneous system is represented by two variables. It should also be noted that the response function itself of the whole system is constructed from those of the component subsystems.

5.3 Nonlocal Kinetic-Energy Functional with Energy Electron Density

5.3.1 Linear Response Function on the Energy Coordinate

This Section will be devoted to describe the linear response function χ_0 of the reference system using the energy coordinate ϵ introduced in Chap. 3. To this end, we start our discussion with an atomic system where the electron distribution is spherically symmetric around the nucleus. In such a system, the electron density wave, considered as a Friedel oscillation, emerges along the radial direction. Importantly, the energy coordinate ϵ uniquely specifies the radial coordinate r, that is, $r = -Z/\epsilon$ with Z being the charge of the nucleus. Thus, ϵ can be an appropriate variable of the response function responsible for reproducing the Friedel oscillations in the electron densities in atoms. By virtue of this good property of ϵ, the coordinate can also be applied to molecular systems. As advocated by Parr and Berk [12], contours of the electron density in a molecule are more or less parallel to those of the bare-nuclear potential of the molecule. Therefore, the density wave due to the Friedel oscillation will arise along the direction perpendicular to the contours of the bare-nuclear potential. It suggests that the six-dimensional response function $\chi_0(\boldsymbol{r}, \boldsymbol{r}')$ can be safely reduced to the two-dimensional function $\chi_0^e(\epsilon, \epsilon')$. The response function χ_0^e defined on the energy coordinate ϵ can be obtained by projecting χ_0 onto ϵ. Explicitly, it is written as

$$\chi_0^e(\epsilon, \epsilon') = \int d\boldsymbol{r} d\boldsymbol{r}' \chi_0(\boldsymbol{r}, \boldsymbol{r}') \delta(\epsilon - v_{\text{def}}(\boldsymbol{r})) \delta(\epsilon' - v_{\text{def}}(\boldsymbol{r}')) \tag{5.50}$$

Accordingly, the third term in the right hand side of Eq. (5.49) can be rewritten as

$$\text{third term:} = -\frac{1}{2}\int d\epsilon d\epsilon' \delta n^e(\epsilon)\, \chi_0^e(\epsilon, \epsilon')^{-1} \delta n^e(\epsilon') \tag{5.51}$$

where δn^e is defined as $\delta n^e(\epsilon) = n^e(\epsilon) - n_0^e(\epsilon)$. The potential $\upsilon_{\text{def}}(\boldsymbol{r})$ is the defining potential and usually the external potential itself is taken as the υ_{def} as discussed in Chap. 3. Note that $\chi_0^e(\epsilon, \epsilon')^{-1}$ expresses the inverse matrix of $\chi_0^e(\epsilon, \epsilon')$. To make the inverse of χ_0^e in Eq. (5.51), we have to prove that χ_0^e is positive semidefinite. The proof can be made basically in the same way shown in Eq. (1.190). We consider Q^e given by

$$Q^e = \int d\epsilon d\epsilon' q^e(\epsilon)\, \chi_0^e(\epsilon, \epsilon')\, q^e(\epsilon') \tag{5.52}$$

with $q^e(\epsilon)$ being an arbitrary real function. Q^e can be transformed as follows:

$$\begin{aligned}
Q^e &= \int d\epsilon d\epsilon' q^e(\epsilon)\, q^e(\epsilon') \\
&\quad \times \int d\boldsymbol{r} d\boldsymbol{r}' \chi_0(\boldsymbol{r}, \boldsymbol{r}')\, \delta(\epsilon - \upsilon_{\text{def}}(\boldsymbol{r}))\, \delta(\epsilon' - \upsilon_{\text{def}}(\boldsymbol{r}')) \\
&= \int d\boldsymbol{r} d\boldsymbol{r}' \chi_0(\boldsymbol{r}, \boldsymbol{r}') \int d\epsilon\, q^e(\epsilon)\, \delta(\epsilon - \upsilon_{\text{def}}(\boldsymbol{r})) \\
&\quad \times \int d\epsilon'\, q^e(\epsilon')\, \delta(\epsilon' - \upsilon_{\text{def}}(\boldsymbol{r}')) \\
&= \int d\boldsymbol{r} d\boldsymbol{r}' q(\boldsymbol{r})\, \chi_0(\boldsymbol{r}, \boldsymbol{r}')\, q(\boldsymbol{r}') \geq 0
\end{aligned}$$

where we use the definition in Eq. (5.61) and $q(\boldsymbol{r})$ is defined as

$$q(\boldsymbol{r}) = \int d\epsilon q^e(\epsilon)\, \delta(\epsilon - \upsilon_{\text{def}}(\boldsymbol{r})) \tag{5.53}$$

Since the number of electrons is supposed to be maintained, the response function $\chi_0^e(\epsilon, \epsilon')$ on the energy coordinate satisfies the following relationship

$$\int d\epsilon' \chi_0^e(\epsilon, \epsilon') = 0 \tag{5.54}$$

in parallel to Eq. (1.186). It means that χ_0^e has an eigenvector with null eigenvalue. Thus, for example the Moore-Penrose pseudo-inverse method must be employed to invert χ_0^e. Equation (5.54) can be proved as follows:

$$
\begin{aligned}
\int d\epsilon' \chi_0^e [n_0] (\epsilon, \epsilon') &= \int d\epsilon' \int d\boldsymbol{r} d\boldsymbol{r}' \chi_0 (\boldsymbol{r}, \boldsymbol{r}') \delta (\epsilon - \upsilon_{\text{def}} (\boldsymbol{r})) \delta (\epsilon' - \upsilon_{\text{def}} (\boldsymbol{r}')) \\
&= \int d\boldsymbol{r} \delta (\epsilon - \upsilon_{\text{def}} (\boldsymbol{r})) \\
&\qquad \times \int d\epsilon' \int d\boldsymbol{r}' \chi_0 (\boldsymbol{r}, \boldsymbol{r}') \delta (\epsilon' - \upsilon_{\text{def}} (\boldsymbol{r}')) \\
&= \int d\boldsymbol{r} \delta (\epsilon - \upsilon_{\text{def}} (\boldsymbol{r})) \int d\boldsymbol{r}' \chi_0 (\boldsymbol{r}, \boldsymbol{r}') \\
&= 0
\end{aligned} \tag{5.55}
$$

The kinetic energy functional $E_{\text{kin}}^e [n]$ using the energy electron density $n^e (\epsilon)$ in the nonlocal part of Eq. (5.49) can be written as

$$
\begin{aligned}
E_{\text{kin}}^e [n] = E_{\text{kin}} [n_0] &+ \int d\boldsymbol{r} \upsilon_{\text{kin}} [n_0] (\boldsymbol{r}) \delta n (\boldsymbol{r}) \\
&- \frac{1}{2} \int d\epsilon d\epsilon' \delta n^e (\epsilon) \chi_0^e (\epsilon, \epsilon')^{-1} \delta n (\epsilon')
\end{aligned} \tag{5.56}
$$

In the following, the efficiency of this equation will be examined for atoms and a molecule.

5.3.2 Applications to Atomic Systems

In this Subsection, we apply Eq. (5.56) to atoms to examine the efficiency and the reliability of the response function $\chi_0^e(\epsilon, \epsilon')$ on the energy coordinate. A focus will also be placed on the problem associated with inverting the response function. To this end, we calculate the electronic kinetic energies in atoms with respect to the variations of the core charges Z_υ [13]. Here, we present the results for the 'pseudo' H and Ar atoms, where the bare Coulomb potentials of the core charges Z_υ^{A} (A = H, Ar) are replaced by the local parts of the corresponding pseudopotentials developed by Bachelet, Hamann and Schlüter (BHS) [14] for these atoms. Explicitly, the local potential dependent only on the radial distance r is given by

$$
\begin{cases}
V_{\text{ps}}^{\text{A}} (r) = V_{\text{BHS\_loc}}^{\text{A}} (r) & \text{A} = \text{H} \\
V_{\text{ps}}^{\text{A}} (r) = V_{\text{BHS\_loc}}^{\text{A}} (r) + \Delta V_{\text{BHS\_s}}^{\text{A}} (r) & \text{A} = \text{Ar}
\end{cases} \tag{5.57}
$$

where $V_{\text{BHS\_loc}}^{\text{A}} (r)$ has the form written as

$$
V_{\text{BHS\_loc}}^{\text{A}} (r) = -\frac{Z_\upsilon^{\text{A}}}{r} \sum_{i=1}^{2} c_i^{\text{A}} \operatorname{erf} \left((\alpha_i^{\text{A}})^{1/2} r \right) \tag{5.58}
$$

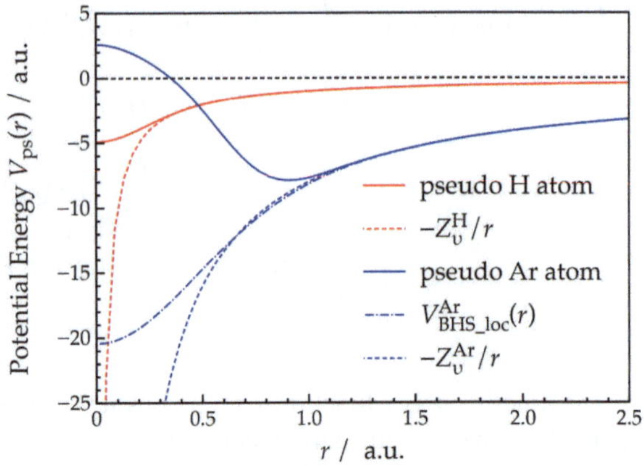

Fig. 5.3 Local potentials $V_{ps}^{H}(r)$ and $V_{ps}^{Ar}(r)$ in Eq. (5.57) for pseudo H and Ar atoms, respectively. $\Delta V_{BHS\_s}^{A}(r)$ for Ar is also plotted. The corresponding Coulomb potentials $-Z_v^{H}/r$ and $-Z_v^{Ar}/r$ are shown for comparisons

The coefficients $\{c_i^A\}$ and the exponents $\{\alpha_i^A\}$ for the atoms A are provided in Table IV in Ref. [14]. The core charges $Z_v^A = 1.0$ and 8.0 for H and Ar, respectively. $\Delta V_{BHS\_s}^{A}(r)$ represents radial behavior of the s component of the nonlocal part of the BHS pseudopotential for the Ar atom and it is expressed as

$$\Delta V_{BHS\_s}^{A}(r) = \sum_{i=1}^{3} \left(P_i^A + r^2 P_{i+3}^A\right) \exp\left(-\alpha_i^A r^2\right) \tag{5.59}$$

The two sets of the coefficients $\{P_i^A\}$ and $\{P_{i+3}^A\}$ are obtained from the reconstruction of the values tabulated in Table IV in Ref. [14]. The addition of the term $\Delta V_{BHS\_s}^{A}(r)$ to the Ar potential is to repel the electrons from the center of the atomic core and to smooth the electron density so that the real-space grid representation works properly. Of course, the potentials in Eq. (5.57) do not realize the valence electron densities of the real H and Ar atoms. Thus, each atom represented by Eq. (5.57) is the pseudo atom for the H or the Ar atom with the same number of valence electrons. However, this does not matter because the comparisons will be made with the KS-DFT calculation for the same pseudo atoms. The potentials defined in Eq. (5.57) are plotted in Fig. 5.3.

The kinetic energy change is computed with respect to the variation of the core charge from Z_v^A to $Z_v^A + \delta q$, where the range of δq is $-0.2 \le \delta q \le 0.2$ in the unit of the elementary charge. For this purpose, the defining potential v_{def} in Eq. (5.61) should be a function of the radial distance r of the electron from the atomic core. Explicitly, the local potential $V_{BHS\_loc}^{A}(r)$ is used as the defining potential $v_{def}(\boldsymbol{r})$ for the energy coordinate ϵ. Of course, when the additional charge δq is to be placed at a position different from the atomic core, the defining potential must be reconstructed accordingly. The system with $\delta q = 0.0$ is regarded as the reference system and its response function is used to construct the kinetic energy functional of Eq. (5.56). The electron densities for the systems with the core charges $Z_v^A + \delta q$ are constructed by the KS-DFT using the real-space grid (RSG) approach. The details of the RSG method itself is provided in Appendix F. For the numerical details of the calculation of the kinetic energy, we refer the reader to Ref. [13]. In implementing Eq. (5.56), the kinetic potential $\upsilon_{\text{kin}}[n_0](\boldsymbol{r})$ of the reference system with the density $n_0(\boldsymbol{r})$ is provided by exploiting Eq. (5.10). That is, $\upsilon_{\text{kin}}[n_0](\boldsymbol{r})$ is given by

$$\upsilon_{\text{kin}}\,[n_0]\,(\boldsymbol{r}) = -\upsilon_{\text{eff}}[n_0]\,(\boldsymbol{r}) + \mu \tag{5.60}$$

The effective potential $\upsilon_{\text{eff}}[n_0]\,(\boldsymbol{r})$ can be determined from the given density n_0 e.g. by the inverse Kohn-Sham method described in Sect. 1.4.5. In the present case, however, $\upsilon_{\text{eff}}[n_0]$ can be obtained directly in the KS-DFT calculation for the reference system ($\delta q = 0$) to determine n_0. The constant shift of the kinetic potential by an amount of the chemical potential μ can be compensated exactly because the relation $\int d\boldsymbol{r}\mu \times (n\,(\boldsymbol{r}) - n_0\,(\boldsymbol{r})) = 0$ holds as long as the densities are being normalized.

Another issue to be discussed for the implementation of Eq. (5.56) is the inversion of the response function $\chi_0^e(\epsilon, \epsilon')$. Suppose that χ_0^e is expressed in the form of the spectral decomposition, thus,

$$\chi_0^e\left(\epsilon, \epsilon'\right) = \sum_i^{N_{\text{grid}}^e} \langle \epsilon |\, \gamma_i\, \rangle\, g_i\, \langle \gamma_i |\, \epsilon' \rangle \tag{5.61}$$

where N_{grid}^e is the number of grids for constructing the energy distribution functions $n^e(\epsilon)$. γ_i and g_i are the eigenvectors and the corresponding eigenvalues of the matrix $\chi_0^e\left(\epsilon, \epsilon'\right)$. Then, the inverted matrix can be formally given by

$$\chi_0^e\left(\epsilon, \epsilon'\right)^{-1} = \sum_i^{N_{\text{grid}}^e} \langle \epsilon |\, \gamma_i\, \rangle\, g_i^{-1}\, \langle \gamma_i |\, \epsilon' \rangle \tag{5.62}$$

Here one meets the same problem as in the inversion of $\chi\,(\boldsymbol{r}, \boldsymbol{r}')$ shown in Eq. (1.191). Since $\chi_0^e\left(\epsilon, \epsilon'\right)$ as well as $\chi\,(\boldsymbol{r}, \boldsymbol{r}')$ are the response functions with null eigenvalues as discussed in Sect. 1.4.1, Eq. (5.62) is divergent. Thus, it is needed to exclude the eigenvector with the null eigenvalue from the summation in Eq. (5.62) to construct the inverted matrix as in the Moore-Penrose inversion method. We also note that the expansion of Eq. (5.61) includes a lot of eigenvectors with effectively null eigenvalues in the expansion of Eq. (5.61). We found that the inclusion of these eigenvectors with small eigenvalues leads to the unphysical oscillations in the resultant $\chi_0^e\left(\epsilon, \epsilon'\right)^{-1}$. Thus, only the several leading terms in the decomposition are used to invert the matrix. This approach is also known as the truncated singular value decomposition (TSVD) method [15] in the context of the inverse Kohn-Sham method. In Fig. 5.4, the response function $\chi_0^e\left(\epsilon, \epsilon'\right)$ of the pseudo Ar atom and its components in Eq. (5.61) with the two largest eigenvalues are also drawn. It is clearly shown in the figure that the response function is dominated by the single eigenvector $|\gamma_1\rangle$ with the largest eigenvalue g_1. The structure of the response function on the energy coordinate is quite simple and it suggests that the increase of the potential in the large energy coordinate ϵ' leads to the increase of the energy electron density in the small energy

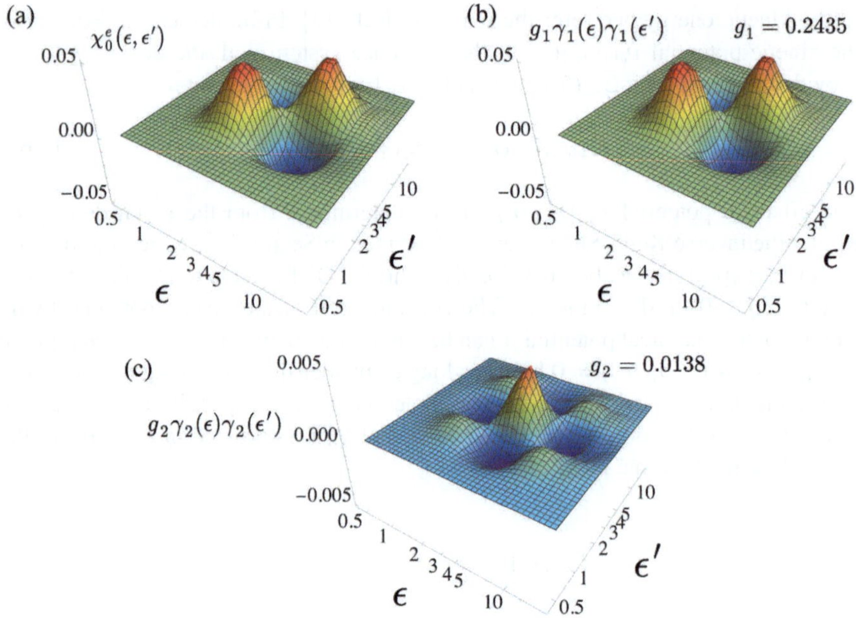

Fig. 5.4 **a** Response function $\chi_0^e(\epsilon, \epsilon')$ for the pseudo Ar atom. **b** The component $g_1\gamma_1(\epsilon)\gamma_1(\epsilon')$ with the largest eigenvalue ($g_1 = 0.2435$) in the expansion of Eq. (5.61). **c** The component $g_2\gamma_2(\epsilon)\gamma_2(\epsilon')$ with the secondary largest eigenvalue ($g_2 = 0.0138$). The axis for the energy ϵ in the unit of a.u. is given in the logarithmic scale. See also Fig. 2 in Ref. [13]

coordinate and its vice versa. The more detailed response can be realized by including the eigenfunction with $g_2 = 0.0138$. However, its contribution to the total response function is found to be negligibly small.

The response function shown in Fig. 5.4a suggests that the density response to the fluctuation of the core charge of Ar is almost bipolar along the radial direction. The actual density responses in the pseudo Ar are visualized in Fig. 5.5a and b for $\delta q = -0.2$ and 0.2, respectively. The reference density n_0 is constructed for the system with $\delta q = 0$. As shown in Fig. 5.5a, the decrease in the core charge induces a rather simple reorganization of the electron density in the Ar. Explicitly, the density distribution near the atomic core decreases and that in the outer region increases. It is thus found that the bipolar structure is formed in the density polarization. The opposite event takes place when the core charge of Ar is increased by $\delta q = 0.2$. These observations are found to be consistent with the structure of the response function $\chi_0^e(\epsilon, \epsilon')$ in Fig. 5.4a for the Ar.

The electronic kinetic energy of the pseudo H atom is shown in Fig. 5.6 as a function of the variation $\delta q(-0.2 \leq \delta q \leq 0.2)$ of the core charge $Z_v^{\mathrm{H}} + \delta q$. It is clearly shown in the figure that the TF energy seriously underestimates the kinetic energies for all the values of δq as compared to that given by the KS-DFT. On the contrary, it is found that the inclusion of the von Weizsäcker correction with

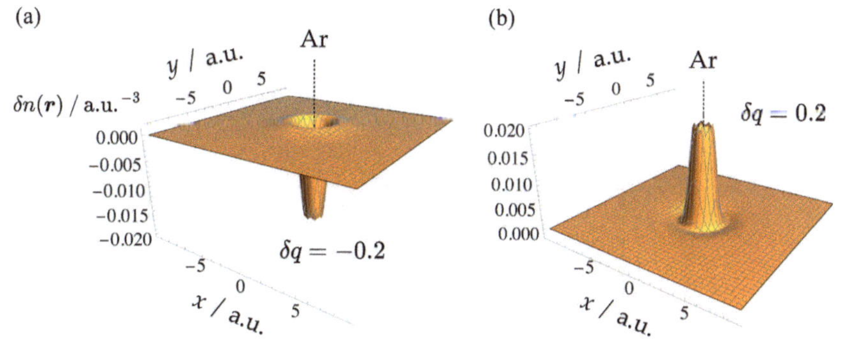

Fig. 5.5 Electron density difference $\delta n(\boldsymbol{r}) = n(\boldsymbol{r}) - n_0(\boldsymbol{r})$ on the xy plane that includes the pseudo Ar atom with core charge $Z_v^{\mathrm{Ar}} + \delta q$. **a** for $\delta q = -0.2$ and **b** for $\delta q = 0.2$ in the unit of the elementary charge. The depletion of the electron density near the atomic core (on the right) is due to the term $\Delta V_{\mathrm{BHS\_s}}^{\mathrm{A}}(r)$ in Eq. (5.59) that was placed to repel the valence electrons from the atomic core

the weight of $\lambda = 1/5$ overshoots the KS-DFT energy. The kinetic energy by the functional TF+1/9W shows the better agreement with the KS-DFT. Remind that the factor TF+1/9W functional reproduces the behavior of the inverse of the Lindhard function $C_{\mathrm{Lind}}(\eta)$ only in the small region of η as shown in Fig. 5.1. The kinetic GGA functional, which is provided in Ref. [16], is explicitly written as

$$E_{\mathrm{kin}}^{\mathrm{GGA}}[n_\sigma] = 2^{\frac{2}{3}} C_{\mathrm{TF}} \int d\boldsymbol{r}\; n_\sigma(\boldsymbol{r})^{\frac{5}{3}} \times \left(1 + \frac{A_1 q_\sigma^2}{1 + A_2 q_\sigma \sinh^{-1} q_\sigma} - \frac{A_3 q_\sigma}{1 + A_4 q_\sigma}\right) \tag{5.63}$$

where q_σ is the inhomogeneity parameter defined in Eq. (1.150). The parameters in Eq. (5.63) are given as $A_1 = 0.0055$, $A_2 = 0.0253$, $A_3 = 0.072$, and $A_4 = 2^{5/3}$ in Ref. [16]. It was demonstrated in Ref. [16] that the GGA functional gives the best performance in calculating the kinetic energies for a set of 77 molecules as compared to various kinetic energy functionals. It is notable that the TF+1/9W functional gives almost the same result as the GGA functional. It is quite impressive that the kinetic energy obtained by Eq. (5.56) shows the nice agreement with the KS-DFT except for the region of small core charges. It is thus found that the functional $E_{\mathrm{kin}}^{e}[n]$ defined in Eq. (5.56) is almost comparable in accuracy with the KS-DFT at least for the pseudo H atom.

In Fig. 5.7, we show the results for the pseudo Ar atom with $Z_v^{\mathrm{Ar}} = 8.0$. As expected we can see the same trend as the pseudo H atom in the results given by the TF model and the TF+λW calculations. It is remarkable that the present work using the functional $E_{\mathrm{kin}}^{e}[n]$ of Eq. (5.56 shows the faithful agreement with the KS-DFT. Actually, the curves obtained by these methods are hardly discernible in the figure, suggesting the reliability and the applicability of the functional to the atomic system. It should be noted, however, that the test calculations considered in this

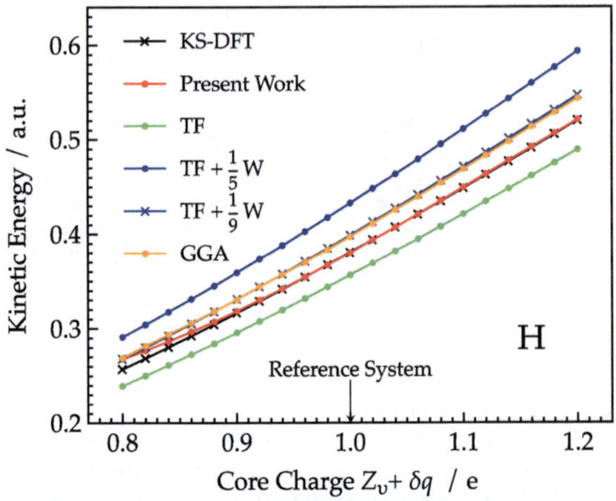

Fig. 5.6 Plots of the kinetic energies of the pseudo H atom ($Z_{\upsilon}^{\rm H} = 1.0$) as functions of the variation of the core charge $Z_{\upsilon}^{\rm H} + \delta q$. The pseudo atom is described with the local potential defined in Eq. (5.57). For the production of these curves, the electron density optimized by KS-DFT with UBLYP functional is used. The legend 'Present Work' refers to the calculation using Eq. (5.56). 'TF' stands for the Thomas-Fermi (TF) energy, and 'TF+λW' indicates the TF energy with the von Weizsäcker correction defined in Eq. (5.3). 'GGA' shows the energy given by the kinetic GGA functional of Eq. (12) in Ref. [16]. See also Fig. 3 in Ref. [13]

Fig. 5.7 Plots of the kinetic energies of the pseudo Ar atom ($Z_{\upsilon}^{\rm Ar} = 8.0$) as functions of the variation of the core charge $Z_{\upsilon}^{\rm Ar} + \delta q$. The other notations are the same as in Fig. 5.6

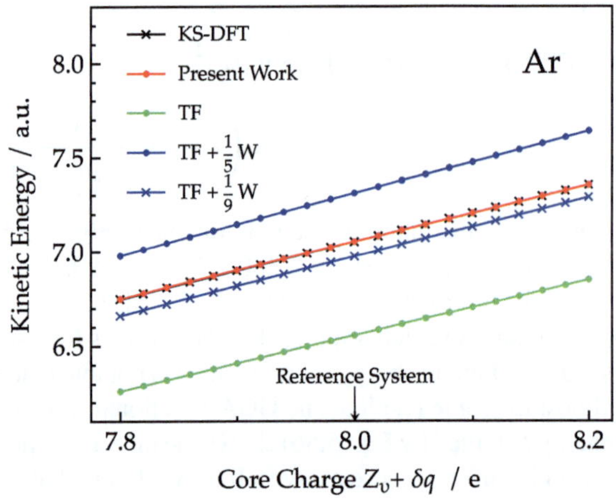

Subsection are practically of no use in general. Thus, in the next Subsection, we consider to describe a covalent bond in a simple molecule. To this end, the concept of the composite response function represented on the energy coordinate will be introduced.

5.3.3 Composite Linear Response Function

The next issue is how to construct the linear response function $\chi_0^e(\epsilon, \epsilon')$ for molecular systems. To make the explanation simple without loss of generality, we consider a system consisting of two subsystems A and B. Suppose that these two subsystems are individually isolated with their internal geometries being fixed at those in the molecular system. The electron densities of the ground states are given by $n_0^{\mathrm{A}}(\boldsymbol{r})$ and $n_0^{\mathrm{B}}(\boldsymbol{r})$, respectively, for the subsystems A and B. Then, a reasonable choice of the *total* density n_0 of the reference system is to take merely the overlap of these densities, thus,

$$n_0(\boldsymbol{r}) = n_0^{\mathrm{A}}(\boldsymbol{r}) + n_0^{\mathrm{B}}(\boldsymbol{r}) \tag{5.64}$$

For the closed shell system, the reference density $n_{0\sigma}(\boldsymbol{r})$ for each spin σ is obtained by

$$n_{0\sigma}(\boldsymbol{r}) = \frac{1}{2}\left(n_0^{\mathrm{A}}(\boldsymbol{r}) + n_0^{\mathrm{B}}(\boldsymbol{r})\right) \tag{5.65}$$

In the following, the subscript σ will be omitted for the sake of notational simplicity. The choice of Eq. (5.64) is completely the same as that of the reference density used in the development of the static correlation functional discussed in Sect. 4.2.1. The response function $\chi_0(\boldsymbol{r}, \boldsymbol{r}')$ for the system that has the density n_0 as its ground state density can be constructed e.g. by the inverse Kohn-Sham method introduced in Sect. 1.4.5. In fact, provided that n_0 is non-interacting v-representable, the corresponding effective potential $v_{\mathrm{eff}}[n_0](\boldsymbol{r})$ can be uniquely determined through the inv-KS. Then, we get the following Schrödinger equations which yield the set of wave functions $\{\varphi_i[n_0]\}$,

$$\left(-\frac{1}{2}\nabla^2 + v_{\mathrm{eff}}[n_0](\boldsymbol{r})\right)\varphi_i[n_0](\boldsymbol{r}) = \epsilon_i^0\varphi_i[n_0](\boldsymbol{r}) \tag{5.66}$$

with $\{\epsilon_i^0\}$ being the eigenvalues of the corresponding wave functions. Of course, it is assumed that the set $\{\varphi_i[n_0]\}$ provides the electron density n_0 in Eq. (5.65). The response function $\chi_0(\boldsymbol{r}, \boldsymbol{r}')$ for the system with n_0 can be constructed using Eq. (1.185) based on the perturbation theory. Explicitly, it can be expressed as

$$\begin{aligned}\chi_0^{\mathrm{full}}[n_0]\left(\boldsymbol{r}, \boldsymbol{r}'\right) = -2\sum_i^{occ}\sum_a^{vir}\frac{1}{\epsilon_a^0 - \epsilon_i^0}\varphi_i[n_0](\boldsymbol{r})\,\varphi_a^*[n_0](\boldsymbol{r}) \\ \times\varphi_a[n_0]\left(\boldsymbol{r}'\right)\varphi_i^*[n_0]\left(\boldsymbol{r}'\right) + c.c.\end{aligned} \tag{5.67}$$

where a are the indices for the virtual orbitals. The implicit dependence of χ_0 on the reference density n_0 is explicitly denoted in Eq. (5.67). The superscript 'full' means that the response function is built from the density n_0 of the whole system. Note that the relation $\int d\boldsymbol{r}'\chi_0^{\mathrm{full}}[n_0]\left(\boldsymbol{r}, \boldsymbol{r}'\right) = 0$ holds as shown in Eq. (1.186). For the evaluation of Eq. (5.49), $E_{\mathrm{kin}}[n_0]$ and $v_{\mathrm{kin}}[n_0](\boldsymbol{r})$ must also be constructed in addition to χ_0. $E_{\text{kin}}[n_0]$ can be obtained as the kinetic energy of the system specified by Eq. (5.66). For the evaluation of $v_{\text{kin}}[n_0](\boldsymbol{r})$, Eq. (5.60) can be exploited.

Now we are ready to compute the orbital-free kinetic energy functional of Eq. (5.49) for the reference electron density n_0 given by Eq. (5.64). Remind, however, that the inv-KS needs to be performed for the preparation of the quantities $E_{\text{kin}}[n_0]$, $v_{\text{kin}}[n_0](\boldsymbol{r})$ and $\chi_0(\boldsymbol{r}, \boldsymbol{r}')$ in Eq. (5.49). In fact, the computational cost for it is larger than that for the KS-DFT calculation for the entire system under consideration. However, it is quite helpful to compare the kinetic energy of Eq. (5.49) with that given by KS-DFT. If the response function $\chi_0^{\text{full}}[n_0]\left(\boldsymbol{r}, \boldsymbol{r}'\right)$ and the kinetic potential $v_{\text{kin}}[n_0](\boldsymbol{r})$, respectively, given by Eqs. (5.67) and (5.60) do not work properly in Eq. (5.49), there will be no hope that $\chi_0^{\text{full}}[n_0]\left(\boldsymbol{r}, \boldsymbol{r}'\right)$ can be safely reduced to the response function $\chi_0^{e,\text{full}}[n_0]\left(\epsilon, \epsilon'\right)$ on the energy coordinate. Later, the real-space grid implementation of the functional will be addressed.

According to Eq. (5.61), the projection of Eq. (5.67) onto the energy coordinate to yield $\chi_0^{e,\text{full}}(\epsilon, \epsilon')$ is explicitly written as

$$\chi_0^{e,\text{full}}[n_0]\left(\epsilon, \epsilon'\right) = \int d\boldsymbol{r}d\boldsymbol{r}'\chi_0^{\text{full}}[n_0]\left(\boldsymbol{r}, \boldsymbol{r}'\right)\delta\left(\epsilon - v_{\text{def}}\left(\boldsymbol{r}\right)\right)\delta\left(\epsilon' - v_{\text{def}}\left(\boldsymbol{r}'\right)\right) \tag{5.68}$$

where $v_{\text{def}}(\boldsymbol{r})$ can be taken as the external potential $v_{\text{ext}}(\boldsymbol{r})$ of the system of interest, that is $v_{\text{def}}(\boldsymbol{r}) = v_{\text{ext}}(\boldsymbol{r})$. With this projection of the response function, there will be some degradation in the description of the kinetic energy, which can be examined by comparing the computational results given by the functional with Eq. (5.68) and that with Eq. (5.67). It is also helpful to recall that the projected response function satisfies the relation $\int d\epsilon'\chi_0^{e,\text{full}}[n_0]\left(\epsilon, \epsilon'\right) = 0$ regardless of the value of ϵ as shown in Eq. (5.54).

So far, no idea has been devised to reduce the computational cost of computing the kinetic energy of Eq. (5.49). As mentioned above, the construction of $\chi_0^{\text{full}}[n_0]\left(\boldsymbol{r}, \boldsymbol{r}'\right)$ itself requires a larger computational cost than the KS-DFT calculation. To avoid this situation, we introduce a function $\chi_0^{\text{cmp}}(\boldsymbol{r}, \boldsymbol{r}')$ given as a 'composite' of the response functions of the subsystems, thus,

$$\chi_0^{\text{cmp}}[n_0^{\text{A}}, n_0^{\text{B}}]\left(\boldsymbol{r}, \boldsymbol{r}'\right) = \chi_0^{\text{A}}[n_0^{\text{A}}]\left(\boldsymbol{r}, \boldsymbol{r}'\right) + \chi_0^{\text{B}}[n_0^{\text{B}}]\left(\boldsymbol{r}, \boldsymbol{r}'\right) \tag{5.69}$$

Accordingly, we also obtain the projected response function,

$$\chi_0^{e,\text{cmp}}[n_0^{\text{A}}, n_0^{\text{B}}]\left(\epsilon, \epsilon'\right) = \chi_0^{e,\text{A}}[n_0^{\text{A}}]\left(\epsilon, \epsilon'\right) + \chi_0^{e,\text{B}}[n_0^{\text{B}}]\left(\epsilon, \epsilon'\right) \tag{5.70}$$

It should be stressed that the projection from $\chi_0^{\text{A}}[n_0^{\text{A}}]\left(\boldsymbol{r}, \boldsymbol{r}'\right)$ to $\chi_0^{e,\text{A}}[n_0^{\text{A}}]\left(\epsilon, \epsilon'\right)$ is performed using the same defining potential as in Eq. (5.68) for the entire system and not for the subsystem A as will be shown explicitly in the next Subsection. The composite response functions defined in Eqs. (5.69) and (5.70) satisfy, respectively, the sum rules in Eqs. (1.186) and (5.54), since each component response function satisfies the sum rule. Due to the property that the composite response functions also preserve the sum rule, the number of electrons in the system is maintained for any perturbation in principle.

Each response function $\chi_0^p[n_0^p](\boldsymbol{r}, \boldsymbol{r}')$ (p = A, B) in Eq. (5.69) is, of course, obtained from the set of wave functions $\{\varphi_{0i}^p\}$ of the isolated subsystem p in the ground state. Note that from the given ground state density $n_0^p(\boldsymbol{r})$, it is possible to obtain the wave function $\{\varphi_{0i}^p\}$ and the corresponding response functions χ_0^p and $\chi_0^{e,p}$, assuming that $n_0^p(\boldsymbol{r})$ is non-interacting υ-representable. Since the component response function $\chi_0^p\left[n_0^p\right](\boldsymbol{r}, \boldsymbol{r}')$ is confined to a spatially limited region enclosing the subsystem p, the computational cost for constructing the composite response function $\chi_0^{e,\mathrm{cmp}}[n_0^{\mathrm{A}}, n_0^{\mathrm{B}}]$ can be suppressed to order N in principle. One might think that Eq. (5.69) or (5.70) does not make sense at least as a response function of the whole system since it completely ignores the nonlocal correlations between the two subsystems. Note, however, that the spatially nonlocal response can be included in the composite response function on the energy coordinate shown in Eq. (5.70). This issue will be examined below by computing the potential energy curve as a function of the bond distance of an H_2 molecule.

5.3.4 Application to Covalent Bond

As a simplest application of the composite response function in Eq. (5.70) to the description of a covalent bond, we consider here an H_2 molecule. To this end, an isolated H atom is taken as the component system for the construction of the reference densities n_0^{A} and n_0^{B} in Eq. (5.64). Throughout the calculations in this Subsection, the electron densities of the component systems are obtained using the nonlocal pseudopotentials. Accordingly, the component response functions $\chi_0^{\mathrm{A}}\left[n_0^{\mathrm{A}}\right](\boldsymbol{r}, \boldsymbol{r}')$ and $\chi_0^{\mathrm{B}}\left[n_0^{\mathrm{B}}\right](\boldsymbol{r}, \boldsymbol{r}')$ in Eq. (5.69) are constructed using the perturbation theory for the subsystems A and B, respectively. The composite response function $\chi_0^{e,\mathrm{cmp}}[n_0^{\mathrm{A}}, n_0^{\mathrm{B}}](\epsilon, \epsilon')$ on the energy coordinate is to be constructed according to Eq. (5.70), that is, the function is given by sum of the projected response functions of the individual subsystems. Note, however, that the computational cost to build the function $\chi_0^{e,p}\left[n_0^p\right](\epsilon, \epsilon')$ is not negligible, because it requires the double loop for the integration with respect to the 3-dimensional space:

$$\chi_0^{e,p}\left[n_0^p\right](\epsilon, \epsilon') = \int_{\Omega_p} d\boldsymbol{r} d\boldsymbol{r}' \chi_0^p\left[n_0^p\right](\boldsymbol{r}, \boldsymbol{r}')\, \delta\left(\epsilon - \upsilon_{\mathrm{def}}(\boldsymbol{r})\right) \delta\left(\epsilon' - \upsilon_{\mathrm{def}}(\boldsymbol{r}')\right) \tag{5.71}$$

where Ω_p represents the spatial region that effectively encloses the wave function of the subsystem p. The problem can be resolved by the method as shown below, although not tested in the present implementation. Suppose that the component response function for subsystem p is being decomposed with the eigenfunctions $\{\phi_i^p(\boldsymbol{r})\}$ and the corresponding eigenvalues $\{\gamma_i^p\}$,

$$\chi_0^p\left[n_0^p\right](\boldsymbol{r},\boldsymbol{r}') = \sum_i \phi_i^p(\boldsymbol{r})\,\gamma_i^p\,\phi_i^p(\boldsymbol{r}') \tag{5.72}$$

A possible way to obtain the decomposition is to first represent $\chi_0^p\left[n_0^p\right]$ in terms of the KS wave functions $\{\varphi_i^p(\boldsymbol{r})\}$, which are given as the solutions of the KS equation for the subsystem p. Then, using the unitary transformation of $\{\varphi_i^p(\boldsymbol{r})\}$ to diagonalize the matrix, the set $\{\phi_i^p(\boldsymbol{r})\}$ can be obtained. By substituting Eq. (5.72) into Eq. (5.71), we get

$$\begin{aligned}
\chi_0^{e,p}\left[n_0^p\right](\epsilon,\epsilon') &= \int_{\Omega_p} d\boldsymbol{r}d\boldsymbol{r}' \sum_i \phi_i^p(\boldsymbol{r})\,\gamma_i^p\,\phi_i^p(\boldsymbol{r}')\,\delta\left(\epsilon - \upsilon_{\mathrm{def}}(\boldsymbol{r})\right)\delta\left(\epsilon' - \upsilon_{\mathrm{def}}(\boldsymbol{r}')\right) \\
&= \sum_i \gamma_i^p \int_{\Omega_p} d\boldsymbol{r}\phi_i^p(\boldsymbol{r})\,\delta\left(\epsilon - \upsilon_{\mathrm{def}}(\boldsymbol{r})\right) \\
&\quad \times \int_{\Omega_p} d\boldsymbol{r}'\phi_i^p(\boldsymbol{r}')\,\delta\left(\epsilon' - \upsilon_{\mathrm{def}}(\boldsymbol{r}')\right) \\
&= \sum_i \phi_i^{e,p}(\epsilon)\,\gamma_i^p\,\phi_i^{e,p}(\epsilon')
\end{aligned} \tag{5.73}$$

It is thus shown that the projected response function can be obtained by projecting the eigenfunctions $\{\phi_i^p(\boldsymbol{r})\}$ to $\{\phi_i^{e,p}(\epsilon)\}$ on the energy coordinate. Then, the two-fold loop for the spatial coordinates $\boldsymbol{r}$ and $\boldsymbol{r}'$ can be reduced to the single loop for $\boldsymbol{r}$ combined with the loop for the truncated subspace of the eigenfunctions $\{\phi_i^p(\boldsymbol{r})\}$. Remind that the defining potential $\upsilon_{\mathrm{def}}(\boldsymbol{r})$ is constructed for the entire system of interest as noted in the previous Subsection, and thus υ_{def} itself has value over the whole system. But, $\chi_0^p\left[n_0^p\right]$ is limited to the spatial region Ω_p enclosing the system p. Therefore, as was described, the total computational cost increases with order N.

For the reference density n_0 given by Eq. (5.64), the corresponding kinetic potential $\upsilon_{\mathrm{kin}}[n_{0\sigma}](\boldsymbol{r})$ in Eq. (5.49) or in Eq. (5.56) is obtained from the effective potential $\upsilon_{\mathrm{eff}}[n_0](\boldsymbol{r})$ by virtue of Eq. (5.60), where $\upsilon_{\mathrm{eff}}[n_0](\boldsymbol{r})$ can be constructed from the inverse KS method for example. In the present calculation for the H_2 molecule, the kinetic potential $\upsilon_{\mathrm{kin}}[n_0](\boldsymbol{r})$ can be determined by a much simpler approach. For a one-electron system or a closed-shell two electron system, the kinetic potential for the electron density $n_{0\sigma}$ with spin σ can be given by

$$\upsilon_{\mathrm{kin}}[n_{0\sigma}](\boldsymbol{r}) = n_{0\sigma}(\boldsymbol{r})^{-1/2}\left(-\frac{1}{2}\nabla^2\right)n_{0\sigma}(\boldsymbol{r})^{1/2} \tag{5.74}$$

In our implementation, it is confirmed that the KS equation with the effective potential $\upsilon_{\mathrm{eff}}[n_{0\sigma}](\boldsymbol{r})$ given by using Eqs. (5.60) and (5.74) yields almost the same electron density as the input density $n_{0\sigma}(\boldsymbol{r})$. Note that the zeroth order term $E_{\mathrm{kin}}[n_0]$ in Eq. (5.49) can also be computed readily. As mentioned in Sect. 5.3.3, we are interested also in the performance of the full linear response function $\chi_0^{\mathrm{full}}[n_0](\boldsymbol{r},\boldsymbol{r}')$ described in Eq. (5.67), although there is no advantage to use it in reducing the computational cost. The function can be represented with its eigenfunctions $\{\phi_i^{\mathrm{full}}\}$ and the corresponding eigenvalues $\{\gamma_i^{\mathrm{full}}\}$

$$\chi_0^{\mathrm{full}}[n_0](\boldsymbol{r}, \boldsymbol{r}') = \sum_i \phi_i^{\mathrm{full}}(\boldsymbol{r})\, \gamma_i^{\mathrm{full}} \phi_i^{\mathrm{full}}(\boldsymbol{r}') \tag{5.75}$$

Its inversion is obtained as

$$\chi_0^{\mathrm{full}}[n_0](\boldsymbol{r}, \boldsymbol{r}')^{-1} = {\sum_i}' \phi_i^{\mathrm{full}}(\boldsymbol{r}) \frac{1}{\gamma_i^{\mathrm{full}}} \phi_i^{\mathrm{full}}(\boldsymbol{r}') \tag{5.76}$$

where $\sum'$ implies the summation excluding the eigenfunctions with the (effectively) zero eigenvalues. Remind that the reference density $n_0(\boldsymbol{r})$ as well as $n(\boldsymbol{r})$ of the interacting system are cylindrically symmetric. Thus, in the summation of Eq. (5.76), we manually excludes the eigenfunctions that cause the polarization to break the cylindrical symmetry of the density around the molecular axis. Consequently, only two eigenfunctions are included in the summation of Eq. (5.76). For the function $\chi_0^{\mathrm{full}}[n_0](\boldsymbol{r}, \boldsymbol{r}')$, its projection $\chi_0^{e,\mathrm{full}}[n_0](\epsilon, \epsilon')$ onto the energy coordinate is computed using Eq. (5.68). It is quite interesting to note that the contribution due to the eigenfunctions breaking the cylindrical symmetry of the density vanishes in principle upon the integration in Eq. (5.68).

Now, we have two variations for the functional in the form of Eq. (5.49), and also for the functional of Eq. (5.56). The only difference in these functionals arises from the treatment of the nonlocal term. These functionals are summarized as

$$\begin{aligned}
E_{\mathrm{kin}}^{\mathrm{full}}[n] &= E_{\mathrm{kin}}^{\mathrm{0th+1st}}[n] - \frac{1}{2}\int d\boldsymbol{r} d\boldsymbol{r}' \delta n(\boldsymbol{r})\, \chi_0^{\mathrm{full}}[n_0](\boldsymbol{r}, \boldsymbol{r}')\, \delta n(\boldsymbol{r}') \\
E_{\mathrm{kin}}^{e,\mathrm{full}}[n] &= E_{\mathrm{kin}}^{\mathrm{0th+1st}}[n] - \frac{1}{2}\int d\epsilon d\epsilon' \delta n^e(\epsilon)\, \chi_0^{e,\mathrm{full}}[n_0](\epsilon, \epsilon')\, \delta n^e(\epsilon') \\
E_{\mathrm{kin}}^{\mathrm{cmp}}[n] &= E_{\mathrm{kin}}^{\mathrm{0th+1st}}[n] - \frac{1}{2}\int d\boldsymbol{r} d\boldsymbol{r}' \delta n(\boldsymbol{r})\, \chi_0^{\mathrm{cmp}}\left[n_0^{\mathrm{A}}, n_0^{\mathrm{B}}\right](\boldsymbol{r}, \boldsymbol{r}')\, \delta n(\boldsymbol{r}') \\
E_{\mathrm{kin}}^{e,\mathrm{cmp}}[n] &= E_{\mathrm{kin}}^{\mathrm{0th+1st}}[n] - \frac{1}{2}\int d\epsilon d\epsilon' \delta n^e(\epsilon)\, \chi_0^{e,\mathrm{cmp}}\left[n_0^{\mathrm{A}}, n_0^{\mathrm{B}}\right](\epsilon, \epsilon')\, \delta n^e(\epsilon')
\end{aligned} \tag{5.77}$$

The kinetic energy functionals from the first to the fourth in Eq. (5.77) include, respectively, the nonlocal terms defined in Eqs. (5.67), (5.68), (5.69) and (5.70). The common term $E_{\mathrm{kin}}^{\mathrm{0th+1st}}[n]$ in Eq. (5.77) represents the sum of the first and the second terms in Eq. (5.56).

These functionals are implemented on the 'Vmol' program based on the real-space grid method and used to describe the covalent bond of an H_2 molecule. To this end, the total energy E_{tot} of the molecule is computed for the variation of the R(H-H) distance. At each R(H-H) distance, the ground state electron density $n(\boldsymbol{r})$, which serves as the input to the four functionals in Eq. (5.77), is obtained from the

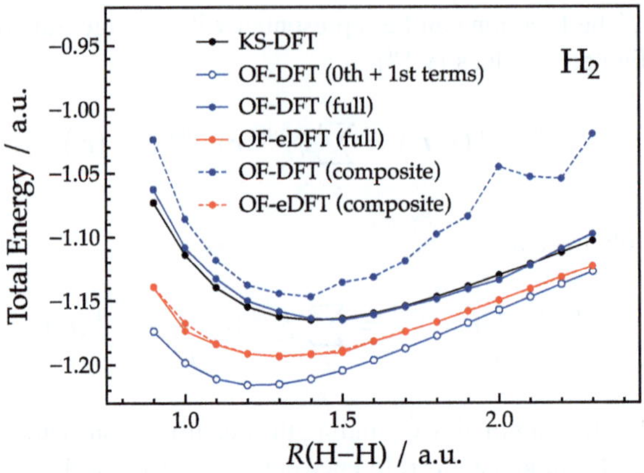

Fig. 5.8 Potential energy curves of an H_2 molecule as functions of the distance R(H-H). The curve with the legend 'KS-DFT' is obtained by the KS-DFT calculation with the BLYP functional. 'OF-DFT(0th+1st terms)' stands for the calculation with the kinetic energy obtained by the first two terms on the right side of Eq. (5.56). The following four OF-DFT calculations ('OF-DFT (full)', 'OF-eDFT (full)', 'OF-DFT (composite)' and 'OF-eDFT (composite)') use the kinetic energy functionals listed in Eq. (5.77), in order from top to bottom

wave functions determined by the KS-DFT with the BLYP functional. The electron density n_0^p of the isolated subsystem p is obtained with the same functional. The interaction energy between the nuclei and electrons is computed by means of a nonlocal pseudopotentials and the KS wave functions. We refer the reader to Ref. [13] for the numerical details. The results are shown in Fig. 5.8. First, it is revealed in the plot with the legend 'OF-DFT(0th+1st terms)' that the calculation with the kinetic energy given by the sum of the zeroth and the first order terms in Eq. (5.56) seriously underestimates the kinetic energy by the KS-DFT, especially in the region of the small R(H-H) distances. It is found, however, that the inclusion of the nonlocal term with the full response function $\chi_0^{\mathrm{full}}[n_0](\boldsymbol{r}, \boldsymbol{r}')$ defined in Eq. (5.67) remarkably improves the potential energy curve (PEC). In fact, it is shown that the 'OF-DFT(full)' calculation reproduces faithfully the PEC given by the KS-DFT calculation. It is, thus, proved that Eq. (5.67) can be regarded as a benchmark of the nonlocal term in the kinetic energy functional. How about the projected response function $\chi_0^{e,\mathrm{full}}[n_0]\left(\epsilon, \epsilon'\right)$ defined in Eq. (5.68)? Unfortunately, the PEC given by 'OF-eDFT (full)' severely deviates from that by the KS-DFT as shown in the plots. This is, of course, due to the degradation of the response function associated with the projection onto the energy coordinate.

Next, we examine the results given by the kinetic energy functionals that include the composite response functions in their nonlocal term. As shown in the graph of 'OF-DFT (composite)', the PEC given by the functional using the composite response function $\chi_0^{\mathrm{cmp}}\left[n_0^{\mathrm{A}}, n_0^{\mathrm{B}}\right]\left(\boldsymbol{r}, \boldsymbol{r}'\right)$ defined in Eq. (5.69) agrees well with that of the KS-DFT in the region of small R(H-H) distance. However, it shows unphysical behavior in the region of the larger bond lengths. This can be attributed to the fact that the two individual response functions $\chi_0^{\mathrm{A}}\left[n_0^{\mathrm{A}}\right]$ and $\chi_0^{\mathrm{B}}\left[n_0^{\mathrm{B}}\right]$ for the subsystems

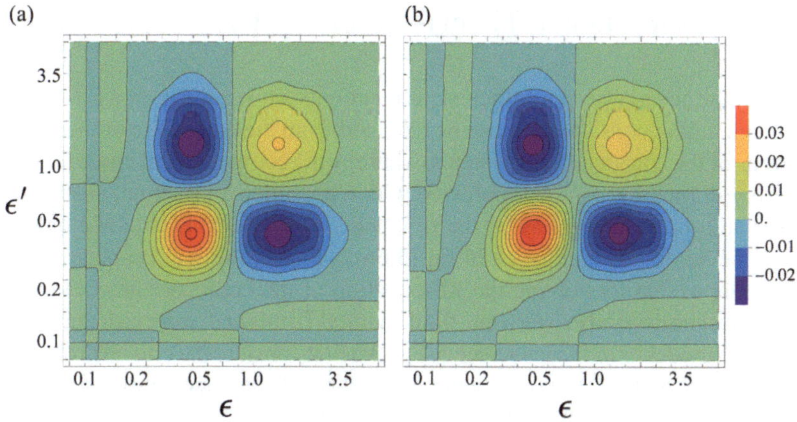

Fig. 5.9 Contour plots of **a** full response function $\chi_0^{e,\text{full}}[n_0](\epsilon, \epsilon')$ in Eq. (5.68) and **b** composite response function $\chi_0^{e,\text{cmp}}[n_0^{\text{A}}, n_0^{\text{B}}](\epsilon, \epsilon')$ in Eq. (5.70) for H_2 molecule with R(H-H) = 1.4 a.u. The axis for the energy ϵ in the unit of a.u. is given in the logarithmic scale. The unit for the color legend is a.u.−2

A and B, respectively, have no correlation with each other. Conversely, the sound behavior of the PEC in the small bond distances is rather surprising.

The most striking is the result of 'OF-eDFT (composite)' given by the functional with the composite response function $\chi_0^{e,\text{cmp}}[n_0^{\text{A}}, n_0^{\text{B}}](\epsilon, \epsilon')$ defined in Eq. (5.70). As shown in Fig. 5.9, the graph is almost coincident with the result 'OF-eDFT (full)' obtained by the functional with the *full* response function $\chi_0^{e,\text{full}}[n_0](\epsilon, \epsilon')$. The agreement is quite surprising since these two response functions are constructed from the different sets of wave functions. In Fig. 5.9, we show the contour plots to make a comparison between the full and the composite response functions on the energy coordinate. It is found that the composite function $\chi_0^{e,\text{cmp}}[n_0^{\text{A}}, n_0^{\text{B}}]$ agrees well with the full function $\chi_0^{e,\text{full}}[n_0]$, although the detailed structures are somewhat different. It is thus proved that the coincidence of the PEC given by $\chi_0^{e,\text{cmp}}[n_0^{\text{A}}, n_0^{\text{B}}]$ with that by $\chi_0^{e,\text{full}}[n_0]$ is a direct consequence of the fact that the composite response function is almost the same as the response function of the whole system. Although its effectiveness is revealed only for the simplest single bond, such the agreement seems to be a good indication to apply the composite response function to larger systems. Anyway, the mechanism underlying the coincidence of $\chi_0^{e,\text{cmp}}[n_0]$ with $\chi_0^{e,\text{full}}[n_0]$ is a matter of great interest and must be elucidated for the further development of the nonlocal response functions on the energy coordinate.

5.4 Reconstruction of Thomas-Fermi Model

Although the accuracy is not sufficient, as mentioned above, the response function of a molecule under consideration can be obtained from the composite of the response functions of the constituent subsystems, whereby the computational cost to construct the response function can be suppressed within the order N. Remind, however, that the first and the second terms on the right side in Eq. (5.56) is obtained only after solving the Schrödinger equation of Eq. (5.66), which requires a computational cost comparable to that for solving the KS equation of the entire system. Obviously, a method must be devised to overcome the difficulty. To this end, we revisit the Thomas-Fermi (TF) model introduced in Sect. 5.1.1. We first consider the crucial problem that arises in applying the TF model to the description of the dissociation of a covalent bond. It is shown below that the source of the problem is essentially the same as that of the static correlation error (SCE) in the local-density approximation (LDA). Recall that the SCE arises when the exchange hole for a spin is split into fragments upon a bond dissociation. It causes a corresponding decrease in the spin density at each atomic site. As a consequence, the absolute of the LDA exchange energy is seriously underestimated, which leads to an overestimation of the dissociation energy of a chemical bond. As discussed in Chap. 4, the problem associated with the delocalization of the electron density for each spin can be resolved by exploiting the energy electron density $n^e(\epsilon)$.

In fact, the TF model also underestimates the kinetic energy of a dissociating covalent bond due to essentially the same mechanism underlying the occurrence of the SCE. Therefore, it is anticipated that the same approach to solve the SCE can also be applied to the problem in the TF model. On the basis of this discussion, in the following Section, the TF kinetic energy functional is reconstructed by means of the coupling-parameter integration scheme. Furthermore, a generalized gradient approximation (GGA) is used to improve the reconstructed TF functional. As a benchmark test, the functional is employed to compute the potential energy curve of the dissociating H_2 molecule. The problem in the functional and the future prospect of the development of the kinetic energy functional will also be addressed.

5.4.1 *Static Correlation Error in Thomas-Fermi Model*

To illustrate the problem of using the Thomas-Fermi (TF) kinetic energy functional to describe the dissociation of a chemical bond, let us consider again the H_2 molecule that has the simplest σ bond. In fact, the discussion on the problem of the TF model is almost parallel to that for the static correlation error (SCE) in the local-density approximation (LDA) in Sect. 4.1. The only differences are the magnitude and the sign of the errors. Figure 5.10 shows the normalized electron density $n_\sigma(\boldsymbol{r})$ for the spin $\sigma(=\alpha$ or $\beta)$ of the infinitely stretched H_2 molecule ($R(\mathrm{H_A}$-$\mathrm{H_B}) = \infty$)). We apply the TF model to the spin polarized (symmetry broken) density $n_{0\sigma}$ shown in

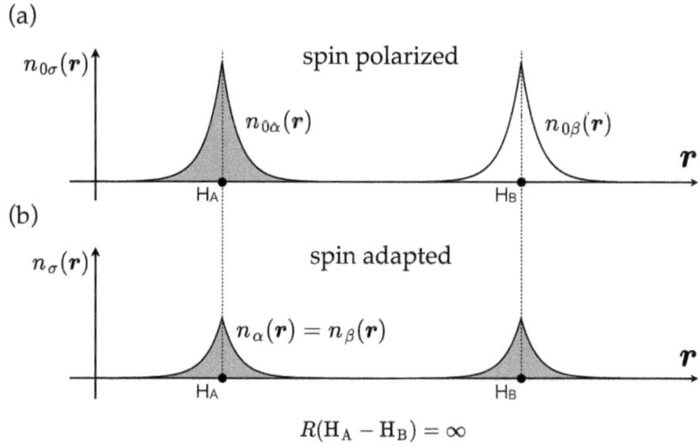

Fig. 5.10 Schematic representation of the electron density n for the spin σ in the H_2 molecule with the infinitely stretched chemical bond. **a** Spin polarized (symmetry broken) density $n_{0\sigma}$ and **b** Spin symmetry adapted density n_{σ}. The axis $\boldsymbol{r}$ represents the electron coordinate

Fig. 5.10a and the spin-symmetry adapted density n_σ in Fig. 5.10b. It is supposed that the total electron density $n_0 = n_{0\alpha} + n_{0\beta}$ is identical to the density $n = n_\alpha + n_\beta$, that is $n_0(\boldsymbol{r}) = n(\boldsymbol{r})$. We first rewrite Eq. (5.1) in the form of the sum of the contributions of the spins. For the densities $n_{0\sigma}$, the TF energy is written as

$$E_{\mathrm{kin}}^{\mathrm{TF}}[\{n_{0\sigma}\}] = 2^{\frac{2}{3}} C_{\mathrm{TF}} \sum_{\sigma} \int d\boldsymbol{r}\; n_{0\sigma}(\boldsymbol{r})^{\frac{5}{3}} \tag{5.78}$$

and similarly, for the densities n_σ, we have

$$E_{\mathrm{kin}}^{\mathrm{TF}}[\{n_{\sigma}\}] = 2^{\frac{2}{3}} C_{\mathrm{TF}} \sum_{\sigma} \int d\boldsymbol{r}\; n_{\sigma}(\boldsymbol{r})^{\frac{5}{3}} \tag{5.79}$$

where $2^{\frac{2}{3}}$ is the spin scaling factor and the coefficient C_{TF} is $\frac{3}{10}\left(3\pi^2\right)^{2/3}$. Since it is assumed that $n_0(\boldsymbol{r}) = n(\boldsymbol{r})$, Eq. (5.79) can be described in terms of the spin polarized densities $n_{0\sigma}$, thus,

$$E_{\mathrm{kin}}^{\mathrm{TF}}[\{n_{\sigma}\}] = 2^{\frac{2}{3}} C_{\mathrm{TF}} \sum_{\sigma} 2 \times \int d\boldsymbol{r} \left(\frac{1}{2} n_{0\sigma}(\boldsymbol{r})\right)^{\frac{5}{3}} \tag{5.80}$$

The multiplication by the factor 2 indicates the identical contributions from the two atoms H_A and H_B for each spin. Then, we get the relation

$$
\begin{aligned}
E_{\rm kin}^{\rm TF}\left[\{n_\sigma\}\right] &= 2^{-\frac{2}{3}} \cdot 2^{\frac{2}{3}} C_{\rm TF} \sum_\sigma \int d\boldsymbol{r}\; n_{0\sigma}\left(\boldsymbol{r}\right)^{\frac{5}{3}} \\
&= 2^{-\frac{2}{3}} \cdot E_{\rm TF}\left[\{n_{0\sigma}\}\right] \\
&= 0.629960\cdots \times E_{\rm TF}\left[\{n_{0\sigma}\}\right]
\end{aligned}
\tag{5.81}
$$

It is thus found that the TF energy for the symmetry adapted densities $\{n_\sigma\}$ is much smaller than that for the symmetry broken densities $\{n_{0\sigma}\}$ by a factor of ~ 0.63. Equation (5.81) reminds us of the discussion for the LDA in Sect. 4.1. Actually, the contents of the discussions are essentially the same. The difference comes only from the fact that the energy density of the TF functional and the LDA scale as powers of $5/3$ and $4/3$, respectively. Anyway, the electronic kinetic energy obtained by the TF functional with the symmetry adapted density leads to a disastrous result in the potential energy curve (PEC) for the H_2 dissociation. That is, the PEC at the dissociation limit is seriously underestimated due to the error of the TF energy. As in the SCE of the LDA, the error cannot be compensated by the GGA corrections to the energy. One might consider that the error can be eliminated by allowing the symmetry broken density in the dissociation. Remind, however, that the spin polarization in the density raises the kinetic energy, as opposed to the exchange energy. Therefore, the self-consistent field (SCF) calculation for the density leads favorably to the spin adapted structure.

The static correlation error (SCE) refers to the failure of the single-configuration wave function to describe the electron correlations associated with the bond dissociation. Hence, the problem of the TF model has no direct relevance to the SCE. In this Section, however, the error of the TF model in describing the bond dissociation is referred to as the SCE. This is solely because the source of the error is essentially the same as that of the SCE of LDA from a numerical point of view.

5.4.2 Coupling-Parameter Integration Scheme

We equivalently reformulate the TF functional using the coupling-parameter integration. To do this, we introduce a *reference* electron density $n_0(\boldsymbol{r})$. With the density n_0, the TF energy $E_{\rm TF}[n]$ for a density n can be written as

$$
E_{\rm kin}^{\rm TF}\left[n\right] = E_{\rm kin}^{\rm TF}\left[n_0\right] + \int_0^1 d\lambda \frac{dE_{\rm kin}^{\rm TF}\left[n_\lambda\right]}{d\lambda}
\tag{5.82}
$$

where λ is the coupling parameter ($0 \le \lambda \le 1$) describing the gradual change of the density from n_0 to $n = n_1$. The above formula is, of course, nothing more than an equation of identity, but it provides a basis for further approximations. The second term of the right hand of Eq. (5.82) can be rewritten in terms of the energy electron density $n_\lambda^e(\epsilon)$, thus,

$$\overline{E}^{e,\mathrm{TF}}_{\mathrm{kin}}[n] = E^{\mathrm{TF}}_{\mathrm{kin}}[n_0] + \int_0^1 d\lambda \frac{dE^{e,\mathrm{TF}}_{\mathrm{kin}}\left[n^e_\lambda\right]}{d\lambda} \tag{5.83}$$

It should be noticed that Eq. (5.83) is completely parallel to Eq. (4.19) for the exchange correlation functional E_{xc}. The energy electron density $n^e_\lambda(\epsilon)$ can be given by the projection defined in Eq. (3.2), thus,

$$n^e_\lambda(\epsilon) = \int d\boldsymbol{r}\, n_\lambda(\boldsymbol{r})\, \delta\left(\epsilon - \upsilon_{\mathrm{def}}(\boldsymbol{r})\right) \tag{5.84}$$

for each spin σ. The potential $\upsilon_{\mathrm{def}}(\boldsymbol{r})$ is the defining potential used in the previous subsection. Note that there are three kinds of different TF functionals $E^{\mathrm{TF}}_{\mathrm{kin}}$, $E^{e,\mathrm{TF}}_{\mathrm{kin}}$, and $\overline{E}^{e,\mathrm{TF}}_{\mathrm{kin}}$ in Eq. (5.83). $E^{e,\mathrm{TF}}_{\mathrm{kin}}$ differs from $E^{\mathrm{TF}}_{\mathrm{kin}}$ since it uses the energy electron density n^e_λ as the input, and $\overline{E}^{e,\mathrm{TF}}_{\mathrm{kin}}$ is different from $E^{e,\mathrm{TF}}_{\mathrm{kin}}$ because it is given by the sum of $E^{\mathrm{TF}}_{\mathrm{kin}}$ and $E^{e,\mathrm{TF}}_{\mathrm{kin}}$. In parallel to Eq. (3.7), the TF energy of the distribution n^e_λ with spin σ is defined as

$$E^{e,\mathrm{TF}}_{\mathrm{kin}}\left[n^e\right] = 2^{\frac{2}{3}} C_{\mathrm{TF}} \int d\epsilon\, \widetilde{n}^e(\epsilon)^{\frac{5}{3}}\, \Omega(\epsilon) \tag{5.85}$$

where $\widetilde{n}^e(\epsilon)$ is the average electron density on the energy coordinate ϵ and defined in Eq. (3.6). $\Omega(\epsilon)$ in Eq. (5.85) is the volume of the spatial region with the energy coordinate ϵ as defined in Eq. (3.4). As discussed also for the LDA energy in Sect. 3.1, it is readily understood that the functional $E^{e,\mathrm{TF}}_{\mathrm{kin}}[n^e]$ gives exactly the same values as the TF energy $E^{\mathrm{TF}}_{\mathrm{kin}}[n]$ provided that the electron density $n(\boldsymbol{r})$ is constant on the hypersurface $S(\epsilon)$ with the coordinate ϵ. Such a situation occurs in an atomic system where the electron density $n(\boldsymbol{r})$ is spherically symmetric around the nucleus.

How should the reference density n_0 be chosen for describing the bond dissociation of H_2? As in the case of the development of the exchange-correlation functional for the static correlation, the density n_0 in Eq. (5.83) is to be provided by the overlap of the electron densities of the isolated atoms H_A and H_B that constitute the molecule of interest. Explicitly, the density $n_{0\sigma}$ for the spin σ is written as

$$n_{0\sigma}(\boldsymbol{r}) = n^{\mathrm{A}}_{0\sigma}(\boldsymbol{r}) + n^{\mathrm{B}}_{0\sigma}(\boldsymbol{r}) \tag{5.86}$$

Actually, the density is already illustrated schematically in Fig. 5.10a for the hydrogen molecule with $R(H_A\text{-}H_B) = \infty$. Thus, the electrons with α and β spins are localized at atoms H_A and H_B, respectively. Note that this choice is equivalent to Eq. (4.12). It is quite important to note here that the TF model combined with a GGA correction shows rather good performances for calculating the kinetic energies of molecules at their equilibrium structures, as demonstrated in Ref. [16] for example. In fact, it was shown in Ref. [16] that the root-mean-square (RMS) error of the kinetic GGA functional given in Eq. (5.63) was found to be only 0.33% for the kinetic energies of 77 molecules. It is thus anticipated that the TF with a GGA functional can reasonably describe the individual kinetic energies of the localized densities $n^{\mathrm{A}}_{0\alpha}$ and $n^{\mathrm{B}}_{0\beta}$ in Eq. (5.86). In other words, the kinetic energy of the electron density at the dissociation limit of H_2 depicted in Fig. 5.10a can be calculated with reasonable accuracy. With this assumption in mind, we consider the second term on the right-hand side of Eq. (5.83). Integrating with respect to the coupling parameter λ, we get

$$\overline{E}_{\text{kin}}^{e,\text{TF}}[n] = E_{\text{kin}}^{\text{TF}}[n_0] + E_{\text{kin}}^{e,\text{TF}}\left[n^e\right] - E_{\text{kin}}^{e,\text{TF}}\left[n_0^e\right] \tag{5.87}$$

where n_1 is denoted with n for simplicity. The form of this equation is completely parallel to that of Eq. (4.13) for the static correlation functional E_{xc}. Here, we assume that the density $n = n_1$ is the spin-adapted density shown in Fig. 5.10b. Then, it means that the coupling-parameter integration describes the energy change $E_{\text{kin}}^{e,\text{relax}}[n^e, n_0^e]$ associated with the density relaxation from n_0 to n. Explicitly, the kinetic relaxation energy is defined as

$$E_{\text{kin}}^{e,\text{relax}}[n^e, n_0^e] = E_{\text{kin}}^{e,\text{TF}}\left[n^e\right] - E_{\text{kin}}^{e,\text{TF}}\left[n_0^e\right] \tag{5.88}$$

For the later references, it is also useful to define the kinetic static correlation energy $\Delta E_{\text{kin}}^{e,\text{SC}}[n^e, n_0]$, which reads

$$E_{\text{kin}}^{e,\text{SC}}[n_0] = E_{\text{kin}}^{\text{TF}}[n_0] - E_{\text{kin}}^{e,\text{TF}}\left[n_0^e\right] \tag{5.89}$$

The advantage of using Eq. (5.83) or equivalently Eq. (5.87) lies in the property of the energy electron density $n^e(\epsilon)$ as discussed in Sect. 4.1. At the dissociation limit, the spin-polarized density $n_{0\alpha}$ in Fig. 5.10a is completely different from the spin-adapted density n_α in Fig. 5.10b. However, as proved in Eq. (4.7), their projections $n_{0\alpha}^e(\epsilon)$ and $n_\alpha^e(\epsilon)$ onto the energy coordinate ϵ are equivalent to each other. The same is true for the electron with spin β. As a result, the relaxation term defined in Eq. (5.88) exactly vanishes and only the first term $E_{\text{kin}}^{\text{TF}}[n_0]$ in Eq. (5.87) remains at the dissociation limit. Provided that the kinetic energies of the electron densities of the independent atoms can be reasonably calculated by the TF model with a GGA correction, which is an assumption made in the previous paragraph, the functional of Eq. (5.87) delivers the correct kinetic energy at the dissociation. We also consider the opposite situation where the bond length is shorter than that of the equilibrium structure. In that case, the first and the third terms on the right-hand side of Eq. (5.87) will almost cancel each other, and the second term will remain to give the correct kinetic energy. Thus, it is expected that the overall profile of the potential energy curve of the H_2 dissociation can be described by a TF+GGA kinetic functional.

Before describing the numerical implementation of Eq. (5.87), we provide the graphs for the reference density $n_{0\alpha}$ and the spin-symmetry adapted density n_α of the H_2 molecules with the bond lengths of 0.9 and 7.0 a.u., in Figs. 5.11 and 5.12, respectively, for reader's references. As shown in Fig. 5.11, at the short bond length ($R(\text{H}_\text{A}\text{-H}_\text{B}) = 0.9$ a.u.), the electron density n_α subject to the spin symmetry has a spatial distribution similar to the reference density $n_{0\alpha}$. On the contrary, at the long bond length ($R(\text{H}_\text{A}\text{-H}_\text{B}) = 7.0$ a.u.), the density n_α completely differs from the reference density $n_{0\alpha}$ as seen in Fig. 5.12. It is quite interesting to observe what

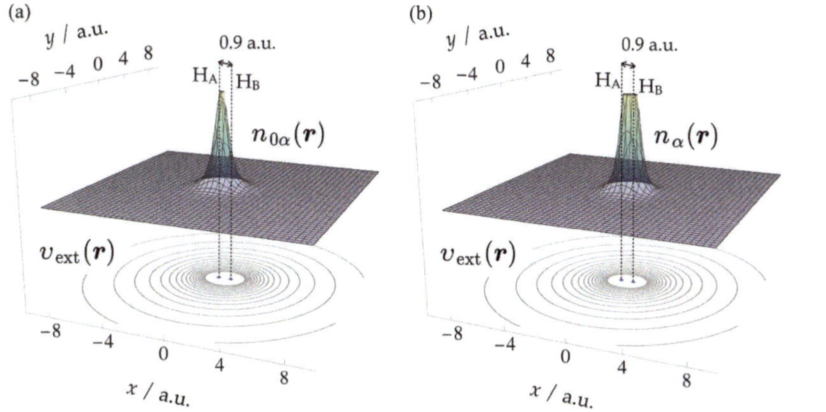

Fig. 5.11 **a** Left: 3-dimensional representation of the reference electron density $n_{0\alpha}(\boldsymbol{r})$ with spin α given as the ground state density of the isolated hydrogen atom placed at the position of H_A in the H_2 molecule with $R(H_A\text{-}H_B) = 0.9$ a.u. Lower: contour plot of the external potential $v_{\text{ext}}(\boldsymbol{r})$ of the molecule. **b** Right: 3-dimensional representation of the spin-symmetry adapted electron density $n_\alpha(\boldsymbol{r})$. Lower: the same external potential $v_{\text{ext}}(\boldsymbol{r})$ as in (**a**). These graphs are plotted for the logarithms $\log(|q| + 1.0)$ of the values q of the density n or potential v_{ext}. The plot ranges are $0 \leq \log(n(\boldsymbol{r}) + 1.0) \leq 0.2$ for the density, and $0 \leq \log(|v_{\text{ext}}(\boldsymbol{r})| + 1.0) \leq 1.0$ for the potential. The electron densities are constructed using 'Vmol' program package based on the real-space grid approach, where the BLYP exchange correlation functional is used

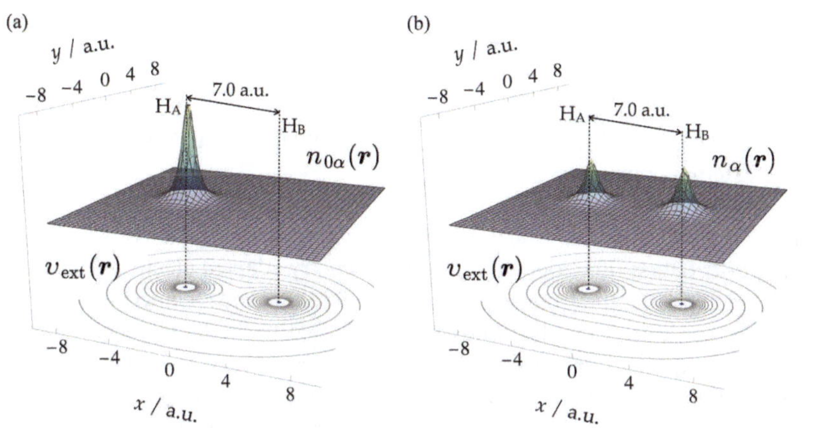

Fig. 5.12 **a** Left: The reference electron density $n_{0\alpha}$ with spin α placed at the position of H_A in the H_2 molecule with $R(H_A\text{-}H_B) = 7.0$ a.u. The shape of the distribution itself is the same as that in Fig. 5.11a. Lower: contour plot of the external potential $v_{\text{ext}}(\boldsymbol{r})$ of the molecule. **b** Right: 3-dimensional representation of the spin-symmetry adapted electron density $n_\alpha(\boldsymbol{r})$ of the molecule. Lower: the same external potential $v_{\text{ext}}(\boldsymbol{r})$ as in (**a**). Other notations are the same as in Fig. 5.11

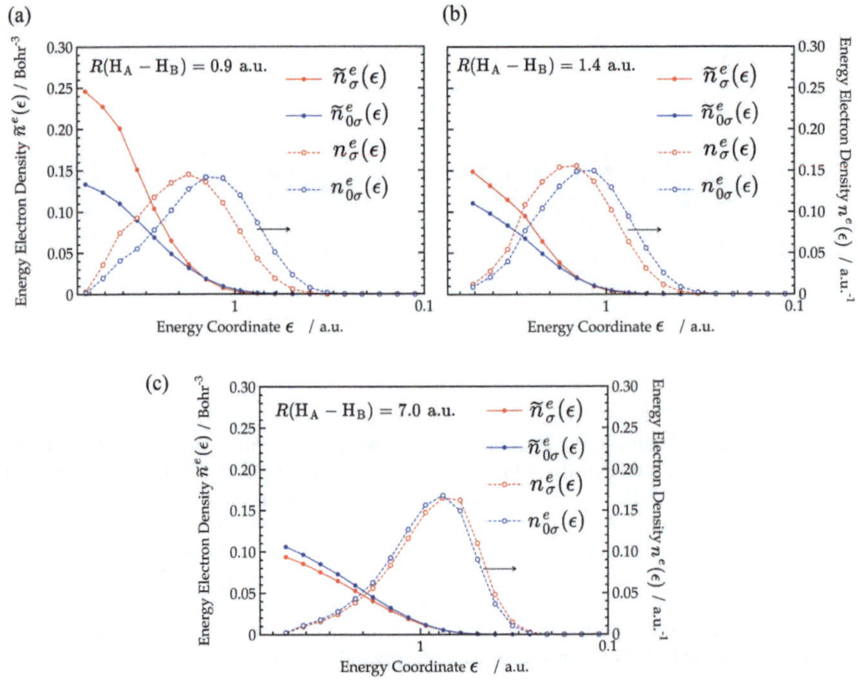

Fig. 5.13 Plots of the energy distribution functions $\widetilde{n}^e_\sigma(\epsilon)$ and $n^e_\sigma(\epsilon)$ for the spin-symmetry adapted electron density with spin σ. $\widetilde{n}^e_{0\sigma}(\epsilon)$ and $n^e_{0\sigma}(\epsilon)$ are for the reference densities. $\widetilde{n}^e$ is the average electron density on the energy coordinate ϵ and is defined in Eq. (3.6). The plots in (**a**), (**b**), and (**c**) are for the H_2 molecule with $R(\mathrm{H_A}\text{-}\mathrm{H_B}) = 0.9$, 1.4, and 7.0 a.u., respectively

happens when these distributions are projected onto the energy coordinate ϵ defined by the external potential $\upsilon_{\mathrm{ext}}(\boldsymbol{r})$.

In Fig. 5.13, the energy distribution functions are presented for the H_2 molecule with $R(\mathrm{H_A}\text{-}\mathrm{H_B}) = 0.9$, 1.4, and 7.0 a.u. When the bond distance is 0.9 a.u., the energy distribution functions for the spin-symmetry adapted electron density $n_\sigma(\boldsymbol{r})$ are distinctly different from those for the reference density $n_{0\sigma}(\boldsymbol{r})$. In fact, it is seen that the peak of the distribution $n^e_\sigma(\epsilon)$ shifted to the large energy coordinate and the average electron density $\widetilde{n}^e_\sigma(\epsilon)$ at the largest energy coordinate is increased significantly. This is due to the fact that the reference density $n_{0\sigma}(\boldsymbol{r})$ relaxes to $n_\sigma(\boldsymbol{r})$ to enjoy the attractive potential of the nuclei. However, when the bond length is extended to 1.4 a.u., the amount of the peak shift becomes modest, and the same is true for the increase in $\widetilde{n}^e_\sigma$ at the left end of the energy coordinate. At the bond length of 7.0 a.u., it is observed that the distribution $n^e_\sigma(\epsilon)$ almost coincides with $n^e_{0\sigma}(\epsilon)$ for the reference density, which shows a clear contrast to the *spatial* electron densities $n_\sigma(\boldsymbol{r})$ and $n_{0\sigma}(\boldsymbol{r})$ presented in Fig. 5.12. Such a property of the energy electron density is quite favorable in the development of the kinetic energy functional to describe the bond dissociation because it guarantees the coincidence of the electronic kinetic energy of the H_2 molecule at the dissociation limit with that of the two isolated hydrogen atoms under any choice of the approximate functional. Thus, the problem of the TF model shown in Eq. (5.81), which arises in the description of the bond dissociation, can be resolved by means of the functional in Eq. (5.87).

5.4.3 Generalized Gradient Approximation

The kinetic energy functional in Eq. (5.87) is described in terms of the TF functional. It is also possible to apply a GGA correction to the functional [17]. This Subsection provides the method in detail. In the GGA kinetic energy functional $E_{\mathrm{kin}}^{\mathrm{GGA}}[n_\sigma]$ for each spin σ, as in the GGA exchange functional, the inhomogeneity parameter q_σ defined in Eq. (3.70) plays a role in general. In parallel to Eq. (1.149), $E_{\mathrm{kin}}^{\mathrm{GGA}}[n_\sigma]$ can be written in the form as

$$E_{\mathrm{kin}}^{\mathrm{GGA}}[n_\sigma] = E_{\mathrm{kin}}^{\mathrm{TF}}[n_\sigma] + 2^{\frac{2}{3}} C_{\mathrm{TF}} \int d\boldsymbol{r}\; n_\sigma(\boldsymbol{r})^{\frac{5}{3}} F_{\mathrm{kin}}(q_\sigma) \tag{5.90}$$

where $F_{\mathrm{kin}}(q_\sigma)$ is the enhancement factor for the GGA functional. A series of the kinetic GGA functionals use the same function forms for $F_{\mathrm{kin}}(q_\sigma)$ as the corresponding exchange GGA functionals. For instance, for the exchange functional by Becke [18], Lee et al. [19] developed a corresponding GGA functional with the new parameters optimized for the kinetic energy. Similarly, referring to the exchange functional by Perdew et al. [20], Tran and Wesołowski [21] proposed an enhancement factor $F_{\mathrm{kin}}^{\mathrm{TW}}(q_\sigma)$ for the GGA kinetic functional in the form written as

$$F_{\mathrm{kin}}^{\mathrm{TW}}(q_\sigma) = \frac{\mu q_\sigma^2}{1 + \dfrac{\mu}{\kappa} q_\sigma^2} \tag{5.91}$$

where the parameters μ and κ were optimized as $\mu = 0.2319$ and $\kappa = 0.8438$. Using this enhancement factor in Eq. (5.90), we get the following TW functional,

$$E_{\mathrm{kin}}^{\mathrm{TW}}[n_\sigma] = E_{\mathrm{kin}}^{\mathrm{TF}}[n_\sigma] + 2^{\frac{2}{3}} C_{\mathrm{TF}} \int d\boldsymbol{r}\; n_\sigma(\boldsymbol{r})^{\frac{5}{3}} F_{\mathrm{kin}}^{\mathrm{TW}}(q_\sigma(\boldsymbol{r})) \tag{5.92}$$

In the present work, we utilize the Tran and Wesołowski (TW) functional modified for the functional of the energy electron density $n_\sigma^e(\epsilon)$. To this end, the inhomogeneity parameter $q_\sigma^e(\epsilon)$ defined on the energy coordinate will be introduced as given in Eqs. (3.71) and (3.72). Using n_σ^e and q_σ^e, the GGA kinetic energy functional of Eq. (5.90) can be transformed to

$$E_{\mathrm{kin}}^{e,\mathrm{GGA}}\left[n_\sigma^e\right] = E_{\mathrm{kin}}^{e,\mathrm{TF}}[n_\sigma^e] + 2^{\frac{2}{3}} C_{\mathrm{TF}} \int d\epsilon\, \widetilde{n}_\sigma^e(\epsilon)^{\frac{5}{3}} F_{\mathrm{kin}}(q_\sigma^e(\epsilon))\Omega(\epsilon) \tag{5.93}$$

Specifically, for the enhancement factor of Eq. (5.91) of the TW functional, we have

$$E_{\mathrm{kin}}^{e,\mathrm{TW}}\left[n_\sigma^e\right] = E_{\mathrm{kin}}^{e,\mathrm{TF}}[n_\sigma^e] + 2^{\frac{2}{3}} C_{\mathrm{TF}} \int d\epsilon\, \widetilde{n}_\sigma^e(\epsilon)^{\frac{5}{3}} F_{\mathrm{kin}}^{\mathrm{TW}}(q_\sigma^e(\epsilon))\Omega(\epsilon) \tag{5.94}$$

By replacing the TF energies in Eq. (5.87) with the GGA counterparts given in Eqs. (5.90) and (5.93), we obtain the GGA version of Eq. (5.87), thus,

$$\overline{E}_{\mathrm{kin}}^{e,\mathrm{GGA}}[n] = E_{\mathrm{kin}}^{\mathrm{GGA}}[n_0] + E_{\mathrm{kin}}^{e,\mathrm{GGA}}\left[n^e\right] - E_{\mathrm{kin}}^{e,\mathrm{GGA}}\left[n_0^e\right] \tag{5.95}$$

Specifically for the TW functional, we have

$$\overline{E}_{\mathrm{kin}}^{e,\mathrm{TW}}[n] = E_{\mathrm{kin}}^{\mathrm{TW}}[n_0] + E_{\mathrm{kin}}^{e,\mathrm{TW}}\left[n^e\right] - E_{\mathrm{kin}}^{e,\mathrm{TW}}\left[n_0^e\right] \tag{5.96}$$

However, in the application of the equation to the dissociation of the H_2 molecule, we employed the modified equation given by

$$\widetilde{E}_{\mathrm{kin}}^{e,\mathrm{TW}}[n] = E_{\mathrm{kin}}^{\mathrm{TF}}[n_0] + E_{\mathrm{kin}}^{e,\mathrm{TW}}\left[n^e\right] - E_{\mathrm{kin}}^{e,\mathrm{TW}}\left[n_0^e\right] \tag{5.97}$$

That is, the first term on right-hand side of Eq. (5.96) is being replaced by the TF functional $E_{\mathrm{kin}}^{\mathrm{TF}}$ for the density n_0. Such an artificial modification was made because it was found in a preliminary calculation that including the GGA correction in the first term erroneously raises the kinetic energy of the molecule. It is speculated that the kinetic static correlation $E_{\mathrm{kin}}^{e,\mathrm{SC}}[n_0]$ defined in Eq. (5.89) already includes the correction given by the GGA functional. Anyway, the modification only causes a constant shift in the kinetic energy by an amount of $E_{\mathrm{kin}}^{\mathrm{TF}}[n_0] - E_{\mathrm{kin}}^{\mathrm{TW}}[n_0]$ and does not change the profile of the potential energy in the case of the H_2 molecule.

5.4.4 Potential Gradient Approximation

In the application of Eq. (5.87) to the dissociation of the H_2 molecule we introduce, in addition to the GGA approach, a new method to correct the TF energy, that is referred to as the potential gradient approximation (PGA). This Subsection is devoted to describe the details of the PGA approach [17].

The magnitude of the gradient $|\nabla n(\boldsymbol{r})|$ of the electron density is, of course, related to the slope of the external potential $v_{\mathrm{ext}}(\boldsymbol{r})$ acting on the electrons. It is therefore quite natural to use the magnitude $|\nabla v_{\mathrm{ext}}(\boldsymbol{r})|$ instead of $|\nabla n(\boldsymbol{r})|$ as the variable to make a correction to the TF model. This constitutes the motivation for developing the PGA approach. To apply the PGA to Eq. (5.87), the average potential gradient $s^e(\epsilon)$ on the energy coordinate ϵ is introduced, which is defined as

$$s^e(\epsilon) = \Omega^{-1}(\epsilon) \int d\boldsymbol{r} \, |\nabla \upsilon_{\text{ext}}(\boldsymbol{r})| \, \delta(\epsilon - \upsilon_{\text{def}}(\boldsymbol{r})) \tag{5.98}$$

where $\upsilon_{\text{def}}(\boldsymbol{r})$ is the defining potential of the energy coordinate. Using this parameter, in parallel to Eq. (5.93), the TF energy can be corrected as,

$$E_{\text{kin}}^{e,\text{PGA}}\left[n_\sigma^e\right] = E_{\text{kin}}^{e,\text{TF}}[n_\sigma^e] + 2^{\frac{2}{3}} C_{\text{TF}} \int d\epsilon \, \widetilde{n}_\sigma^e(\epsilon)^{\frac{5}{3}} F_{\text{kin}}^{\text{PGA}}(s^e(\epsilon)) \Omega(\epsilon) \tag{5.99}$$

where the enhancement factor $F_{\text{kin}}^{\text{PGA}}(s^e)$ for the PGA is assumed to have the form similar to that for the GGA approach. In the present application, we employed the PBE functional as the template of the PGA enhancement factor, which reads

$$F_{\text{kin}}^{\text{PGA}}(s^e) = \frac{\beta(s^e)^2}{1 + \gamma(s^e)^2} \tag{5.100}$$

The fitting parameters β and γ in the enhancement factor $F_{\text{kin}}^{\text{PGA}}$ are roughly determined to be $\beta = \gamma = 4.0 \times 10^{-3}$. Because the major purpose of the present application is to examine the applicability of the PGA method to the kinetic energy functional, the fine tuning of the parameters is not performed. We notice that the important property of the function in Eq. (5.100) is that it increases monotonically with increasing potential gradient s^e as in the GGA functional.

In parallel to Eq. (5.97), we obtain the kinetic energy of Eq. (5.87) with the PGA correction written as

$$\widetilde{E}_{\text{kin}}^{e,\text{PGA}}[n] = E_{\text{kin}}^{\text{TF}}[n_0] + E_{\text{kin}}^{e,\text{PGA}}\left[n^e\right] - E_{\text{kin}}^{e,\text{PGA}}\left[n_0^e\right] \tag{5.101}$$

Note that the first term on the right-hand side is evaluated by the TF model as in the GGA functional of Eq. (5.97).

Enhancement Factor of a Hydrogenic Atom

Here we consider explicitly the relation between the potential gradient s^e and the corresponding enhancement factor $F_{\text{kin}}^{\text{H}}(s^e)$ for a hydrogenic atom where an electron is bound to a nucleus of charge Z. As shown in Eq. (2.50), the normalized ground state electron density n_0^{H} for the atom is written as

$$\begin{cases} n_0^{\text{H}}(r) & = \frac{\alpha(Z)^3}{8\pi} \exp(-\alpha(Z) r) \\ \alpha(Z) & = 2Z \end{cases} \tag{5.102}$$

where r is the distance between the electron and the nucleus. Since the potential for the electron is merely given by $-Z/r$, the magnitude of the potential gradient $s(r)$ is proportional to Z. Thus the gradient at r can be varied by changing the charge Z. For the density $n_0^{\text{H}}(r)$ of Eq. (5.102), the kinetic energy density $U_{\text{kin}}[n_0^{\text{H}}]$ at r is calculated as

$$U_{\text{kin}}\left[n_0^{\text{H}}\right](r) = \frac{1}{2}\left(\frac{dn_0^{\text{H}}(r)^{\frac{1}{2}}}{dr}\right)^2 = \frac{1}{8}\alpha^2 A(\alpha)\exp(-\alpha r) \tag{5.103}$$

where $A(\alpha)$ stands for the normalization factor $\frac{\alpha(Z)^3}{8\pi}$ in Eq. (5.102). On the other hand, the TF kinetic energy density $U_{\text{kin}}^{\text{TF}}$ is given by

$$U_{\text{kin}}^{\text{TF}}\left[n_0^{\text{H}}\right](r) = 2^{\frac{2}{3}}\frac{3}{10}\left(3\pi^2\right)^{\frac{2}{3}} A(\alpha)^{\frac{5}{3}}\exp\left(-\frac{5}{3}\alpha r\right) \tag{5.104}$$

Then, the enhancement factor $F_{\text{kin}}^{\text{H}}(r)$ is obtained as

$$F_{\text{kin}}^{\text{H}}(r) = \frac{U_{\text{kin}}\left[n_0^{\text{H}}\right](r)}{U_{\text{kin}}^{\text{TF}}\left[n_0^{\text{H}}\right](r)} = \frac{5}{3}(6\pi)^{-\frac{2}{3}}\exp\left(\frac{2}{3}\alpha r\right) \tag{5.105}$$

It is shown that the enhancement factor increases monotonically with increasing $\alpha = 2Z > 0$ for a given r. Thus, the PGA enhancement factor $F_{\text{kin}}^{\text{PGA}}(s^e)$ in Eq. (5.100) is consistent with the behavior of the function $F_{\text{kin}}^{\text{H}}(r)$ of Eq. (5.105). As the two atoms in the hydrogen molecules approach each other, the absolute value of the potential gradient in the molecule increases. Equation (5.100) allows to describe the resultant increase in the kinetic energy enhancement factor. However, $F_{\text{kin}}^{\text{PGA}}(s^e)$ does not realize the increase of the enhancement factor with increasing r when α is fixed. We also note that the real enhancement factor $F_{\text{kin}}^{\text{H}}(r)$ of Eq. (5.105) becomes less than 1 for small r. In fact, $F_{\text{kin}}^{\text{H}}(0)$ is found to be $0.235314\cdots$, which indicates that the kinetic energy density at $r = 0$ is much smaller than that given by the TF model regardless of the value of Z. This property cannot be realized in either the GGA or the PGA functionals.

Applications to Bond Dissociation of an H_2 molecule

In the following, we apply the kinetic functionals in Eqs. (5.92), (5.87), (5.97), and (5.101) to the calculations of the dissociation curve of the H_2 molecule to make comparisons. In these functionals, only Eq. (5.92) uses the normal electron density $n(\boldsymbol{r})$ as an argument, while the remaining three functionals employ the energy distribution $n^e(\epsilon)$. In the following, we refer to these kinetic energy functionals by shorthand notations for convenience. The correspondences are as follows: Eq. (5.92) → 'TW', Eq. (5.87) → 'eTF', Eq. (5.97) → 'eTW', and Eq. (5.101) → 'ePGA'. The character 'e' indicates that the functional uses the energy electron density n^e as an argument.

The electron density $n(\boldsymbol{r})$ used as an argument of these functionals is that determined by the spin-restricted KS-DFT calculation using the real-space grids. The total electronic KS energy of the system can be expressed as

$$E_{\text{tot}}^{\text{KS}}[\{\varphi_i\}] = E_{\text{kin}}^{\text{KS}}[\{\varphi_i\}] + E_{\text{H}}[n] + E_{xc}[n] + E_{\text{ext}}^{\text{ps}}[\{\varphi_i\}] \tag{5.106}$$

where the first term on the right-hand side is the KS(non-interacting) kinetic energy described with the set of KS wave functions $\{\varphi_i\}$, the second is the Hartree energy for the electron density n, the third stands for the exchange correlation energy, and the fourth is the interaction between the electrons and the nuclei. Since we use the real-space grids, the last term is calculated using the *nonlocal* pseudopotentials and hence the KS wave functions are needed. The energy E_{xc} is evaluated using the BLYP functional [18, 22]. The total energies evaluated with the kinetic energy functionals are obtained by replacing the KS kinetic energy $E_{\rm kin}^{\rm KS}[\{\varphi_i\}]$ with the corresponding functionals. Specifically, for the functional 'TW' defined in Eq. (5.92), the total energy is given by

$$E_{\rm tot}^{\rm TW}[\{\varphi_i\}] = E_{\rm kin}^{\rm TW}[n] + E_{\rm H}[n] + E_{xc}[n] + E_{\rm ext}^{\rm ps}[\{\varphi_i\}] \tag{5.107}$$

For the remaining kinetic energy functionals indicated by 'eTF', 'eTW', and 'ePGA', the static correlation energy $E_{\rm sc}[n_0^e]$ evaluated with Eq. (4.16) is also included in $\overline{E}_{xc}$. To be more precise, $\overline{E}_{xc}[n]$ is expressed by sum of the two contributions,

$$\overline{E}_{xc}[n] = \overline{E}_x^{e,\rm LDA}\left[n^e\right] + E'_{\rm BLYP}[n] \tag{5.108}$$

where the exchange LDA energy $\overline{E}_x^{e,\rm LDA}[n^e]$ includes the static correlation and is calculated with the following equation

$$\overline{E}_x^{e,\rm LDA}\left[n^e\right] = E_x^{\rm LDA}[n_0] + \left(E_x^{e,\rm LDA}\left[n^e\right] - E_x^{e,\rm LDA}\left[n_0^e\right]\right) \tag{5.109}$$

The energy $E_x^{e,\rm LDA}[n^e]$ is explicitly defined in Eq. (3.7) and $E'_{\rm BLYP}[n]$ in Eq. (5.108) is obtained by subtracting the LDA energy from the BLYP energy for the normal density n, thus,

$$E'_{\rm BLYP}[n] = E_{\rm BLYP}[n] - E_x^{\rm LDA}[n] \tag{5.110}$$

Employing the functional of Eq. (5.108), we construct the functional $E_{\rm tot}^{\rm A}[\{\varphi_i\}]$ for the total energy

$$E_{\rm tot}^{\rm A}[\{\varphi_i\}] = E_{\rm kin}^{\rm A}[n] + E_{\rm H}[n] + \overline{E}_x^{e,\rm LDA}\left[n^e\right] + E'_{\rm BLYP}[n] + E_{\rm ext}^{\rm ps}[\{\varphi_i\}] \tag{5.111}$$

where the superscript A stands for the notations, 'eTF', 'eTW', and 'ePGA'. As noted above, these notations indicate the kinetic energy functionals in Eqs. (5.87), (5.97), and (5.101), respectively. The reference density n_0 used in the kinetic energy functionals is the same as that used in the exchange functional $\overline{E}_x^{e,\rm LDA}[n]$ and is constructed according to Eq. (5.86). In summary, we use the functionals for the total energy defined in Eqs. (5.106), (5.107), and (5.111) to calculate the potential energies for the bond dissociation of the H_2 molecule. And in the application of Eq. (5.111), three kinds of the kinetic energy functionals using the energy electron density n^e are adopted for comparisons.

Next we briefly provide the numerical details of the calculations. The KS wave function and the density of the H_2 molecule are enclosed in a cubic cell with the size $L = 18.354$ a.u. The real-space cell is discretized by 64 equally spaced grids along each direction. As a result, the grid width becomes $h = 0.28579$ a.u. The nonlocal pseudopotential with the Kleinman and Bylander form is adopted for the H atoms to evaluate the energy $E^{\text{ps}}_{\text{ext}}[\{\varphi_i\}]$. To describe the detailed structure of the pseudopotential, the double grid technique by Ono and Hirose [23] is employed, where the width d of the dense grid is set at $d = h/5$. To define the energy coordinate ϵ for the construction of the distribution $n^e(\epsilon)$, the local part of the pseudopotential developed by the Bachelet et al. [14] for the H atom is utilized. Each coarse grid is uniformly discretized by 5 grids to increase the number of sampling points for constructing the energy distribution functions, where the values on the dense grids are evaluated by the fourth-order polynomial interpolation.

Before showing the results given by the kinetic energy functionals, we examine the effects of the following two issues on the description of the total energy. The first concern is the effect of using the energy distribution function obtained by projecting the electron density $n(\boldsymbol{r})$ onto the energy coordinate on the LDA exchange energy. Although the effect has been already studied e.g. in Fig. 3.11 in Chap. 3, the comparison is made for the sake of clarity. The second issue is the effect of using the static correlation functional for the exchange energy on the energy profile. Elucidation of these two matters is important to extract the pure effects of the kinetic energy functionals. Specifically for the first issue, we consider the KS energy $E^{\text{KS-I}}_{\text{tot}}$ given by the following functional,

$$E^{\text{KS-I}}_{\text{tot}}[\{\varphi_i\}] = E^{\text{KS}}_{\text{kin}}[\{\varphi_i\}] + E_{\text{H}}[n] + E^{e,\text{LDA}}_{x}\left[n^e\right] + E'_{\text{BLYP}}[n] + E^{\text{ps}}_{\text{ext}}[\{\varphi_i\}] \tag{5.112}$$

And for the second issue, we introduce the functional $E^{\text{KS-II}}_{\text{tot}}$,

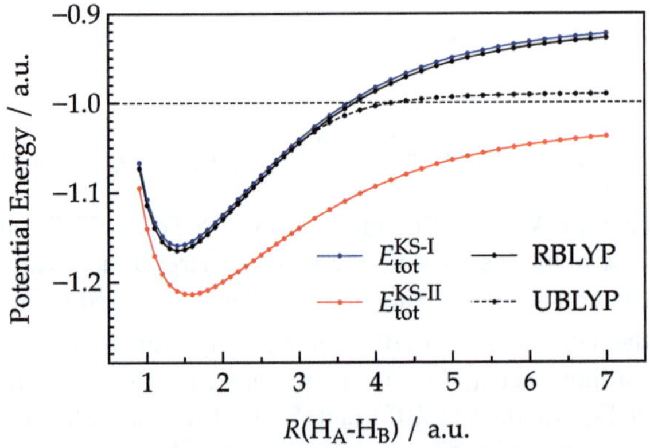

Fig. 5.14 Potential energy curves calculated using the functionals $E^{\text{KS-I}}_{\text{tot}}$ and $E^{\text{KS-II}}_{\text{tot}}$ shown, respectively, in Eqs. (5.112) and (5.113). The results given by the spin-restricted and -unrestricted Kohn-Sham DFT (KS-DFT) calculations (Eq. (5.106)) with the BLYP functional are also shown for comparisons. For the calculations with the functionals $E^{\text{KS-I}}_{\text{tot}}$ and $E^{\text{KS-II}}_{\text{tot}}$, the electron density optimized by the RBLYP calculation is employed as an argument

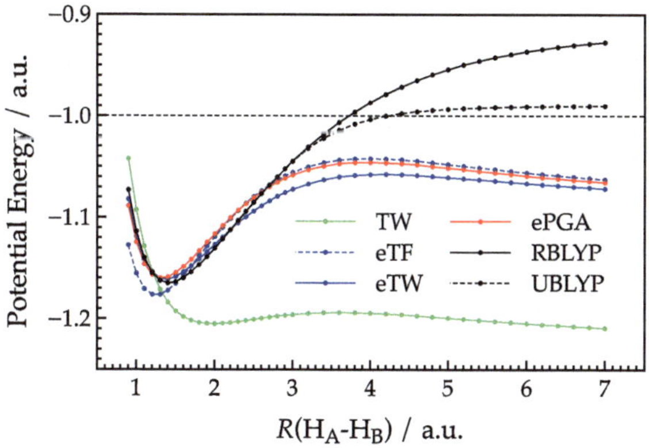

Fig. 5.15 Potential energy curves calculated using the functionals with the shorthand notations; 'TW', 'eTF', 'eTW', and 'ePGA'. The correspondences between the notations and the equations used for the total and kinetic energies are summarized in Table 5.1. The electron density obtained by the KS-DFT (RBLYP) calculation is used as an argument for all the kinetic energy functionals

$$E_{\text{tot}}^{\text{KS-II}}\left[\{\varphi_i\}\right] = E_{\text{kin}}^{\text{KS}}\left[\{\varphi_i\}\right] + E_{\text{H}}[n] + \overline{E}_x^{e,\text{LDA}}\left[n^e\right] + E'_{\text{BLYP}}[n] + E_{\text{ext}}^{\text{ps}}\left[\{\varphi_i\}\right] \quad (5.113)$$

Equation (5.112) differs from Eq. (5.106) in the point that the LDA exchange energy in the BLYP functional is evaluated with the energy electron density $n^e(\epsilon)$. And the difference between $E_{\text{tot}}^{\text{KS-I}}$ of Eq. (5.112) and $E_{\text{tot}}^{\text{KS-II}}$ of Eq. (5.113) is that the latter includes the static correlation $E_{\text{sc}}[n_0^e] = E_x^{\text{LDA}}[n_0] - E_x^{e,\text{LDA}}\left[n_0^e\right]$ in the exchange energy as shown in Eq. (5.109). In Fig. 5.14, we make comparisons for the potential energy curves (PECs) given by these functionals. In consistent with the PECs in Fig. 3.11, the PEC by the functional $E_{\text{tot}}^{\text{KS-I}}$ shows a faithful agreement with that given by the RBLYP calculation over the whole range of the bond distances in the graph. It is thus confirmed that the projection of $n(\boldsymbol{r})$ onto the energy coordinate does not degrade the LDA exchange energy. It is shown, however, $E_{\text{tot}}^{\text{KS-I}}$ and RBLYP overshoot the correct energy in the dissociation limit evaluated by the UBLYP calculation. On the contrary, the functional $E_{\text{tot}}^{\text{KS-II}}$, which includes the static correlation $E_{\text{sc}}[n_0^e]$, underestimates the PEC, especially in the intermediate bond distances as already shown in Fig. 4.4.

Next, the KS kinetic energy $E_{\text{kin}}^{\text{KS}}\left[\{\varphi_i\}\right]$ in the functional $E_{\text{tot}}^{\text{KS-II}}$ is replaced by the orbital-free functionals to make comparisons. The results are shown in Fig. 5.15. We refer the readers to Table 5.1 to specify the equations corresponding to the notations in the graph. At first sight, it is worth noting that the functional indicated by 'TW' shows rather good agreement with the R(U)BLYP calculation in the range of the small bond distances. However, when the bond length becomes larger than the equilibrium nuclear distance $R_{\text{eq}} = 1.4$ a.u., the functional is found to give erroneously low energies, which cannot be compensated by the static correlation error associated with the LDA exchange energy. As discussed in Sect. 5.4.1, such an error

Table 5.1 Correspondence table between the shorthand notations and the sets of the total energy functional E_{tot} and the kinetic energy functional $E_{\text{kin}}^{\text{A}}$ included in E_{tot}. The term 'DFT' stands for the conventional DFT, while 'eDFT' represents the DFT using the energy electron density n^e as an argument

Theory	Notation	$E_{\text{kin}}^{\text{A}}$	E_{tot}
DFT	TW	Eq. (5.92)	Eq. (5.107)
eDFT	eTF	Eq. (5.87)	Eq. (5.111)
	eTW	Eq. (5.97)	
	ePGA	Eq. (5.101)	

in the kinetic energy is, of course, due to the fact that the electron density $n_\sigma(\boldsymbol{r})$ for each spin severely delocalizes upon the bond dissociation. It is found, however, that the inclusion of the static correlation in the kinetic energy greatly improves the profile of the potential energy, as shown in the graph indicated with 'eTF'. Recall that the static correlation (SC) is negatively overestimated in Eq. (5.109), as seen in Fig. 5.14. From Figs. 5.14 and 5.15, it can be seen that the overestimation of the SC in the exchange-correlation energy is favorably compensated by the overestimation of the SC in the kinetic energy with the opposite sign. As a result, the potential profile of 'eTF' in the region of the intermediate bond lengths agrees well with that of the KS-DFT calculation. In the region of the small inter-nuclear distances, on the other hand, it is shown that the potential energies of 'eTF' are remarkably lower than those of the KS-DFT calculations. In the calculation with the notation 'eTW', a GGA correction [21] is being adapted to the kinetic energy functional in 'eTF'. It is shown in the graph that the energy profile in the region of the small bond lengths is clearly improved by the 'eTW' functional. The result given by the functional 'ePGA' with the potential gradient approximation also shows the similar trend as that of the 'eTW' calculation, suggesting the effectiveness of the PGA. The mechanism underlying the increase in the kinetic energy due to the PGA correction can be easily understood as follows. The approach of the two nuclei leads to an increase in the gradient of the Coulomb potential formed by the nuclei, which results in an increase in the PGA enhancement factor in Eq. (5.100). The applicability of the PGA to other molecules needs to be examined in the future works. Lastly, we also make a remark on the lowering of the potential energies of 'eTF', 'eTW', and 'ePGA' in the region of the long bond distances. As noted above, for the calculations with these functionals, the electron density obtained by the RBLYP calculation is used. Then, the *total* electron density of RBLYP will be somewhat delocalized as compared with that of the UBLYP since the static correlation cannot be incorporated in the RBLYP calculation. As a consequence, the kinetic energies by the functionals 'eTF', 'eTW', and 'ePGA' will be erroneously lowered. In fact, it can be shown (not presented in Fig. 5.15) that the potential energies at the dissociation limit are certainly increased by employing the spin density given by half the total electron density of UBLYP.

References

1. L.H. Thomas, Proc. Camb. Phil. Soc. **23**, 541 (1927)
2. L.W. Wang, M.P. Teter, Phys. Rev. B **45**(23), 13196 (1992)
3. Y.A. Wang, E.A. Carter, *Chapter 5 in Theoretical Methods in Condensed Phase Chemistry (Progress in Theoretical Chemistry and Physics)*, vol. 5 (Kluwer, Dordrecht, 2000)
4. T.A. Wesolowski, Y.A. Wang (eds.), *Recent Progress in Orbital Free Density Functional Theory (Recent Advances in Computational Chemistry)* (World Scientific, Singapore, 2013)
5. W. Kohn, L.J. Sham, Phys. Rev. **140**(4A), A1133 (1965)
6. L. Hung, E.A. Carter, Chem. Phys. Lett. **475**, 163 (2009)
7. L. Hung, C. Huang, I. Shin, G.S. Ho, V.L. Ligneres, E.A. Carter, Comput. Phys. Commun. **181**, 2208 (2010)
8. E. Fermi, Z. Phys. **48**, 73 (1928)
9. R.G. Parr, W. Yang, *Density-Functional Theory of Atoms and Molecules* (Oxford University Press, New York, 1989)
10. C.F. von Weizsacker, Z. Phys. **96**, 431 (1935)
11. J. Lindhard, Kgl. Danske Videnskab. Selskab Mat.-fys. Medd. **28** (1954)
12. R.G. Parr, A. Berk, in *Chemical Applications of Atomic and Molecular Electrostatic Potentials*, ed. by P. Politzer, D.G. Truhlar (Plenum Press, 1981), pp. 51–62
13. H. Takahashi, Int. J. Quantum Chem. **122**(20), e26969 (2022)
14. G.B. Bachelet, D.R. Hamann, M. Schlüter, Phys. Rev. B **26**(8), 4199 (1982)
15. A.J.C. Felipe A. Bulat, Tim Heaton-Burgess, W. Yang, J. Chem. Phys. **127**, 174101(9) (2007)
16. A.J. Thakkar, Phys. Rev. A **46**(11), 6920 (1992). https://doi.org/10.1103/PhysRevA.46.6920
17. H. Takahashi, Electron. Struct. **7**(2), 025003 (2025)
18. A.D. Becke, Phys. Rev. A **38**(6), 3098 (1988)
19. H. Lee, C. Lee, R.G. Parr, Phys. Rev. A **44**(1), 768 (1991)
20. J.P. Perdew, K. Burke, M. Ernzerhof, Phys. Rev. Lett. **77**(18), 3865 (1996)
21. F. Tran, T.A. Wesołowski, Int. J. Quantum Chem. **89**, 441 (2002)
22. C. Lee, W. Yang, R.G. Parr, Phys. Rev. B **37**(2), 785 (1988)
23. T. Ono, K. Hirose, Phys. Rev. Lett. **82**(25), 5016 (1999)

Appendix A
Local Spin Density Approximation

In this Appendix, the local density approximation (LDA) for the spin compensated total density n is modified to the equation for the density n_σ for spin σ to obtain the local spin density approximation (LSDA). The same treatment can also be applied to the Thomas-Fermi (TF) model. As shown below, the modification of the equation due to the change of the argument from n to n_σ is just the multiplication of a spin scaling factor. This appendix provides a note on how to determine the scaling factors for the LDA and the TF models. It is important to note that the factors are not derived from any physical model, but is merely determined so that the energy of the spin-compensated density can be realized.

For the *total* electron density $n(\boldsymbol{r})$ without a spin polarization, the LDA exchange energy $E_x^{\mathrm{LDA}}[n]$ is given by

$$E_x^{\mathrm{LDA}}[n] = -C_x^{\mathrm{LDA}} \int d\boldsymbol{r}\ n(\boldsymbol{r})^{\frac{4}{3}} \tag{A.1}$$

where C_x^{LDA} is the constant for the exchange energy of the homogeneous electron gas (HEG) and is summarized below. Our purpose is to obtain the equation $E_{x\sigma}^{\mathrm{LDA}}[n_\sigma]$ for the spin σ so that $E_{x\sigma}^{\mathrm{LDA}}$ is consistent with Eq. (A.1) for the spin-compensated density. Since it is assumed that the density is spin-compensated in Eq. (A.1), it can be rewritten as

$$\begin{aligned} E_x^{\mathrm{LDA}}[n] &= -C_x^{\mathrm{LDA}} \int d\boldsymbol{r}\ (2n_\sigma(\boldsymbol{r}))^{\frac{4}{3}} \\ &= -2^{\frac{4}{3}} C_x^{\mathrm{LDA}} \int d\boldsymbol{r}\ n_\sigma(\boldsymbol{r})^{\frac{4}{3}} \\ &= -2 \cdot 2^{\frac{1}{3}} C_x^{\mathrm{LDA}} \int d\boldsymbol{r}\ n_\sigma(\boldsymbol{r})^{\frac{4}{3}} \end{aligned} \tag{A.2}$$

Since the relation

$$E_x^{\mathrm{LDA}}[n] = 2E_{x\sigma}^{\mathrm{LDA}}[n_\sigma] \tag{A.3}$$

H. Takahashi, Density-Functional Theory for Electrons,
https://doi.org/10.1007/978-981-95-1229-4

Table A.1 List of the values of the constants C_x^{LDA} and C_{TF} for the LDA exchange and the TF energies, respectively. The corresponding spin scaling factors are also presented

Energy	Notation	Value	Factor
Exchange (LDA)	C_x^{LDA}	$\frac{3}{4}\left(\frac{3}{\pi}\right)^{\frac{1}{3}} = 0.738559$	$2^{\frac{1}{3}} = 1.25992$
Kinetic (TF)	C_{TF}	$\frac{3}{10}\left(3\pi^2\right)^{\frac{2}{3}} = 2.87123$	$2^{\frac{2}{3}} = 1.58740$

must be satisfied for the density with no spin polarization, we get the LSDA written as

$$E_{x\sigma}^{\mathrm{LDA}}\left[n_\sigma\right] = -2^{\frac{1}{3}} C_x^{\mathrm{LDA}} \int d\boldsymbol{r}\; n_\sigma\left(\boldsymbol{r}\right)^{\frac{4}{3}} \tag{A.4}$$

where $2^{\frac{1}{3}}$ is the spin scaling factor for the exchange energy. Note that the use of the term 'spin scaling factor' is not common in the literature. However, the factor is always referred to as that term throughout this book.

In parallel to Eq. (A.2), for the kinetic energy $E_{\mathrm{kin}}^{\mathrm{TF}}[n]$ of the TF model, we have

$$\begin{aligned} E_{\mathrm{kin}}^{\mathrm{TF}}\left[n\right] &= C_{\mathrm{TF}} \int d\boldsymbol{r}\; \left(2n_\sigma\left(\boldsymbol{r}\right)\right)^{\frac{5}{3}} \\ &= 2 \cdot 2^{\frac{2}{3}} C_{\mathrm{TF}} \int d\boldsymbol{r}\; n_\sigma\left(\boldsymbol{r}\right)^{\frac{5}{3}} \\ &= 2E_{\mathrm{kin}\sigma}^{\mathrm{TF}}\left[n_\sigma\right] \end{aligned} \tag{A.5}$$

where $E_{\mathrm{kin}\sigma}^{\mathrm{TF}}$ for spin σ is defined as

$$E_{\mathrm{kin}\sigma}^{\mathrm{TF}}\left[n_\sigma\right] = 2^{\frac{2}{3}} C_{\mathrm{TF}} \int d\boldsymbol{r}\; n_\sigma\left(\boldsymbol{r}\right)^{\frac{5}{3}} \tag{A.6}$$

Thus, the spin scaling factor for the TF model is found to be $2^{\frac{2}{3}}$. The values of the constants C_x^{LDA} and C_{TF} are summarized in Table A.1.

Appendix B
Lagrange's Multiplier Method

Lagrange's multiplier method is often employed in the main text for the minimiza tions of functions or functionals under some constraints. Here let us review the method with an emphasis on its mathematical aspect. We start our discussion with the consideration of the simplest example. Let $F(x, y)$ be a real function of the real numbers x and y. The objective is to obtain the local extremum (minimum or maximum) of the function F with a constraint specified by an equation

$$\varphi(x, y) = 0 \tag{B.1}$$

Suppose that the extremum of F is attained at a point $P_0(x_0, y_0)$. Then, the small variation (dx, dy) from P_0 of x and y that fulfills the constraint Eq. (B.1) is given by

$$d\varphi = \left.\frac{\partial\varphi}{\partial x}\right|_{P_0} dx + \left.\frac{\partial\varphi}{\partial y}\right|_{P_0} dy = 0 \tag{B.2}$$

And we also have the same relation for F, thus,

$$dF = \left.\frac{\partial F}{\partial x}\right|_{P_0} dx + \left.\frac{\partial F}{\partial y}\right|_{P_0} dy = 0 \tag{B.3}$$

It is of essential importance to note that the variation vector (dx, dy) is common to Eqs. (B.2) and (B.3). Thus, the gradient vector ∇F shares its direction with that of φ, that is,

$$\left(\frac{\partial F}{\partial x}, \frac{\partial F}{\partial y}\right) = k\left(\frac{\partial\varphi}{\partial x}, \frac{\partial\varphi}{\partial y}\right) \tag{B.4}$$

where the subscripts P_0 are omitted for the sake of simplicity and k is a real number representing the ratio of the length of the gradient vector ∇F to that of φ at P_0. It is

H. Takahashi, Density-Functional Theory for Electrons,
https://doi.org/10.1007/978-981-95-1229-4

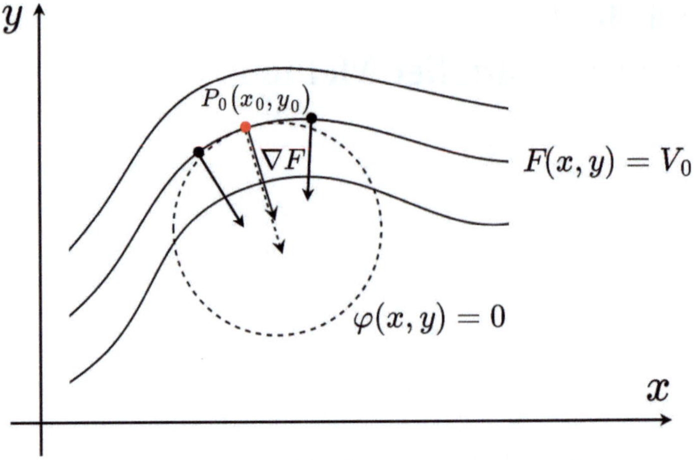

Fig. B.1 Three real curves represent contours of the function $F(x, y)$. Three dots are placed on the contour line of the value V_0. The three arrows with real lines represent the gradient vectors ∇F on the points. At the point P_0, the function F takes the extremum under the constraint of $\varphi(x, y) = 0$ indicated by a circle drawn with a broken line. At P_0, the direction of the gradient vector ∇F coincides with that of $\nabla \varphi$ (an arrow with a broken line)

recognized in Fig. B.1 that P_0 is the point where the gradient vector ∇F coincides with the gradient vector $\left(\frac{\partial \varphi}{\partial x}, \frac{\partial \varphi}{\partial y}\right)$. Equation (B.4) consists of two independent equations,

$$\begin{cases} \frac{\partial F}{\partial x} = k\frac{\partial \varphi}{\partial x} \\ \frac{\partial F}{\partial y} = k\frac{\partial \varphi}{\partial y} \end{cases} \tag{B.5}$$

To ensure that these two equation hold simultaneously, we introduce a new function $L(x, y)$ defined as

$$L(x, y) = F(x, y) - k\varphi(x, y) \tag{B.6}$$

where k is Lagrange's multiplier. Then, we impose the extremum conditions for L without constraints,

$$\begin{cases} \frac{\partial L(x,y)}{\partial x} = 0 \\ \frac{\partial L(x,y)}{\partial y} = 0 \end{cases} \tag{B.7}$$

It is readily recognized that these equations are equivalent to Eqs. (B.5).

With the above discussion kept in mind, we next consider a general case. To this end, let the function F be a function of the n real variables, that is, $F(x_1, x_2, \cdots, x_n)$. It is assumed that there are m constraints for the n variables, thus,

$$\varphi_a(x_1, x_2, \cdots, x_n) = 0 \qquad (a = 1, 2, \cdots, m) \tag{B.8}$$

It is supposed that these constraints are fulfilled at the point P_0. Then, for the small variation $\delta \boldsymbol{x} = (dx_1, dx_2, \cdots, dx_n)$ from P_0, the following relations hold,

$$\frac{\partial \varphi_a}{\partial x_1} dx_1 + \frac{\partial \varphi_a}{\partial x_2} dx_2 + \cdots + \frac{\partial \varphi_a}{\partial x_n} dx_n = 0 \qquad (a = 1, 2, \cdots, m) \tag{B.9}$$

We define the vector $\boldsymbol{v}_a$ for each constraint,

$$\boldsymbol{v}_a = \left(\frac{\partial \varphi_a}{\partial x_1}, \frac{\partial \varphi_a}{\partial x_2}, \cdots, \frac{\partial \varphi_a}{\partial x_n} \right) \qquad (a = 1, 2, \cdots, m) \tag{B.10}$$

Each gradient vector satisfies the orthogonalization condition with the variation vector $\delta \boldsymbol{x}$,

$$\boldsymbol{v}_a \cdot \delta \boldsymbol{x} = 0 \qquad (a = 1, 2, \cdots, m) \tag{B.11}$$

Now we suppose that the function F achieves an extremum at P_0, which reads

$$dF = \frac{\partial F}{\partial x_1} dx_1 + \frac{\partial F}{\partial x_2} dx_2 + \cdots + \frac{\partial F}{\partial x_n} dx_n = 0 \tag{B.12}$$

or, equivalently,

$$\nabla F \cdot \delta \boldsymbol{x} = 0 \tag{B.13}$$

This relation holds as a consequence of the relations shown in Eqs. (B.11), implying that the gradient vector ∇F is expressed by a linear combination of the vectors $\{\boldsymbol{v}_a\}$,

$$\nabla F = \sum_a^m k_a \boldsymbol{v}_a \tag{B.14}$$

with k_a the coefficient for the vector $\boldsymbol{v}_a$. In parallel to the discussion for the function with two variables, we introduce the function L, thus,

$$L(x_1, x_2, \cdots, x_n) = F(x_1, x_2, \cdots, x_n) - \sum_a^m k_a \varphi_a (x_1, x_2, \cdots, x_n) \tag{B.15}$$

where k_a are Lagrange's multipliers. The extremum of F under the constraints can be achieved by searching the extremum of L without the constraints, which can be represented as

$$\frac{\partial L(x_1, x_2, \cdots, x_n)}{\partial x_i} = 0 \qquad (i = 1, 2, \cdots, n) \tag{B.16}$$

Appendix C
Potential-functional Theory

The potential-functional theory (PFT) dual to the density-functional theory (DFT) was advocated by Yang et al. [1] PFT provides a theoretical basis for the optimized-effective potential (OEP) method described in Sect. 1.4. In this appendix, the potential functional specific to the HF energy E_0^{HF} will be formulated, whereby the formal equivalence of the HF-OEP and the inverse Kohn-Sham (inv-KS) method will be discussed. The Hohenberg-Kohn (HK) functional F^{HK} plays a decisive role in the formulation of the general PFT [1]. Thus, for the construction of the PF specific to the HF ground state energy E_0^{HF}, it is also needed to develop the HK functional $F_{\text{HF}}^{\text{HK}}$ for the HF electron density as an argument. In the following, we first prove the HK theorem for the HF density in completely parallel to the original HK theorem. Then, the PFT for the HF energy will be formulated.

C.1 Hohenberg-Kohn Theorem for Hartree-Fock Electron Density

In the proof of the HK theorem given in Sect. 1.3.1, we consider a full Hamiltonian $\hat{\mathbf{H}}$ of an interacting system with N electrons. As shown in Eq. (1.68), $\hat{\mathbf{H}}$ is associated with an external potential $\hat{\upsilon}_{\text{ext}}$, while the full Hamiltonian $\hat{\mathbf{H}}'$ in Eq. (1.70) includes another external potential $\hat{\upsilon}'_{\text{ext}}$. The major difference of the proof of the HK theorem for the HF energy from that of the original theorem is that the wave functions are confined within a subset of single Slater determinants Ψ_{SD}, and hence it is not possible to specify the eigenfunction for the Hamiltonian $\hat{\mathbf{H}}$ or $\hat{\mathbf{H}}'$ within the subset. Instead of assuming that the wave function is the eigenfunction for the Hamiltonian, we suppose that a Slater determinant Ψ_{SD} minimizes the expectation value $\langle \Psi_{\text{SD}} | \hat{\mathbf{H}} | \Psi_{\text{SD}} \rangle$, which leads to E_0^{HF}. Similarly, it is supposed that Ψ'_{SD} minimizes $\langle \Psi'_{\text{SD}} | \hat{\mathbf{H}}' | \Psi'_{\text{SD}} \rangle$ providing $E_0'^{\text{HF}}$.

H. Takahashi, Density-Functional Theory for Electrons,
https://doi.org/10.1007/978-981-95-1229-4

With these preparations, the proof using the 'proof by contradiction' can be made in parallel to the original HK theorem. To do this we assume that the different external potentials $\hat{\upsilon}_{\text{ext}}$ and $\hat{\upsilon}'_{\text{ext}}$ yield the same HF electron density n_0^{HF}. That is, Ψ_{SD} and Ψ'_{SD} provides the same electron density n_0^{HF}. Of course, it is presumed that $\hat{\upsilon}_{\text{ext}}$ differs from $\hat{\upsilon}'_{\text{ext}}$ more than a constant shift and that the wave functions are not degenerated. Then, from the variational principle, the following inequality holds,

$$\begin{aligned} E_0^{\text{HF}} = \langle \Psi_{\text{SD}} | \hat{\mathbf{H}} | \Psi_{\text{SD}} \rangle < \langle \Psi'_{\text{SD}} | \hat{\mathbf{H}} | \Psi'_{\text{SD}} \rangle &= E_0'^{\text{HF}} + \langle \Psi'_{\text{SD}} | \hat{\mathbf{H}} - \hat{\mathbf{H}}' | \Psi'_{\text{SD}} \rangle \\ &= E_0'^{\text{HF}} + \int d\boldsymbol{r} n_0^{\text{HF}}(\boldsymbol{r}) \left(\upsilon_{\text{ext}}(\boldsymbol{r}) - \upsilon'_{\text{ext}}(\boldsymbol{r}) \right) \end{aligned} \tag{C.1}$$

In deriving the last equality, we use the assumption mentioned above. For the energy $E_0'^{\text{HF}}$, we get a similar inequality

$$\begin{aligned} E_0'^{\text{HF}} = \langle \Psi'_{\text{SD}} | \hat{\mathbf{H}}' | \Psi'_{\text{SD}} \rangle < \langle \Psi_{\text{SD}} | \hat{\mathbf{H}}' | \Psi_{\text{SD}} \rangle &= E_0 + \langle \Psi_{\text{SD}} | \hat{\mathbf{H}}' - \hat{\mathbf{H}} | \Psi_{\text{SD}} \rangle \\ &= E_0^{\text{HF}} + \int d\boldsymbol{r} n_0^{\text{HF}}(\boldsymbol{r}) \left(\upsilon'_{\text{ext}}(\boldsymbol{r}) - \upsilon_{\text{ext}}(\boldsymbol{r}) \right) \end{aligned} \tag{C.2}$$

Sum of these Eqs. (C.1) and (C.2) leads to a contradiction,

$$E_0^{\text{HF}} + E_0'^{\text{HF}} < E_0'^{\text{HF}} + E_0^{\text{HF}} \tag{C.3}$$

Therefore, the HF electron density n_0^{HF} uniquely determines the corresponding external potential $\hat{\upsilon}_{\text{ext}}$ and also the Slater determinant Ψ_{SD}.

Now the HK functional $F_{\text{HF}}^{\text{HK}}[n]$ for the HF ground state density n as an argument can be defined. $F_{\text{HF}}^{\text{HK}}[n]$ is explicitly defined as

$$F_{\text{HF}}^{\text{HK}}[n] = \langle \Psi_{\text{SD}}[n] | \hat{\mathbf{T}} + \hat{\mathbf{V}}_{ee} | \Psi_{\text{SD}}[n] \rangle \tag{C.4}$$

where $\Psi_{\text{SD}}[n]$ represents the Slater determinant corresponding to the density n. It is quite important to note that the domain of definition of the functional $F_{\text{HF}}^{\text{HK}}[n]$ is not the υ-representable density, but the 'HF non-interacting' υ-representable density. Otherwise, Ψ_{SD} cannot be identified for the density n. For later discussion, it is useful to define the HK functional $E_{\text{HF}}^{\text{HK}}[n]$ for the HF total energy, that is,

$$E_{\text{HF}}^{\text{HK}}[n] = F_{\text{HF}}^{\text{HK}}[n] + \int d\boldsymbol{r}\, n(\boldsymbol{r})\, \upsilon_{\text{ext}}(\boldsymbol{r}) \tag{C.5}$$

Then, the HF ground state energy E_0^{HF} can be given by

$$E_0^{\text{HF}} = \min_{n \to \text{HF non-int.}} E_{\text{HF}}^{\text{HK}}[n] \tag{C.6}$$

where the search of the electron density n is constrained within the set of HF non-interacting υ-representable densities.

C.2 Potential Functional for Hartree-Fock Energy

The following discussions to formulate the potential functionals $E_{\mathrm{PF}}^{\mathrm{HF}}[w_{\mathrm{ext}}]$ and $E_{\mathrm{PF0}}^{\mathrm{HF}}[w_{\mathrm{KS}}]$ are parallel to those for the functionals developed by Yang et al. [1] $E_{\mathrm{PF}}^{\mathrm{HF}}[w_{\mathrm{ext}}]$ represents the PF for an external potential $w_{\mathrm{ext}}(\boldsymbol{r})$, while $E_{\mathrm{PF0}}^{\mathrm{HF}}[w_{\mathrm{KS}}]$ stands for the PF for a KS (effective) potential w_{KS} of a non-interacting reference system. As shown in Eq. (C.5) the HK functional $E_{\mathrm{HF}}^{\mathrm{HK}}[n]$ for the HF ground state energy is specific to an external potential υ_{ext}. Thus, the subscript υ_{ext} will be attached to the functionals hereafter to avoid confusions. By exploiting the functional $E_{\mathrm{HF},\upsilon_{\mathrm{ext}}}^{\mathrm{HK}}[n]$, the potential functional $E_{\mathrm{PF},\upsilon_{\mathrm{ext}}}^{\mathrm{HF}}[w_{\mathrm{ext}}]$ is defined as

$$E_{\mathrm{PF},\upsilon_{\mathrm{ext}}}^{\mathrm{HF}}[w_{\mathrm{ext}}] = E_{\mathrm{HF},\upsilon_{\mathrm{ext}}}^{\mathrm{HK}}\left[n_{w_{\mathrm{ext}}}\right] \tag{C.7}$$

Note that the density $n_{w_{\mathrm{ext}}}$ is built from the Slater determinant $\Psi_{\mathrm{SD}}[w_{\mathrm{ext}}]$ that minimizes the expectation value of the full Hamiltonian with the external potential $w_{\mathrm{ext}}(\boldsymbol{r})$. It is easy to see that the HF ground state energy $E_{0,\upsilon_{\mathrm{ext}}}^{\mathrm{HF}}$ can be obtained by the minimization with respect to the external potential w_{ext}

$$E_{0,\upsilon_{\mathrm{ext}}}^{\mathrm{HF}} = \min_{w_{\mathrm{ext}}} E_{\mathrm{PF},\upsilon_{\mathrm{ext}}}^{\mathrm{HF}}[w_{\mathrm{ext}}] \tag{C.8}$$

Of course, the minimum is obtained when $w_{\mathrm{ext}} = \upsilon_{\mathrm{ext}}$ is attained in the minimization.

Next we consider a non-interacting system specified by the Kohn-Sham (KS) equation. In the following, the subscript 'KS' can be equivalently replaced by 'eff', that is, an abbreviation of the term 'effective'. Suppose that n_{KS} is a non-interacting υ-representable density and $w_{\mathrm{KS}}(\boldsymbol{r})$ is the effective potential that uniquely corresponds to n_{KS}. Then, the KS equation can be written as

$$\left(-\frac{1}{2}\nabla^2 + w_{\mathrm{KS}}[n_{\mathrm{KS}}](\boldsymbol{r})\right)\varphi_i(\boldsymbol{r}) = \epsilon_i\varphi_i(\boldsymbol{r}) \tag{C.9}$$

where n_{KS} is given by $n_{\mathrm{KS}}(\boldsymbol{r}) = \sum_i^N \varphi_i(\boldsymbol{r})\varphi_i^*(\boldsymbol{r})$ with N the number of electrons in the system. Assuming that the HF ground state density n_0^{HF} is non-interacting υ-representable, the energy $E_{0,\upsilon_{\mathrm{ext}}}^{\mathrm{HF}}$ can be obtained by the minimization with respect to the density n_{KS}

$$E_{0,\upsilon_{\mathrm{ext}}}^{\mathrm{HF}} = \min_{n_{\mathrm{KS}}} E_{\mathrm{HF},\upsilon_{\mathrm{ext}}}^{\mathrm{HK}}[n_{\mathrm{KS}}] \tag{C.10}$$

where the minimum is achieved when n_{KS} coincides with n_0^{HF}. The potential functional $E_{\mathrm{PF0},\upsilon_{\mathrm{ext}}}^{\mathrm{HF}}[w_{\mathrm{KS}}]$ can then be defined as

$$E_{\mathrm{PF0},\upsilon_{\mathrm{ext}}}^{\mathrm{HF}}[w_{\mathrm{KS}}] = E_{\mathrm{HF},\upsilon_{\mathrm{ext}}}^{\mathrm{HK}}[n_{\mathrm{KS}}] \tag{C.11}$$

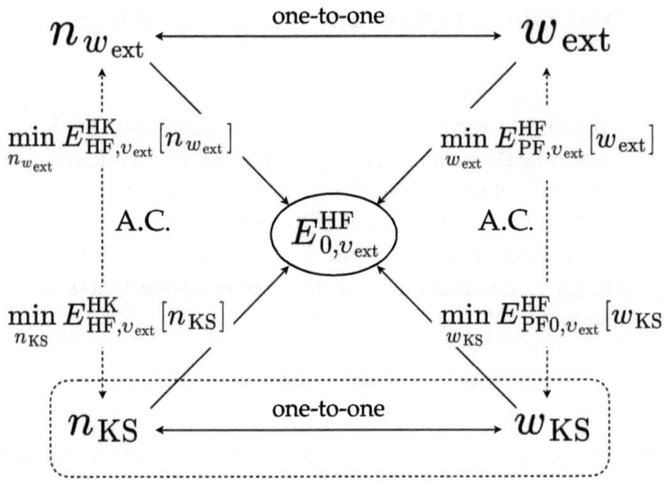

Fig. C.1 Potential functional (right) dual to the density functional (left) for the Hartree-Fock (HF) total energy. The upper line with arrows on the both ends represents the one-to-one mapping between the electron density $n_{w_{\rm ext}}$ and the external potential $w_{\rm ext}$, while the lower line stands for the one-to-one mapping between the non-interacting v-representable density $n_{\rm KS}$ and the Kohn-Sham (KS) effective potential $w_{\rm KS}$. Each arrow pointing the HF ground state energy $E^{\rm HF}_{0,v_{\rm ext}}$ represents the minimization of the potential functional or density functional. 'A.C.' stands for adiabatic connection between the interacting and the non-interacting systems. We refer readers also to Fig. 1 in Ref. [1]

where the density $n_{\rm KS}$ is uniquely derived from the KS potential $w_{\rm KS}$ given as an argument by virtue of the one-to-one mapping. It should be noted that the functional $E^{\rm HF}_{{\rm PF0},v_{\rm ext}}[w_{\rm KS}]$ is different from the functional $E^{\rm HF}_{{\rm PF},v_{\rm ext}}[w_{\rm ext}]$ in the point that the former employs the effective potential $w_{\rm KS}$ as an argument and the latter uses the external potential. Then, it is possible to obtain the energy $E^{\rm HF}_{0,v_{\rm ext}}$ through the minimization

$$E^{\rm HF}_{0,v_{\rm ext}} = \min_{w_{\rm KS}} E^{\rm HF}_{{\rm PF0},v_{\rm ext}}[w_{\rm KS}] \tag{C.12}$$

The minimum energy is achieved when the KS potential $w_{\rm KS}$ becomes the potential which provides the density $n_{\rm KS} = n_0^{\rm HF}$ corresponding to the external potential $v_{\rm ext}$. In Fig. C.1, we summarize the relationship between the density functional and the potential functional for the HF ground state energy $E^{\rm HF}_{0,v_{\rm ext}}$.

Appendix D
Thomas-Fermi Kinetic Energy Functional

The kinetic energy (KE) functional given by Thomas and Fermi [2, 3] for the homogeneous electron gas (HEG) served as the basis for the subsequent functional developments. In this Appendix we provide the derivation of the explicit functional form. The followings are the detailed derivation of Thomas-Fermi (TF) model almost in parallel to those provided in the textbook by Parr and Yang [4] First, we assume that the non-interacting homogeneous electron gas with density n_0 is enclosed in a cubic box with the periodic boundary condition. Then, the 1st-order density matrix $n_1\left(\boldsymbol{r}_1, \boldsymbol{r}_2\right)$ can be expressed in terms of the plane wave basis given by $\Omega^{-1/2}\exp(i\boldsymbol{k}_i \cdot \boldsymbol{r})$ where Ω is the volume of the cubic cell with size L, namely, $\Omega = L^3$ and $\boldsymbol{k}_i$ represents a wave number vector. Provided that γ_i is the occupation number of the ith basis, the density matrix n_1 is expressed as

$$n_1\left(\boldsymbol{r}_1, \boldsymbol{r}_2\right) = \Omega^{-1} \sum_i \gamma_i \exp\left(i\boldsymbol{k}_i \cdot \left(\boldsymbol{r}_1 - \boldsymbol{r}_2\right)\right) \tag{D.1}$$

Assuming the closed shell structure of the electrons in the cell, the occupation numbers γ_i are 2 and 0 for the occupied and unoccupied basis functions, respectively. For later discussions, it is important to note that the momentum of the wave function identified by $\boldsymbol{k}_i$ is discrete since it is enclosed within a cell with the finite size. Then, a wave number vector $\boldsymbol{k}_i$ can be related to the set of integers (n_x, n_y, n_z) $(n_x, n_y,$ and $n_z = 0, \pm 1, \pm 2, \cdots)$, thus,

$$k_{ix} = \frac{2\pi}{L}n_x,\ k_{iy} = \frac{2\pi}{L}n_y,\ k_{iz} = \frac{2\pi}{L}n_z \tag{D.2}$$

H. Takahashi, Density-Functional Theory for Electrons,
https://doi.org/10.1007/978-981-95-1229-4

where k_{ix} is the x component of the vector $\boldsymbol{k}_i$. The summation over i in Eq. (D.1) can be replaced by that over (n_x, n_y, n_z), which reads

$$n_1(\boldsymbol{r}_1, \boldsymbol{r}_2) = \Omega^{-1} \sum_{n_x, n_y, n_z} \gamma_{n_x, n_y, n_z} \exp\left(i\frac{2\pi}{L}(n_x x_1 + n_y y_1 + n_z z_1)\right) \times \exp\left(-i\frac{2\pi}{L}(n_x x_2 + n_y y_2 + n_z z_2)\right) \quad \text{(D.3)}$$

where it is assumed that $\boldsymbol{r}_1 = (x_1, y_1, z_1)$ and $\boldsymbol{r}_2 = (x_2, y_2, z_2)$. When the number of the occupied wave functions are very large, it is possible to replace the summation with an integration with respect to $d\boldsymbol{n} = dn_x dn_y dn_z$,

$$n_1(\boldsymbol{r}_1, \boldsymbol{r}_2) = 2\Omega^{-1} \int_{\Gamma} d\boldsymbol{n} \exp\left(i\frac{2\pi}{L}(n_x x_1 + n_y y_1 + n_z z_1)\right) \times \exp\left(-i\frac{2\pi}{L}(n_x x_2 + n_y y_2 + n_z z_2)\right) \quad \text{(D.4)}$$

Note that the integration shall be performed within the region Γ where $\gamma_{n_x, n_y, n_z} = 2$ holds. Now we change the variable in the integration from $\boldsymbol{n}$ to $\boldsymbol{k}$. Since $k_x = \frac{2\pi}{L} n_x$ holds, we get the relation

$$d\boldsymbol{n} = \frac{L^3}{(2\pi)^3} d\boldsymbol{k} \quad \text{(D.5)}$$

Then, we obtain the expression of the density matrix using the vector $\boldsymbol{k}$,

$$n_1(\boldsymbol{r}_1, \boldsymbol{r}_2) = \frac{1}{4\pi^3} \int_{\omega} d\boldsymbol{k} \exp(i\boldsymbol{k} \cdot (\boldsymbol{r}_1 - \boldsymbol{r}_2)) \quad \text{(D.6)}$$

where ω is the region of $\boldsymbol{k}$ defined by $\omega := \{\boldsymbol{k} \in \omega : 0 \leq |\boldsymbol{k}| \leq k_F\}$ with k_F being the Fermi wave number. For the homogeneous electron gas (HEG) with the density $n(\boldsymbol{r})$ is specified using k_F, thus,

$$n(\boldsymbol{r}) = n_1(\boldsymbol{r}, \boldsymbol{r}) = \frac{1}{4\pi^3} \frac{4}{3} \pi k_F(\boldsymbol{r})^3 = \frac{1}{3\pi^2} k_F(\boldsymbol{r})^3 \quad \text{(D.7)}$$

Actually this relation can be obtained from a much simpler discussion. It is assumed for a spatial region with a volume ΔV, that the momentum eigenfunctions up to k_F are doubly occupied. Then, the electron density n of the cell with the volume ΔV can be evaluated as

$$n = \frac{1}{h^3} \frac{2}{\Delta V} \int_0^{k_F} dk\, 4\pi k^2 \times \Delta V = \frac{1}{\hbar^3} \frac{k_F^3}{3\pi^2} \quad \text{(D.8)}$$

where the use of atomic units is avoided for clarity of discussion and the factor $1/h^3$ means dividing by the unit volume of the phase space. Now we are ready to perform the integration of Eq. (D.6). The integration is quite common and useful, and it leads to a famous equation that can be related to the spherical Bessel function of the first kind. For the given vectors $\boldsymbol{r}_1$ and $\boldsymbol{r}_2$ in Eq. (D.6), the vector $\boldsymbol{s}$ defined by

$$\boldsymbol{s} = \boldsymbol{r}_1 - \boldsymbol{r}_2 \tag{D.9}$$

is taken as the z axis of the $\boldsymbol{k}$ vector. Then, the integration in Eq. (D.6) can be rewritten in terms of the spherical coordinates for $\boldsymbol{k}$,

$$\begin{aligned} n_1(\boldsymbol{r}_1, \boldsymbol{r}_2) &= \frac{1}{4\pi^3} \int_{\omega} d\boldsymbol{k} \exp\left(i\boldsymbol{k} \cdot (\boldsymbol{r}_1 - \boldsymbol{r}_2)\right) \\ &= \frac{1}{4\pi^3} \int_0^{k_F} dk\, k^2 \int_0^{2\pi} d\phi \int_0^{\pi} d\theta \ \sin\theta \exp\left(iks\cos\theta\right) \end{aligned} \tag{D.10}$$

where s is the magnitude of the vector $\boldsymbol{s}$ and k is that of $\boldsymbol{k}$. By substituting $t = \cos\theta$, the equation can be further processed as

$$\begin{aligned} n_1(\boldsymbol{r}_1, \boldsymbol{r}_2) &= \frac{1}{2\pi^2} \int_0^{k_F} dk\, k^2 \int_{-1}^{1} dt \ \exp(ikst) \\ &= \frac{1}{2\pi^2} \int_0^{k_F} dk\, k^2 \int_{-1}^{1} dt \ (iks)^{-1} \left(\exp(ikst)\right)' \\ &= \frac{1}{\pi^2} \int_0^{k_F} dk \ \frac{k}{s} \sin(ks) \\ &= \frac{1}{\pi^2 s^3} \left(\sin(sk_F) - sk_F \cos(sk_F)\right) \\ &= \frac{k_F^3}{\pi^2} \frac{\sin p - p\cos p}{p^3} = 3n \frac{\sin p - p\cos p}{p^3} \end{aligned} \tag{D.11}$$

In the above equations, we introduce the variable $p(\boldsymbol{r}) = sk_F(\boldsymbol{r})$ and Eq. (D.7) is used. From Eq. (D.11), by the way, the exchange hole function $n_x^{\text{HEG}}(\boldsymbol{r}_1)$ at $\boldsymbol{r}_1$ of the HEG can be readily obtained as

$$n_x^{\text{HEG}}(\boldsymbol{r}_1) = -\frac{1}{n(\boldsymbol{r}_1)} \left| n_1(\boldsymbol{r}_1, \boldsymbol{r}_2) \right|^2 = -9n \left(\frac{\sin p - p\cos p}{p^3} \right)^2 \tag{D.12}$$

Using this equation, it is also possible to derive the formula of the local density approximation (LDA) for the exchange energy. We refer readers to Ref. [4] for the detailed derivation.

Here we define the vector $\boldsymbol{r}$ in Eq. (D.7), thus,

$$\boldsymbol{r} = \frac{1}{2}\left(\boldsymbol{r}_1 + \boldsymbol{r}_2\right) \tag{D.13}$$

The coordinates $\boldsymbol{s}$ and $\boldsymbol{r}$ defined, respectively, by Eqs. (D.9) and (D.13) provides the coordinates equivalent to $\boldsymbol{r}_1$ and $\boldsymbol{r}_2$.

The kinetic energy at $\boldsymbol{r}_1$ can be expressed using the 1st-order density matrix as

$$T\left[n_1\right]\left(\boldsymbol{r}_1\right) = -\frac{1}{2}\left[\nabla_{\boldsymbol{r}_1}^2 n_1\left(\boldsymbol{r}_1, \boldsymbol{r}_2\right)\right]_{\boldsymbol{r}_2=\boldsymbol{r}_1} \tag{D.14}$$

The density matrix in Eq. (D.11) for HEG is substituted to Eq. (D.14) to obtain the explicit form of the TF model. To this end, the coordinate $\boldsymbol{r}$ and $\boldsymbol{s}$ are expressed with their components,

$$\boldsymbol{r} = \left(r_x, r_y, r_z\right) = \frac{1}{2}\left(x_1 + x_2, y_1 + y_2, z_1 + z_2\right) \tag{D.15}$$

$$\boldsymbol{s} = \left(s_x, s_y, s_z\right) = \left(x_1 - x_2, y_1 - y_2, z_1 - z_2\right) \tag{D.16}$$

where it is assumed that $\boldsymbol{r}_1 = (x_1, y_1, z_1)$ and similarly for $\boldsymbol{r}_2$. Hence, the differential operator $\frac{\partial}{\partial x_1}$ can be expressed as

$$\frac{\partial}{\partial x_1} = \frac{\partial r_x}{\partial x_1}\frac{\partial}{\partial r_x} + \frac{\partial s_x}{\partial x_1}\frac{\partial}{\partial s_x} = \frac{1}{2}\frac{\partial}{\partial r_x} + \frac{\partial}{\partial s_x} \tag{D.17}$$

and for the operator $\frac{\partial^2}{\partial x_1^2}$ we get

$$\frac{\partial^2}{\partial x_1^2} = \frac{1}{4}\frac{\partial^2}{\partial r_x^2} + \frac{\partial}{\partial r_x}\frac{\partial}{\partial s_x} + \frac{\partial^2}{\partial s_x^2} \tag{D.18}$$

Applying the same procedures to the variables y_1 and z_1, the Laplacian $\nabla_{\boldsymbol{r}_1}^2$ in Eq. (D.14) can be summarized as

$$\nabla_{\boldsymbol{r}_1}^2 = \frac{1}{4}\nabla_{\boldsymbol{r}}^2 + \nabla_{\boldsymbol{r}}\nabla_{\boldsymbol{s}} + \nabla_{\boldsymbol{s}}^2 \tag{D.19}$$

Below we consider the individual operations of these three terms on the density matrix in Eq. (D.11). The x-component of the gradient $\nabla_{\boldsymbol{r}} n_1(\boldsymbol{r}_1, \boldsymbol{r}_2)$ is given by

$$\begin{aligned}\frac{\partial}{\partial r_x} n_1\left(\boldsymbol{r}_1, \boldsymbol{r}_2\right) = {} & 3\left(\frac{\partial}{\partial r_x} n\left(\boldsymbol{r}\right)\right) \times \frac{\sin p - p\cos p}{p^3} \\ & + 3n\left(\boldsymbol{r}\right)\frac{\partial p\left(\boldsymbol{r}\right)}{\partial r_x}\frac{\partial}{\partial p}\frac{\sin p - p\cos p}{p^3}\end{aligned} \tag{D.20}$$

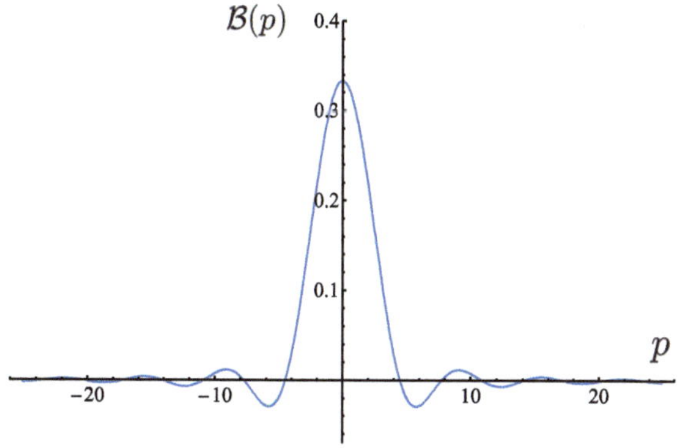

Fig. D.1 Function $\mathcal{B}(p)$ defined in Eq. (D.21), which describes the behavior of the density matrix $n_1(\boldsymbol{r}_1, \boldsymbol{r}_2)$ of the homogeneous electron gas as function of $s = |\boldsymbol{r}_1 - \boldsymbol{r}_2|$

Now we define the function $\mathcal{B}(p)$ in the density matrix for notational simplicity

$$\mathcal{B}(p) = \frac{\sin p - p \cos p}{p^3} \tag{D.21}$$

The function $\mathcal{B}(p)$ is drawn in Fig. D.1 for reader's references.

The first and second derivatives of $\mathcal{B}(p)$ become

$$\mathcal{B}'(p) = \frac{\left(p^2 - 3\right)\sin p + 3p\cos p}{p^4} \tag{D.22}$$

$$\mathcal{B}''(p) = \frac{\cos p}{p^2} - \frac{5\sin p}{p^3} + \frac{12(\sin p - p\cos p)}{p^5} \tag{D.23}$$

When $s \to 0$, p goes to zero by definition. And the functions, $\mathcal{B}(0)$, $\mathcal{B}'(0)$, and $\mathcal{B}''(0)$ are obtained as

$$\begin{cases} \mathcal{B}(0) & = \frac{1}{3} \\ \mathcal{B}'(0) & = 0 \\ \mathcal{B}''(0) & = -\frac{1}{15} \end{cases} \tag{D.24}$$

These values can be obtained by applying the Taylor expansions for sin and cos functions around $p = 0$ in Eqs. (D.21)–(D.23). Anyway, the second derivative of $n_1(\boldsymbol{r}_1, \boldsymbol{r}_2)$ with respect to r_x becomes

$$\begin{aligned} \frac{\partial^2}{\partial r_x^2} n_1(\boldsymbol{r}_1, \boldsymbol{r}_2) &= 3\left(\frac{\partial^2}{\partial r_x^2} n(\boldsymbol{r})\right)\mathcal{B}(p) + 3\left(\frac{\partial}{\partial r_x} n(\boldsymbol{r})\right)\frac{\partial p}{\partial r_x}\mathcal{B}'(p) \\ &+ 3\left(\frac{\partial}{\partial r_x} n(\boldsymbol{r})\right)\frac{\partial p(\boldsymbol{r})}{\partial r_x}\mathcal{B}'(p) + 3n(\boldsymbol{r})\frac{\partial^2 p(\boldsymbol{r})}{\partial r_x^2}\mathcal{B}'(p) \\ &+ 3n(\boldsymbol{r})\left(\frac{\partial p(\boldsymbol{r})}{\partial r_x}\right)^2 \mathcal{B}''(p) \end{aligned} \tag{D.25}$$

Note that the derivatives $\frac{\partial p(\boldsymbol{r})}{\partial r_x}$ and $\frac{\partial^2 p(\boldsymbol{r})}{\partial r_x^2}$ in Eq. (D.25) can be transformed to

$$\frac{\partial p(\boldsymbol{r})}{\partial r_x} = s \frac{\partial k_F(\boldsymbol{r})}{\partial r_x} = s \cdot \frac{1}{3}\left(3\pi^2\right)^{1/3} n(\boldsymbol{r})^{-2/3} \tag{D.26}$$

and

$$\begin{aligned} \frac{\partial^2 p(\boldsymbol{r})}{\partial r_x^2} &= -s \cdot \frac{2}{9}\left(3\pi^2\right)^{1/3} n(\boldsymbol{r})^{-5/3}\left(\frac{\partial n(\boldsymbol{r})}{\partial r_x}\right)^2 \\ &\quad + s \cdot \frac{1}{3}\left(3\pi^2\right)^{1/3} n(\boldsymbol{r})^{-2/3} \frac{\partial^2 n(\boldsymbol{r})}{\partial r_x^2} \end{aligned} \tag{D.27}$$

Thus, it is shown that the terms in Eq. (D.25) include the first or second derivative of the density $n(\boldsymbol{r})$ with respect to r_x. Since the density matrix is for the HEG, all the terms in Eq. (D.25) vanish. Actually, the same rule also applies to the second term in Eq. (D.19). Therefore, only the third term in Eq. (D.19) remains in the operation of the Laplacian. To be precise, the third term ∇_s^2 is written as

$$\nabla_s^2 = \frac{\partial^2}{\partial s_x^2} + \frac{\partial^2}{\partial s_y^2} + \frac{\partial^2}{\partial s_z^2} \tag{D.28}$$

Since the density matrix of HEG is only dependent on the magnitude s of the vector $\boldsymbol{s}$, The Laplacian ∇_s^2 can be reduced to

$$\nabla_s^2 = \frac{\partial^2}{\partial s^2} + \frac{2}{s}\frac{\partial}{\partial s} \tag{D.29}$$

The operation of the second derivative in the Laplacian on the density matrix becomes

$$\begin{aligned} \left.\frac{\partial^2}{\partial s^2} n_1(\boldsymbol{r}_1, \boldsymbol{r}_2)\right|_{s=0} &= \left.\frac{\partial}{\partial s}\left(\frac{\partial}{\partial s} 3n(\boldsymbol{r})\mathcal{B}(p)\right)\right|_{s=0} \\ &= \left.3n(\boldsymbol{r})\frac{\partial}{\partial s}\left(\frac{\partial p}{\partial s}\mathcal{B}'(p)\right)\right|_{s=0} \\ &= \left.3n(\boldsymbol{r})k_F\frac{\partial}{\partial s}\mathcal{B}'(p)\right|_{s=0} \\ &= \left.3n(\boldsymbol{r})k_F^2\,\mathcal{B}''(p)\right|_{s=0} \\ &= -\frac{1}{5}\left(3\pi^2\right)^{\frac{2}{3}} n(\boldsymbol{r})^{\frac{5}{3}} \end{aligned} \tag{D.30}$$

The subscript $s = 0$ means that s goes to zero after the operation of the Laplacian is performed. This procedure corresponds to the manipulation of $\boldsymbol{r}_2 = \boldsymbol{r}_1$ in Eq. (D.14).

We also note that the last relation in Eq. (D.24) is utilized in deriving the last equality Eq. (D.30). Next, we consider the second term on the right hand side of Eq. (D.29). It reads

$$\begin{aligned}\left.\frac{2}{s}\frac{\partial}{\partial s}n_1(\boldsymbol{r}_1,\boldsymbol{r}_2)\right|_{s=0} &= 6n(\boldsymbol{r})\left.\frac{1}{s}\frac{\partial p}{\partial s}\mathcal{B}'(p)\right|_{s=0}\\ &= 6n(\boldsymbol{r})\,k_F^2\left.\frac{1}{p}\frac{\left(p^2-3\right)\sin p+3p\cos p}{p^4}\right|_{s=0}\\ &= -\frac{2}{5}\left(3\pi^2\right)^{\frac{2}{3}}n(\boldsymbol{r})^{\frac{5}{3}}\end{aligned} \tag{D.31}$$

In deriving the last equality of Eq. (D.31), the Taylor expansions for the trigonometric functions are utilized. Finally, by summing up the contributions from Eqs. (D.30) and (D.31), we obtain the TF model of the kinetic energy, thus,

$$\begin{aligned}T_{\mathrm{TF}}[n](\boldsymbol{r}) &= -\frac{1}{2}\left.\nabla_s^2 n_1(\boldsymbol{r}_1,\boldsymbol{r}_2)\right|_{s=0}\\ &= -\frac{1}{2}\left(\frac{\partial^2}{\partial s^2}+\frac{2}{s}\frac{\partial}{\partial s}\right)\left.n_1(\boldsymbol{r}_1,\boldsymbol{r}_2)\right|_{s=0}\\ &= \frac{3}{10}\left(3\pi^2\right)^{\frac{2}{3}}n(\boldsymbol{r})^{\frac{5}{3}}\end{aligned} \tag{D.32}$$

This is the electronic kinetic energy at $\boldsymbol{r}$ of the HEG with density $n = n(\boldsymbol{r})$. Applying Eq. (D.32) to an inhomogeneous electrons with density $n(\boldsymbol{r})$, we get the approximate kinetic energy functional $E_{\mathrm{kin}}^{\mathrm{TF}}$ based on the TF model, thus,

$$E_{\mathrm{kin}}^{\mathrm{TF}}[n] = C_{\mathrm{TF}}\int d\boldsymbol{r}\, n(\boldsymbol{r})^{\frac{5}{3}} \tag{D.33}$$

with $C_{\mathrm{TF}} = \frac{3}{10}\left(3\pi^2\right)^{\frac{2}{3}}$.

Appendix E
Lindhard Function

The linear response function of the homogeneous electron gas was first formulated by Lindhard [5] and it plays an important role in the discussion of Chap. 5. Especially, in deriving the Wang-Teter functional in Sect. 5.1.2, the Lindhard function $\widetilde{\chi}_{\text{Lind}}(\boldsymbol{q})$ is directly utilized. Here we provide the derivation of the function in detail. To do this, suppose that a small perturbation $\upsilon(\boldsymbol{r})$ with a wave vector $\boldsymbol{q}$ is applied to the electron gas. The perturbative potential $\upsilon(\boldsymbol{r})$ can be expressed as

$$\upsilon(\boldsymbol{r}) = \overline{\upsilon}(\boldsymbol{q}) \exp(i\boldsymbol{q} \cdot \boldsymbol{r}) + \overline{\upsilon}(\boldsymbol{q})^* \exp(-i\boldsymbol{q} \cdot \boldsymbol{r}) \tag{E.1}$$

Note that $\upsilon(\boldsymbol{r})$ is a real function by definition and it represents a monotonic wave with the wave length $\frac{2\pi}{|\boldsymbol{q}|}$. Using the first order perturbation theory, the perturbed wave function $\varphi_{\boldsymbol{k}}^{(1)}(\boldsymbol{r})$ for the wave function $\varphi_{\boldsymbol{k}}(\boldsymbol{r})$ with a wave vector $\boldsymbol{k}$ is given by

$$\varphi_{\boldsymbol{k}}^{(1)}(\boldsymbol{r}) = \varphi_{\boldsymbol{k}}(\boldsymbol{r}) + \sum_{\boldsymbol{k}' \neq \boldsymbol{k}} \frac{E_{\boldsymbol{k}'\boldsymbol{k}}}{\epsilon_{\boldsymbol{k}} - \epsilon_{\boldsymbol{k}'}} \varphi_{\boldsymbol{k}'}(\boldsymbol{r}) \tag{E.2}$$

$\varphi_{\boldsymbol{k}}(\boldsymbol{r})$ is the normalized zeroth order wave function and assumed to be a plane wave with wave vector $\boldsymbol{k}$ confined within a periodic cell with length L, which reads

$$\varphi_{\boldsymbol{k}}(\boldsymbol{r}) = L^{-\frac{3}{2}} \exp(i\boldsymbol{k} \cdot \boldsymbol{r}) \tag{E.3}$$

The corresponding eigenvalues $\epsilon_{\boldsymbol{k}}$ is given by

$$\epsilon_{\boldsymbol{k}} = \frac{1}{2} |\boldsymbol{k}|^2 \tag{E.4}$$

The matrix elements $E_{\boldsymbol{k}'\boldsymbol{k}}$ in Eq. (E.2) is written as

H. Takahashi, Density-Functional Theory for Electrons,
https://doi.org/10.1007/978-981-95-1229-4

$$E_{k'k} = L^{-3} \int d\boldsymbol{r} \exp\left(-i\boldsymbol{k}' \cdot \boldsymbol{r}\right) \left(\overline{v}(\boldsymbol{q}) \exp(i\boldsymbol{q} \cdot \boldsymbol{r}) + \overline{v}(\boldsymbol{q})^* \exp(-i\boldsymbol{q} \cdot \boldsymbol{r})\right) \times \exp(i\boldsymbol{k} \cdot \boldsymbol{r}) \tag{E.5}$$

After the integration, we obtain

$$E_{k'k} = \overline{v}(\boldsymbol{q})\, \delta_{k',q+k} + \overline{v}(\boldsymbol{q})^*\, \delta_{k',k-q} \tag{E.6}$$

Inserting this result to Eq. (E.2), we get

$$\varphi_{\boldsymbol{k}}^{(1)}(\boldsymbol{r}) = L^{-\frac{3}{2}} \left(\exp(i\boldsymbol{q} \cdot \boldsymbol{r}) + \frac{2\overline{v}(\boldsymbol{q})}{|\boldsymbol{k}|^2 - |\boldsymbol{k}+\boldsymbol{q}|^2} \exp(i(\boldsymbol{k}+\boldsymbol{q}) \cdot \boldsymbol{r}) + \frac{2\overline{v}(\boldsymbol{q})^*}{|\boldsymbol{k}|^2 - |\boldsymbol{k}-\boldsymbol{q}|^2} \exp(i(\boldsymbol{k}-\boldsymbol{q}) \cdot \boldsymbol{r}) \right) \tag{E.7}$$

Then, the density variation $\delta n_{\boldsymbol{q}}(\boldsymbol{r})$ can be expressed as

$$\delta n_{\boldsymbol{q}}(\boldsymbol{r}) = \sum_{\boldsymbol{k}} f_k \left(\left|\varphi_{\boldsymbol{k}}^{(1)}(\boldsymbol{r})\right|^2 - |\varphi_k(\boldsymbol{r})|^2 \right) \tag{E.8}$$

with f_k being the occupation number for the wave function with the wave vector $\boldsymbol{k}$. Inserting the expression of Eqs. (E.7)–(E.8), and considering up to the first order variation in the wave function, $\delta n_{\boldsymbol{q}}(\boldsymbol{r})$ is explicitly given by

$$\begin{aligned}
\delta n_{\boldsymbol{q}}(\boldsymbol{r}) = L^{-3} \sum_{\boldsymbol{k}} f_k \Bigg\{ & \exp(-i\boldsymbol{k} \cdot \boldsymbol{r}) \left(\frac{2\overline{v}(\boldsymbol{q})}{|\boldsymbol{k}|^2 - |\boldsymbol{k}+\boldsymbol{q}|^2} \exp(i(\boldsymbol{k}+\boldsymbol{q}) \cdot \boldsymbol{r}) + \frac{2\overline{v}(\boldsymbol{q})^*}{|\boldsymbol{k}|^2 - |\boldsymbol{k}-\boldsymbol{q}|^2} \exp(i(\boldsymbol{k}-\boldsymbol{q}) \cdot \boldsymbol{r}) \right) \\
& + \exp(i\boldsymbol{k} \cdot \boldsymbol{r}) \left(\frac{2\overline{v}(\boldsymbol{q})^*}{|\boldsymbol{k}|^2 - |\boldsymbol{k}+\boldsymbol{q}|^2} \exp(-i(\boldsymbol{k}+\boldsymbol{q}) \cdot \boldsymbol{r}) + \frac{2\overline{v}(\boldsymbol{q})}{|\boldsymbol{k}|^2 - |\boldsymbol{k}-\boldsymbol{q}|^2} \exp(-i(\boldsymbol{k}-\boldsymbol{q}) \cdot \boldsymbol{r}) \right) \Bigg\} \\
= 2L^{-3} \sum_{\boldsymbol{k}} & \left(\frac{f_k}{|\boldsymbol{k}|^2 - |\boldsymbol{k}+\boldsymbol{q}|^2} + \frac{f_k}{|\boldsymbol{k}|^2 - |\boldsymbol{k}-\boldsymbol{q}|^2} \right) \times (\overline{v}(\boldsymbol{q}) \exp(i\boldsymbol{q} \cdot \boldsymbol{r}) + c.c.)
\end{aligned} \tag{E.9}$$

By definition, the linear response function $\chi_0(\boldsymbol{q})$ of the homogeneous electron gas (HEG) is written as

$$\chi_0(\boldsymbol{q}) = 2L^{-3} \sum_{\boldsymbol{k}} \left(\frac{f_k}{|\boldsymbol{k}|^2 - |\boldsymbol{k}+\boldsymbol{q}|^2} + \frac{f_k}{|\boldsymbol{k}|^2 - |\boldsymbol{k}-\boldsymbol{q}|^2} \right) \tag{E.10}$$

It is important to notice that the second term in Eq. (E.10) is identical to the first one. In fact, it can be readily recognized that the second term represents the wave vector $\boldsymbol{k}$ with the opposite direction to the first, and its summation over $\boldsymbol{k}$ gives exactly the same contribution as the first. Thus, Eq. (E.10) can be reduced to

$$\chi_0(\boldsymbol{q}) = 4L^{-3}\sum_{\boldsymbol{k}} \frac{f_{\boldsymbol{k}}}{|\boldsymbol{k}|^2 - |\boldsymbol{k}+\boldsymbol{q}|^2} \tag{E.11}$$

The discrete summation over the wave vectors $\boldsymbol{k}$ in Eq. (E.11) is to be performed over the 3-dimensional grid points specified with a set of integers (n_x, n_y, n_z) as shown in Eq. (D.2). Then, using the relation of Eq. (D.5), the summation of Eq. (E.11) can be replaced by the integration with respect to $\boldsymbol{k}$, thus,

$$\chi_0(\boldsymbol{q}) = \frac{1}{2\pi^3}\int d\boldsymbol{k}\frac{f_{\boldsymbol{k}}}{|\boldsymbol{k}|^2 - |\boldsymbol{k}+\boldsymbol{q}|^2} \tag{E.12}$$

To perform the integration, we introduce the spherical coordinate (r, θ, ϕ) where the direction of the vector $\boldsymbol{q}$ is set to the z axis, and the angle between the vector $\boldsymbol{k}$ and $\boldsymbol{q}$ is taken as the variable θ. Then, the integration can be expressed as

$$\begin{aligned}\chi_0(\boldsymbol{q}) &= 2\cdot\frac{1}{2\pi^3}\int_0^{k_F} dk \int_0^{2\pi} d\phi \int_0^{\pi} d\theta \frac{k^2\sin\theta}{-q^2 - 2kq\cos\theta} \\ &= \frac{1}{\pi^2}\int_0^{k_F} dk \int_0^{\pi} d\theta \frac{k^2\sin\theta}{-q^2 - 2kq\cos\theta}\end{aligned} \tag{E.13}$$

where k_F is the Fermi wave number and the factor 2 on the right side of the first equality represents the contributions from α and β spins. Introducing a variable $x = \cos\theta$ in Eq. (E.13), we get

$$\begin{aligned}\chi_0(\boldsymbol{q}) &= \frac{1}{\pi^2}\int_0^{k_F} dk \int_{-1}^{1} dx \frac{k^2}{-q^2 - 2kqx} \\ &= \frac{1}{\pi^2}\int_0^{k_F} dk \left(-\frac{k^2}{2kq}\right)\left(\ln\left|-q^2 - 2kq\right| - \ln\left|-q^2 + 2kq\right|\right) \\ &= -\frac{1}{\pi^2}\int_0^{k_F} dk \frac{k}{2q}\ln\left|\frac{q+2k}{q-2k}\right|\end{aligned} \tag{E.14}$$

In deriving this equation, we use a formula

$$\int \frac{1}{a+bx} dx = \frac{1}{b} \ln |a+bx| \tag{E.15}$$

The integration with respect to k in Eq. (E.14) can be performed using a formula

$$\int dx\ x \ln (a+bx) = \frac{1}{2}\left(x^2 - \frac{a^2}{b^2}\right) \ln (a+bx) - \frac{1}{2}\left(\frac{x^2}{2} - \frac{a}{b}x\right) \tag{E.16}$$

It reads

$$\begin{aligned}
\chi_0(\boldsymbol{q}) &= \frac{1}{2\pi^2 q} \int_0^{k_F} dk\ k \ln \left|\frac{q-2k}{q+2k}\right| \\
&= \frac{1}{2\pi^2 q}\left[\frac{1}{2}\left(x^2 - \frac{q^2}{4}\right) \ln \left|\frac{q-2x}{q+2x}\right| - \frac{1}{2}qx\right]_0^{k_F} \\
&= \frac{1}{2\pi^2 q}\left[\frac{1}{2}\left(k_F^2 - \frac{q^2}{4}\right) \ln \left|\frac{\eta-1}{\eta+1}\right| - \frac{1}{2}qk_F\right] \\
&= \frac{1}{\pi^2}\left[\frac{1}{2}k_F\left(\frac{1-\eta^2}{4\eta}\right) \ln \left|\frac{\eta-1}{\eta+1}\right| - \frac{1}{4}k_F\right] \\
&= -\frac{k_F}{2\pi^2}\left[\frac{1}{2} + \left(\frac{1-\eta^2}{4\eta}\right) \ln \left|\frac{1+\eta}{1-\eta}\right|\right]
\end{aligned}$$

where we use the variable η defined by $\eta = \frac{q}{2k_F}$. Finally, we must consider the contribution due to the wave vector $-\boldsymbol{q}$. Obviously, as shown in Eq. (E.10), $-\boldsymbol{q}$ gives the same contribution as $\boldsymbol{q}$. Thus, we get the Lindhard function $\widetilde{\chi}_{\text{Lind}}(\boldsymbol{q})$ introduced in Sect. 5.1.1,

$$\widetilde{\chi}_{\text{Lind}}(\boldsymbol{q}) = -\frac{k_F}{\pi^2}\left[\frac{1}{2} + \left(\frac{1-\eta^2}{4\eta}\right) \ln \left|\frac{\eta+1}{\eta-1}\right|\right] \tag{E.17}$$

Note that the function is utilized in Sect. 5.1.2 to construct the Wang-Teter nonlocal kinetic energy functional.

Appendix F
Real-Space Grid Method

The results of our calculations shown in this book were all provided by conducting the software 'Vmol' for the Kohn-Sham density-functional theory (KS-DFT). The implementation of the Fortran source code was based on the real-space grid (RSG) formalism proposed by Chelikowsky [6, 7]. The 'Vmol' software was extended to the hybrid QM/MM(quantum mechanical/molecular mechanical) method and so far been extensively utilized to various simulations for solutions and biological systems[8–20]. We refer readers to the review of Ref. [21] for the various techniques in the RSG method, and to the book of Ref. [22] for the applications to the transport properties of electrons. This Appendix describes the details of the RSG method used in the 'Vmol' code. To this end, the KS equation shown in Eqs. (1.121) and (1.122) is presented below,

$$\left(-\frac{1}{2}\nabla^2 + \upsilon_{\rm H}[n](\boldsymbol{r}) + \overline{\upsilon}_{xc}[n](\boldsymbol{r})\right)\varphi_i(\boldsymbol{r}) + \int d\boldsymbol{r}'\upsilon_{\rm ps}\left(\boldsymbol{r},\boldsymbol{r}'\right)\varphi_i\left(\boldsymbol{r}'\right) = \epsilon_i\varphi_i(\boldsymbol{r}) \tag{F.1}$$

where $n(\boldsymbol{r})$ represents the electron density at the electron coordinate $\boldsymbol{r}$, and $\{\varphi_i\}(i = 1, 2, \cdots, N/2)$ constitute the set of one-electron wave functions that provide the density of N electrons. Note that the effective potential in Eq. (F.1) differs from Eq. (1.122) in the point that the local external potential $\upsilon_{\rm ext}(\boldsymbol{r})$ in Eq. (1.122) is being replaced by the nonlocal pseudopotential $\upsilon_{\rm ps}\left(\boldsymbol{r},\boldsymbol{r}'\right)$. This is due to the fact that the RSG is not able to describe the rapid behaviors of the bare Coulomb potentials of the nuclei and the resultant electronic wave functions. The use of the pseudopotentials is essential in the RSG approach except for the case where the adaptive grids [21, 23] are employed. In the following description, it is supposed that the uniform grids are placed in the rectangular cell.

H. Takahashi, Density-Functional Theory for Electrons,
https://doi.org/10.1007/978-981-95-1229-4

F.1 Finite Difference Method for Kinetic Energy Operator

The Laplacian in the kinetic energy operator in Eq. (F.1) is described with the 4th-order finite difference approach in the RSG method. Explicitly, in the Kth-order finite difference method, the operation of the Laplacian to the one-electron wave function $\varphi_i(\boldsymbol{r})$ along the x axis is approximated as

$$\frac{\partial^2}{\partial x^2}\varphi_i\left(x_p, y_q, z_r\right) = \frac{1}{h^2}\sum_{k=-K}^{K} C_k \varphi_i\left(x_{p+k}, y_q, z_r\right) + O(h^{(2K+2)}) \tag{F.2}$$

where h is the grid width, and the set of the numbers x_p, y_q, and z_r specifies the grid coordinate $\boldsymbol{r}$ in the real-space cell. The set of the integers (p, q, r) represents the index attached to a grid in the cell. The coefficients $C_k (k = -K, -K+1, \cdots, K)$ are tabulated for various K in Table I in Ref. [7]. In producing the results given in this book, we typically use the order $K = 4$. Explicitly, for $K = 4$, the coefficients C_k are given as follows, $C_0 = -\frac{205}{72}$, $C_{\pm 1} = \frac{8}{5}$, $C_{\pm 2} = -\frac{1}{5}$, $C_{\pm 3} = \frac{8}{315}$, and $C_{\pm 4} = -\frac{1}{560}$. The coefficients can be obtained by consulting the Taylor expansions of the wave function φ_i. To see in more detail, we consider a one-dimensional function $f(x)$ defined on the grid points with the width h. The Taylor expansion of $f(x)$ around the grid with the coordinate x_0 to obtain $f(x_0 + h)$ can be expressed as

$$f(x_0 + h) = f(x_0) + f^{(1)}(x_0)\, h + \frac{1}{2} f^{(2)}(x_0)\, h^2 + \frac{1}{3!} f^{(3)}(x_0)\, h^3 + O\left(h^4\right) \tag{F.3}$$

And for $f(x_0 - h)$, we have

$$f(x_0 - h) = f(x_0) - f^{(1)}(x_0)\, h + \frac{1}{2} f^{(2)}(x_0)\, h^2 - \frac{1}{3!} f^{(3)}(x_0)\, h^3 + O\left(h^4\right) \tag{F.4}$$

The sum of these equations leads to the most primitive approximation to the quantity $\nabla^2 f(x)|_{x_0}$, which reads,

$$\nabla^2 f(x)\Big|_{x_0} = \frac{1}{h^2}\left(f(x_0 + h) - 2f(x_0) + f(x_0 - h)\right) + O\left(h^4\right) \tag{F.5}$$

Thus, for $K = 1$ in Eq. (F.2), we get the coefficients $C_0 = -2$ and $C_{\pm 1} = 1$. To eliminate the 4th-order term in the Taylor expansion, we further consider the expansions for $f(x_0 + 2h)$ and $f(x_0 - 2h)$, thus,

$$\begin{aligned} f(x_0 + 2h) = {} & f(x_0) + f^{(1)}(x_0)\, 2h + \frac{1}{2} f^{(2)}(x_0)\,(2h)^2 + \frac{1}{3!} f^{(3)}(x_0)\,(2h)^3 \\ & + \frac{1}{4!} f^{(4)}(x_0)\,(2h)^4 + \frac{1}{5!} f^{(5)}(x_0)\,(2h)^5 + O\left(h^6\right) \end{aligned} \tag{F.6}$$

and

$$f(x_0 - 2h) = f(x_0) - f^{(1)}(x_0)\,2h + \frac{1}{2} f^{(2)}(x_0)(2h)^2 - \frac{1}{3!} f^{(3)}(x_0)(2h)^3 + \frac{1}{4!} f^{(4)}(x_0)(2h)^4 - \frac{1}{5!} f^{(5)}(x_0)(2h)^5 + O\left(h^6\right) \tag{F.7}$$

Then, the coefficients for $K = 2$ in Eq. (F.2) can be determined so that the h^4 terms in the equation vanish, which leads to $C_0 = -\frac{5}{2}$, $C_{\pm 1} = \frac{4}{3}$, and $C_{\pm 2} = -\frac{1}{12}$. The coefficients for the larger K can also be obtained through the same manner. It was shown in Fig. 1 in Ref. [7] that the use of K larger than 4 hardly improves the eigenvalues of the electrons in the Oxygen atom, regardless of the value of h.

F.2 Hartree Potential

Construction of the Hartree potential $\upsilon_{\mathrm{H}}[n](\boldsymbol{r})$ in Eq. (F.1) is one of the major issues in the implementation of the real-space grid method. In fact, devices need to be developed to construct the potential. As shown in Eq. (1.110), $\upsilon_{\mathrm{H}}[n](\boldsymbol{r})$ can be explicitly written as

$$\upsilon_{\mathrm{H}}[n](\boldsymbol{r}) = \int d\boldsymbol{r}' \frac{n\left(\boldsymbol{r}'\right)}{|\boldsymbol{r} - \boldsymbol{r}'|} \tag{F.8}$$

When the system is supposed to be periodic, the potential, given in the form of the convolution integral, can be easily obtained through the Fourier transform. The Hartree potential $\widetilde{\upsilon}_{\mathrm{H}}[n](\boldsymbol{k})$ defined on the wave number vector $\boldsymbol{k}$ can be expressed as

$$\begin{aligned}\widetilde{\upsilon}_{\mathrm{H}}[n](\boldsymbol{k}) &= \int d\boldsymbol{r} \exp(-i\boldsymbol{k}\cdot\boldsymbol{r}) \int d\boldsymbol{r}' \frac{n\left(\boldsymbol{r}'\right)}{|\boldsymbol{r} - \boldsymbol{r}'|} \\ &= \widetilde{n}(\boldsymbol{k})\,\widetilde{\upsilon}_c(\boldsymbol{k})\end{aligned} \tag{F.9}$$

where $\widetilde{n}(\boldsymbol{k})$ and $\widetilde{\upsilon}_c(\boldsymbol{k})$ are the Fourier transforms of the electron density and the Coulomb potential, respectively. Explicitly, $\widetilde{\upsilon}_c$ is defined as

$$\widetilde{\upsilon}_c(\boldsymbol{k}) = \int d\boldsymbol{r} \exp(-i\boldsymbol{k}\cdot\boldsymbol{r}) \frac{1}{|\boldsymbol{r}|} \tag{F.10}$$

It is well known that the Fourier transform of the Coulomb potential becomes $\frac{4\pi}{|\boldsymbol{k}|^2}$. We provide its derivation for reader's references. By introducing the spherical coordinate (r, θ, ϕ) to perform the integration in Eq. (F.10), we have

$$\begin{aligned}\widetilde{\upsilon}_c(\boldsymbol{k}) &= \int_0^{2\pi} d\phi \int_0^{\pi} d\theta \sin\theta \int_0^{\infty} dr\, r^2 \exp(-ikr\cos\theta) \frac{1}{r} \\ &= 2\pi \int_0^{\pi} d\theta \int_0^{\infty} dr\, r \sin\theta \exp(-ikr\cos\theta)\end{aligned} \tag{F.11}$$

Introducing the variable t defined as $t = -\cos\theta$, we obtain

$$\begin{aligned}
\widetilde{v}_c(\boldsymbol{k}) &= 2\pi \int_{-1}^{1} dt \int_{0}^{\infty} dr\ r \exp(ikrt) \\
&= 2\pi \int_{0}^{\infty} dr\ r \frac{1}{ikr} (\exp(ikr) - \exp(-ikr)) \\
&= 2\pi \int_{0}^{\infty} dr\ \frac{2}{k} (\sin(kr))
\end{aligned} \tag{F.12}$$

Thus, it is shown that the integral does not converge. To avoid this problem, we introduce Yukawa potential, a short-range potential, described as

$$V_{\mathrm{Y}}(r) = \exp(-\lambda r)\frac{1}{r} \tag{F.13}$$

where λ is a real positive number. It is readily confirmed that the potential is short range. We substitute $V_{\mathrm{Y}}(r)$ instead of $1/r$ in Eq. (F.10) and then, the limit of $\lambda \to 0$ will be taken after the integration, which leads to the result for the Coulomb potential since $\exp(-\lambda r)$ goes to 1 at the limit of $\lambda \to 0$. The integration in Eq. (F.12) can then be rewritten as,

$$\begin{aligned}
\widetilde{v}_c(\boldsymbol{k}) &= 2\pi \lim_{\lambda\to 0} \int_{-1}^{1} dt \int_{0}^{\infty} dr\ r \exp(-\lambda r) \exp(ikrt) \\
&= 2\pi \lim_{\lambda\to 0} \int_{0}^{\infty} dr\ r \exp(-\lambda r) \frac{\exp(ikr) - \exp(-ikr)}{ikr} \\
&= 2\pi \lim_{\lambda\to 0} \int_{0}^{\infty} dr \frac{\exp((ik-\lambda) r) - \exp(-(ik+\lambda) r)}{ik} \\
&= 2\pi \lim_{\lambda\to 0} \frac{1}{ik}\left(\frac{-1}{ik-\lambda} - \frac{-1}{-(ik+\lambda)}\right) \\
&= 2\pi \lim_{\lambda\to 0} \frac{2}{k^2+\lambda^2} \\
&= \frac{4\pi}{k^2}
\end{aligned} \tag{F.14}$$

From Eq. (F.9), the Hartree potential in the reciprocal space can be given by

$$\widetilde{\upsilon}_{\mathrm{H}}[n](\boldsymbol{k}) = \widetilde{n}(\boldsymbol{k}) \frac{4\pi}{|\boldsymbol{k}|^2} \tag{F.15}$$

This equation is majorly employed in the calculations using the plane wave basis for the system with periodicity. Equation (F.15) can also be implemented in the real-space grid calculations. Importantly, the real-space grid implementation allows to construct the Hartree potential for both the periodic and non-periodic systems. In the following, we address the methods of implementing the Hartree potential separately for (A) periodic and (B) non-periodic systems.

F.2.1 Periodic System

For the periodic systems, the implementation of the Hartree potential $\upsilon_{\mathrm{H}}[n](\boldsymbol{r})$ on the real-space grid is rather straightforward. We suppose that the real-space grids $\boldsymbol{s}$ are specified by

$$\boldsymbol{s} = \left(\frac{s_x}{N_x}L_x, \frac{s_y}{N_y}L_y, \frac{s_z}{N_z}L_z\right) \tag{F.16}$$

where N_α ($\alpha = x, y,$ and z) are the numbers of the grids along the axes of the rectangular cell and L_α are the sizes of the cell. The integer s_α for each grid is within the range of

$$-\frac{N_\alpha}{2} \le s_\alpha \le \frac{N_\alpha}{2} - 1 \tag{F.17}$$

for $\alpha = x, y,$ and z. Correspondingly, the grid $\boldsymbol{k}$ in the reciprocal space is expressed as

$$\boldsymbol{k} = 2\pi\left(\frac{k_x}{L_x}, \frac{k_y}{L_y}, \frac{k_z}{L_z}\right) \tag{F.18}$$

where k_α is within the region

$$-\frac{N_\alpha}{2} \le k_\alpha \le \frac{N_\alpha}{2} - 1 \tag{F.19}$$

The electron density $\widetilde{n}$ on the grid (k_x, k_y, k_z) in the reciprocal space is explicitly given by

$$\widetilde{n}\left(k_x, k_y, k_z\right) = \frac{h^3}{L_x L_y L_z} \sum_{s_x} \sum_{s_y} \sum_{s_z} n\left(s_x, s_y, s_z\right) \exp\left(-i\frac{2\pi k_x}{N_x} s_x\right) \times \exp\left(-i\frac{2\pi k_y}{N_y} s_y\right) \exp\left(-i\frac{2\pi k_z}{N_z} s_z\right) \tag{F.20}$$

where h represents the grid width. The summations with respect to s_α in the equation can be performed using a standard FFT library such as FFTW [24, 25]. Note that the factor $\frac{h^3}{L_x L_y L_z}$ can be reduced to M^{-1} with the definition of $M = N_x \times N_y \times N_z$. We obtain the Hartree potential $\widetilde{\upsilon}_{\mathrm{H}}[n]\left(k_x, k_y, k_z\right)$ in the reciprocal space using Eq. (F.15) and then transform it to the real space through the backward Fourier transform to get the Hartree potential in the real space, thus,

$$\upsilon_{\mathrm{H}}[n]\left(s_x, s_y, s_z\right) = \sum_{k_x} \sum_{k_y} \sum_{k_z}{}' \, 4\pi \frac{\widetilde{n}\left(k_x, k_y, k_z\right)}{k_x^2 + k_y^2 + k_z^2} \exp\left(i\frac{2\pi k_x}{N_x} s_x\right) \times \exp\left(i\frac{2\pi k_y}{N_y} s_y\right) \exp\left(i\frac{2\pi k_z}{N_z} s_z\right) \tag{F.21}$$

where the notation $\sum'$ indicates that the term of $\boldsymbol{k}^2 = 0$ is excluded from the summation. The exclusion of the $\boldsymbol{k}^2 = 0$ term does not matter when the charge neutrality of the system is maintained, since the effect is completely compensated by the same term with the opposite charges. Thus, the implementation of the Hartree potential on the real-space grid is rather straightforward when a FFT library is available. The computational cost for the FFT is $2 \times \Omega \log \Omega$ for the forward and the backward Fourier transforms. Later in this paragraph, a remark will be made on the parallelization of the FFT. More detailed discussion was provided in Ref. [20].

F.2.2 Non-periodic System

For the non-periodic system, three approaches have been implemented on the 'Vmol' code. That is, 1. Method by Bernett and Landman, 2. Poisson solver using FFT, and 3. Poisson solver using conjugate gradient method. In the following, these methods are listed and their outlines are provided.

1. Method by Bernett and Landman
   Bernett and Landman proposed a method [26] to use Eq. (F.15) with a modification to eliminate the periodicity. The point of their approach is to represent the Coulomb potential assuming the homogeneity of the charge within the real-space grids, which leads to

$$u_c\left(\boldsymbol{s}'\right)\Big|_{|\boldsymbol{s}|\neq 0} = \frac{1}{|\boldsymbol{s}'|} \times h^3 \tag{F.22}$$

where s' represents the grid points in the real-space cell as will be described below. Importantly, at the singularity of the potential that occurs at the grid specified by $(s'_x, s'_y, s'_z) = (0, 0, 0)$, the Coulomb potential is integrated within the grid of size h [7], which becomes

$$\begin{aligned} u_c(s)|_{|s|=0} &= \int_{-h/2}^{h/2} dx \int_{-h/2}^{h/2} dy \int_{-h/2}^{h/2} dz \frac{1}{\sqrt{x^2+y^2+z^2}} \\ &= -h^2 \left(\frac{\pi}{2} + 3 \ln \left(\frac{\sqrt{3}-1}{\sqrt{3}+1} \right) \right) \end{aligned} \tag{F.23}$$

Thus, the bare Coulomb potential with non-periodicity is first realized on the real-space grids, and then it is transformed to the reciprocal space to get the Coulomb potential $\widetilde{u}_c(\boldsymbol{k}')$. Another notable feature in the method by Barnett and Landman is that the real-space cell for s' is twice as large as the original cell for the KS-DFT. Explicitly, the range of s'_α ($\alpha = x, y,$ and z) is given as

$$-N_\alpha \le s'_\alpha \le N_\alpha - 1 \tag{F.24}$$

instead of Eq. (F.17). Accordingly, the number of grids in the reciprocal cell is also increased by a factor of two. The wave number vector $\boldsymbol{k}'$ in the larger cell is represented as

$$\boldsymbol{k}' = 2\pi \left(\frac{k'_x}{2L_x}, \frac{k'_y}{2L_y}, \frac{k'_z}{2L_z} \right) \tag{F.25}$$

which allows to include a component with a longer wave length. The corresponding range of k'_α becomes,

$$-N_\alpha \le k'_\alpha \le N_\alpha - 1 \tag{F.26}$$

After the FFT to get $\widetilde{u}_u(\boldsymbol{k})$, the result must be multiplied by $(8M)^{-1}$. Then, the Hartree potential $\upsilon_H[n](s')$ without periodicity can be obtained through the backward transform, thus,

$$\begin{aligned} \upsilon_H[n]\left(s'_x, s'_y, s'_z\right) &= \sum_{k'_x} \sum_{k'_y} \sum_{k'_z} \widetilde{n}\left(k'_x, k'_y, k'_z\right) \widetilde{u}_c\left(k'_x, k'_y, k'_z\right) \\ &\times \exp\left(i \frac{2\pi k'_x}{N_x} s'_x\right) \exp\left(i \frac{2\pi k'_y}{N_y} s'_y\right) \exp\left(i \frac{2\pi k'_z}{N_z} s'_z\right) \end{aligned} \tag{F.27}$$

In our early stages of the application of the 'Vmol' code, Eq. (F.27) was mostly utilized, where the reliability and applicability of the method were fully confirmed [8, 10–13]. Note, however, that the computational cost of the FFT is increased by a factor of 8 log 8 due to the introduction of the double-sized real-space cell to realize the non-periodicity.

2. Poisson Solver Using FFT
The Hartree potential $\upsilon_{\mathrm{H}}[n](\boldsymbol{r})$ in Eq. (F.8) can be written in the form of the Poisson equation, thus,

$$\nabla^2 \upsilon_{\mathrm{H}}[n](\boldsymbol{r}) = -4\pi n(\boldsymbol{r}) \tag{F.28}$$

Let us first consider a periodic system for a moment. Actually, for the periodic system, Eq. (F.15) can be derived from the Poisson equation. The Hartree potential with a periodicity L is expanded in the Fourier series,

$$\upsilon_{\mathrm{H}}[n](\boldsymbol{r}) = \sum_{k_x,k_y,k_z=-\infty}^{\infty} \widetilde{\upsilon}_{\mathrm{H}}[n](\boldsymbol{k}) \exp(i\boldsymbol{k}\cdot\boldsymbol{r}) \tag{F.29}$$

Note that the wave number vector $\boldsymbol{k}$ is expressed by Eq. (F.18), since the potential has the periodicity L. Applying the Laplacian from the left, we get

$$\begin{aligned}\nabla^2 \upsilon_{\mathrm{H}}[n](\boldsymbol{r}) &= \sum_{k_\alpha=-\infty}^{\infty} -\frac{4\pi^2}{L^2}\left(k_x^2+k_y^2+k_z^2\right) \widetilde{\upsilon}_{\mathrm{H}}[n](\boldsymbol{k}) \exp(i\boldsymbol{k}\cdot\boldsymbol{r}) \\ &= \sum_{k_\alpha=-\infty}^{\infty} -|\boldsymbol{k}|^2\, \widetilde{\upsilon}_{\mathrm{H}}[n](\boldsymbol{k}) \exp(i\boldsymbol{k}\cdot\boldsymbol{r})\end{aligned} \tag{F.30}$$

where $\alpha = x, y$, and z. Consulting the components in the Fourier series expansions of the right side of Eq. (F.28) and that of Eq. (F.30), we obtain Eq. (F.15), thus,

$$\widetilde{\upsilon}_{\mathrm{H}}[n](\boldsymbol{k}) = \widetilde{n}(\boldsymbol{k}) \frac{4\pi}{|\boldsymbol{k}|^2} \tag{F.31}$$

Now we go back to the issue of the potential $\upsilon_{\mathrm{H}}[n](\boldsymbol{r})$ for the non-periodic system. We consider to solve the Poisson equation with the Dirichlet boundary condition. The Laplacian in Eq. (F.28) will be approximated by the finite difference method as discussed in the previous paragraph for the kinetic energy operator. In the following discussion, we consider the problem along one dimension for the sake of simplicity. The extension to the three dimension is straightforward. The operation of the approximated Laplacian on the potential along the x axis is written as

$$\nabla_x^2 n(x_j) \cong \frac{\upsilon_{\mathrm{H}}[n](x_{j+1}) - 2\upsilon_{\mathrm{H}}[n](x_j) + \upsilon_{\mathrm{H}}[n](x_{j-1})}{h^2} \tag{F.32}$$

where x_j represents jth grid on the x axis of the real-space cell and h is the grid width. Thus, the Laplacian approximated with the finite-difference method can be expressed with a tridiagonal matrix $\mathcal{L}$. For the Dirichlet boundary condition specified by $v_{\mathrm{H}}(x)|_{\mathrm{BD}} = 0$, the discrete sine transformation (DST) is useful. As shown by Schumann and Sweet [27], the Laplacian $\mathcal{L}$ has the eigenvectors l_k

$$l_k = 2\sin\left(\frac{\pi}{N_x}\left(j + \frac{1}{2}\right)(k+1)\right) \tag{F.33}$$

where N_x is the number of grids on the x axis. The DST using these basis is referred to as DST of type II (DST-II). Note that these eigenvectors are for the 'staggered' grids. The eigenvalue λ_k can be obtained by applying matrix $\mathcal{L}$ to l_k at ith grid, which reads,

$$\begin{aligned}
\mathcal{L}l_k &= \frac{2}{h^2}\sin\left(\frac{\pi}{N_x}\left(i + \frac{3}{2}\right)(k+1)\right) - \frac{4}{h^2}\sin\left(\frac{\pi}{N_x}\left(i + \frac{1}{2}\right)(k+1)\right) \\
&\quad + \frac{2}{h^2}\sin\left(\frac{\pi}{N_x}\left(i - \frac{1}{2}\right)(k+1)\right) \\
&= \frac{4}{h^2}\sin\left(\frac{\pi}{N_x}\left(i + \frac{1}{2}\right)(k+1)\right)\cos\left(\frac{\pi}{N_x}(k+1)\right) \\
&\quad - \frac{4}{h^2}\sin\left(\frac{\pi}{N_x}\left(i + \frac{1}{2}\right)(k+1)\right) \\
&= \frac{4}{h^2}\sin\left(\frac{\pi}{N_x}\left(i + \frac{1}{2}\right)(k+1)\right)\left(\cos\left(\frac{\pi}{N_x}(k+1)\right) - 1\right) \\
&= \frac{2}{h^2}\sin\left(\frac{\pi}{N_x}\left(i + \frac{1}{2}\right)(k+1)\right)\left(-4\sin^2\left(\frac{\pi}{2N_x}(k+1)\right)\right) \\
&= \left(-\frac{4}{h^2}\sin^2\left(\frac{\pi}{2N_x}(k+1)\right)\right) l_k
\end{aligned}$$

Thus, the eigenvalue λ_k for the eigenvector l_k is given by

$$\lambda_k = -\frac{4}{h^2}\sin^2\left(\frac{\pi}{2N_x}(k+1)\right) \tag{F.34}$$

Using the eigenvectors l_k, the matrix $\mathcal{L}$ can be represented with the spectral decomposition,

$$\mathcal{L} = \sum_{k=-N_x/2}^{N_x/2-1} |l_k\rangle\, \lambda_k\, \langle l_k| \tag{F.35}$$

Then, its inverse is given by

$$\mathcal{L}^{-1} = \sum_{k=-N_x/2}^{N_x/2-1} |l_k\rangle\, \lambda_k^{-1}\, \langle l_k| \tag{F.36}$$

In general, the Poisson equation with the finite difference method can be represented as

$$\mathcal{L}\boldsymbol{v} = \boldsymbol{f} \tag{F.37}$$

Then, the potential $\boldsymbol{v}$ can be obtained as

$$\boldsymbol{v} = \mathcal{L}^{-1}\boldsymbol{f} \tag{F.38}$$

Using the decomposition in Eq. (F.36), we obtain

$$\boldsymbol{v} = \sum_{k=-N_x/2}^{N_x/2-1} |l_k\rangle \lambda_k^{-1} \langle l_k| \boldsymbol{f}\rangle \tag{F.39}$$

Note that $\langle l_k| \boldsymbol{f}\rangle$ represents $\boldsymbol{f}_k$ in the reciprocal space.

Now we consider the Hartree potential $\upsilon_{\mathrm{H}}(\boldsymbol{r})$ in 3-dimensional space [20]. The dependence of the potential on the density n will not be shown hereafter for the sake of simplicity. The potential $\overline{\upsilon}_{\mathrm{H}}$ specified by the index (i_x, i_y, i_z) on the staggered grid is given by

$$\overline{\upsilon}_{\mathrm{H}}\left(i_x, i_y, i_z\right) = \upsilon_{\mathrm{H}}\left(\left(i_x + \frac{1}{2}\right)h, \left(i_y + \frac{1}{2}\right)h, \left(i_z + \frac{1}{2}\right)h\right) \tag{F.40}$$

And, similarly, for the density $\overline{n}\left(i_x, i_y, i_z\right)$ is defined as

$$\overline{n}\left(i_x, i_y i_z\right) = n\left(\left(i_x + \frac{1}{2}\right)h, \left(i_y + \frac{1}{2}\right)h, \left(i_z + \frac{1}{2}\right)h\right) \tag{F.41}$$

From the result given in Eq. (F.34), the eigenvalues $\Lambda(k_x, k_y, k_z)$ of the Laplacian $\mathcal{L}$ in the 3-dimensional space approximated with the finite different scheme are given by

$$\begin{aligned}\Lambda\left(k_x, k_y, k_z\right) = &-\frac{4}{h^2}\sin^2\left(\frac{\pi}{2N_x}(k_x+1)\right) - \frac{4}{h^2}\sin^2\left(\frac{\pi}{2N_y}(k_y+1)\right)\\ &-\frac{4}{h^2}\sin^2\left(\frac{\pi}{2N_z}(k_z+1)\right)\end{aligned} \tag{F.42}$$

The DST of type III (DST-III) is applied to the right hand side of Eq. (F.28), that is, $-4\pi\overline{n}\left(i_x, i_y i_z\right)$, to get the density ρ in the reciprocal space

$$\rho = \mathcal{F}_{\text{DST-III}}\left[-4\pi\overline{n}\right] \tag{F.43}$$

where $\mathcal{F}_{\text{DST-III}}$ stands for the DST-III transform. In the DST-III transform, the eigenvectors

$$\widetilde{l}_j = 2\sin\left(\frac{\pi}{N_x}(j+1)\left(k+\frac{1}{2}\right)\right) \tag{F.44}$$

are employed. Then, the Hartree potential $\overline{\upsilon}_{\text{H}}$ in the real space can be obtained through backward transform with DST of type II, thus,

$$\overline{\upsilon}_{\text{H}} = \mathcal{F}_{\text{DST-II}}\left[\Lambda^{-1} \odot \rho\right] \tag{F.45}$$

where the operation denoted by $\Lambda^{-1} \odot \rho$ stands for the Hadamard product defined as

$$\left[\Lambda^{-1} \odot \rho\right]\left(k_x, k_y, k_z\right) \equiv \Lambda\left(k_x, k_y, k_z\right)\rho\left(k_x, k_y, k_z\right) \tag{F.46}$$

For the details of the DST-II and DST-III transforms, we refer readers to Refs. [20, 27].

Another issue to be considered is the boundary problem, because the Hartree potential υ_{H} does not sufficiently converge to zero at the boundary in many cases. We provide here the method proposed in Ref. [20]. The major part of the method, which was originally developed by Ricker [28], is tricky but easy to implement. Suppose that the actual boundary potential at $x = 0$ is not zero, but can be evaluated by a certain method as will be described below. The approximate boundary potential is denoted as $\upsilon_{\text{BD}}^x(0, y, z)$. It is also supposed as a matter of course that the electron density n is almost converged to zero at the boundary, that is, $n(0, y, z) \approx 0$. Using the Poisson equation expressed by the finite difference method, the potential at $x = -1/2h$ is approximately given by

$$\upsilon_{\text{H}}\left(-\frac{1}{2}h, y, z\right) \approx 2\upsilon_{\text{BD}}^x(0, y, z) - \upsilon_{\text{H}}\left(\frac{1}{2}h, y, z\right) \tag{F.47}$$

On the other hand, the second derivative of the potential at $x = 1/2h$ is approximately given by the finite difference method, thus,

$$\frac{\partial^2}{\partial x^2}\upsilon_{\text{H}}\left(\frac{1}{2}h, y, z\right) \approx \frac{1}{h^2}\left(\upsilon_{\text{H}}\left(-\frac{1}{2}h, y, z\right) - 2\upsilon_{\text{H}}\left(\frac{1}{2}h, y, z\right)\right.$$
$$\left. + \upsilon_{\text{H}}\left(\frac{3}{2}h, y, z\right)\right) \tag{F.48}$$

By substituting Eq. (F.47) into this equation, we get

$$\frac{\partial^2}{\partial x^2}\upsilon_{\mathrm{H}}\left(\frac{1}{2}h, y, z\right) \approx \frac{1}{h^2}\left(-\upsilon_{\mathrm{H}}\left(\frac{1}{2}h, y, z\right) - 2\upsilon_{\mathrm{H}}\left(\frac{1}{2}h, y, z\right)\right.$$
$$\left.+\upsilon_{\mathrm{H}}\left(\frac{3}{2}h, y, z\right)\right) + \frac{2}{h^2}\upsilon_{\mathrm{BD}}^{x}(0, y, z) \quad \text{(F.49)}$$

According to the Poisson equation, the right side of this equation is equal to $-4\pi n(1/2h, y, z)$. This suggests that the shifted electron density n' defined by

$$n'\left(\frac{1}{2}h, y, z\right) = n\left(\frac{1}{2}h, y, z\right) + \frac{\upsilon_{\mathrm{BD}}^{x}(0, y, z)}{2\pi h^2} \quad \text{(F.50)}$$

allows the use of the Poisson solver with the Dirichlet boundary with $\upsilon_{\mathrm{H}}(0, y, z) = 0$. The potential $\upsilon_{\mathrm{BD}}^{x}(0, y, z)$ at the boundary can be approximately evaluated by the Coulomb potential of the point charges placed at the nuclei, thus,

$$\upsilon_{\mathrm{BD}}^{x}(0, y, z) = \sum_{\mathrm{A}} \frac{q_A}{|\boldsymbol{r} - \boldsymbol{R}_{\mathrm{A}}|} \quad \text{(F.51)}$$

where $\boldsymbol{r}$ is supposed to be on the boundary $x = 0$ and $\boldsymbol{R}_{\mathrm{A}}$ is the position vector of nucleus A. The charge q_{A} is the electron population assigned to the nucleus A. There have been several methods to evaluate the charge q_{A}. One way is to determine the set of $\{q_{\mathrm{A}}\}$ through a least-square fittings so that $\{q_{\mathrm{A}}\}$ realizes the electrostatic potential formed by the electron density. Thus, the set $\{q_{\mathrm{A}}\}$ and the resultant Hartree potential $\upsilon_{\mathrm{H}}(\boldsymbol{r})$ are determined self-consistently. Another approach is to utilize the fuzzy cell proposed by Becke [29].

In fact, in the Vmol code, the Poisson solver using the FFT technique is used to calculate the Hartree-Fock (HF) exchange potential rather than the Hartree potential [20] itself. How can it be used for HF exchange energy? The operation of the HF exchange operator $\hat{K}_\sigma$ for the spin σ on the wave function $\varphi_{i\sigma}$ is written as

$$\hat{K}_\sigma \varphi_{i\sigma}(\boldsymbol{r}) = \int d\boldsymbol{r}' \langle \boldsymbol{r}| \hat{K}_\sigma |\boldsymbol{r}'\rangle\langle \boldsymbol{r}'| \varphi_{i\sigma}\rangle$$
$$= \sum_{k}^{occ} \int d\boldsymbol{r}' \frac{\varphi_{k\sigma}(\boldsymbol{r})\varphi_{k\sigma}^{*}(\boldsymbol{r}')}{|\boldsymbol{r} - \boldsymbol{r}'|}\varphi_{i\sigma}(\boldsymbol{r}') \quad \text{(F.52)}$$

We define the potential υ_{ki} for the orbitals pair φ_k and φ_i as

$$\upsilon_{ki}(\boldsymbol{r}) = \int d\boldsymbol{r}' \frac{\varphi_{k\sigma}^{*}(\boldsymbol{r}')\varphi_{i\sigma}(\boldsymbol{r}')}{|\boldsymbol{r} - \boldsymbol{r}'|} \quad \text{(F.53)}$$

Then, the operation of $\hat{K}_\sigma$ can be rewritten as

$$\hat{K}_{\sigma}\varphi_{i\sigma}(\boldsymbol{r}) = \sum_{k}^{occ} \upsilon_{ki}(\boldsymbol{r})\,\varphi_{k\sigma}(\boldsymbol{r}) \tag{F.54}$$

Thus, it can be seen that the calculation of the HF exchange potential consists of the Hartree potentials of the products of the orbital pairs. The Poisson solver based on the FFT technique can be utilized to obtain the potential in Eq. (F.53) for constructing the HF exchange energy. It is of critical importance to parallelize the 3-dimensional FFT for the application of the hybrid DFT to large systems. The technical details for the parallel implementation is provided in Ref. [20].

3. Poisson Solver Using Conjugate Gradient Method
In our approach using 'Vmol' the Hartree potential $\upsilon_{\mathrm{H}}[n](\boldsymbol{r})$ in Eq. (F.1) for a large system is solved using a Poisson solver based on the conjugate gradient (CG) method. To do this, the matrix representation of the Laplacian approximated by the finite difference scheme (Eq. (F.32)) is denoted as $\mathcal{L}$. Then, the Poisson equation can be represented as

$$\mathcal{L}\boldsymbol{x} = -4\pi\boldsymbol{b} \tag{F.55}$$

where $\boldsymbol{x}$ is the vector representing the potential $\upsilon_{\mathrm{H}}[n](\boldsymbol{r})$ and vector $\boldsymbol{b}$ stands for the total electron density $n(\boldsymbol{r})$. With the vector $\boldsymbol{x}^{\dagger}$ defined as the transpose of $\boldsymbol{x}$, we introduce the scalar quantity $Q[\boldsymbol{x}]$, thus,

$$Q[\boldsymbol{x}] = \frac{1}{2}\boldsymbol{x}^{\dagger}\mathcal{L}\boldsymbol{x} + 4\pi\boldsymbol{x}^{\dagger}\boldsymbol{b} \tag{F.56}$$

The minimization of $Q[\boldsymbol{x}]$ with respect to $\boldsymbol{x}$ leads to the functional derivative, thus,

$$\frac{\delta Q[\boldsymbol{x}]}{\delta\boldsymbol{x}} = \mathcal{L}\boldsymbol{x} + 4\pi\boldsymbol{b} = 0 \tag{F.57}$$

This is equivalent to the solution of the Poisson equation. Thus, the Poisson equation can be solved iteratively by the standard conjugate gradient (CG) method. In our implementation, the steepest descent (SD) method is used in the early stages of the convergence, and then switched to CG procedure to expedite the convergence.

The parallel implementation of the Poisson solver using SD combined with the CG procedure requires the *local* communications among the processors by means of the MPI (message passing interface) [30], while the Poisson solver with the parallelized FFT mostly uses the *collective* communications. Thus, the comparison between the solvers using the CG and the FFT is a matter of significance to perform the large-scale parallel computations efficiently. This issue was discussed in detail in Ref. [20] for the water clusters $(H_2O)_{19}$ and $(H_2O)_{63}$ computed using the 64-nodes parallel machine invoking 8 or 16 cores on each node.

F.3 Pseudopotential

F.3.1 Basic Concept

Since the real-space grid (RSG) or the plane wave basis is not capable of realizing the steep behaviors of the bare Coulomb potentials near the nuclei, the use of the pseudopotential is essential to implement the KS-DFT using the uniform grids. Note, however, that the use of the adaptive mesh [21] makes it possible to perform the KS-DFT calculations without pseudopotentials. The important characteristic of the pseudopotential lies in its nonlocality, which complicates the numerical implementations. We first review the basis of the general construction of the pseudopotential to see how the nonlocality arises. What we want to do is to obtain the modified external potential υ'_{ext} for an atom that includes the effective potential due to the core electrons, and as a consequence to yield the smooth 'nodeless' wave functions $\{\overline{\varphi}^i_{\text{val}}\}$ for the valence electrons. For this purpose, we consider separately the KS equations for the core and valence electrons, thus,

$$\hat{H}_{\text{KS}} \left|\varphi^k_{\text{core}}\right\rangle = \epsilon^k_{\text{core}} \left|\varphi^k_{\text{core}}\right\rangle \tag{F.58}$$

$$\hat{H}_{\text{KS}} \left|\varphi^i_{\text{val}}\right\rangle = \epsilon^i_{\text{val}} \left|\varphi^i_{\text{val}}\right\rangle \tag{F.59}$$

Since $\{\varphi^k_{\text{core}}\}$ and $\{\varphi^i_{\text{val}}\}$ are the eigenfunctions of the KS Hamiltonian $\hat{H}_{\text{KS}}$, the overlaps between the wave functions in theses sets are zero, that is,

$$\left\langle\varphi^k_{\text{core}} \middle| \varphi^i_{\text{val}}\right\rangle = 0 \tag{F.60}$$

Thus, in general, the wave functions $\{\varphi^i_{\text{val}}\}$ have oscillatory behaviors around the atomic cores to ensure the orthonormality with $\{\varphi^k_{\text{core}}\}$. However, the smooth valence wave functions $\{\overline{\varphi}^i_{\text{val}}\}$ have the overlaps with the core wave functions $\{\varphi^k_{\text{core}}\}$. We suppose that $\left|\varphi^i_{\text{val}}\right\rangle$ can be constructed by subtracting the component $\left|\varphi^k_{\text{core}}\right\rangle$ from $\left|\overline{\varphi}^i_{\text{val}}\right\rangle$, which reads,

$$\left|\varphi^i_{\text{val}}\right\rangle = \left|\overline{\varphi}^i_{\text{val}}\right\rangle - \sum_k \left|\varphi^k_{\text{core}}\right\rangle \left\langle\varphi^k_{\text{core}}\middle| \overline{\varphi}^i_{\text{val}}\right\rangle \tag{F.61}$$

Since $\left|\varphi^k_{\text{core}}\right\rangle$ and $\left|\varphi^i_{\text{val}}\right\rangle$ are defined explicitly in Eqs. (F.58) and (F.59), respectively, Eq. (F.61) can be regarded as the definition of the smooth wave functions $\{\overline{\varphi}^i_{\text{val}}\}$. Applying the Hamiltonian $\hat{H}_{\text{KS}}$ from the left of Eq. (F.61), we obtain,

$$\hat{H}_{\mathrm{KS}}\left(\left|\overline{\varphi}_{\mathrm{val}}^{i}\right\rangle-\sum_{k}\left|\varphi_{\mathrm{core}}^{k}\right\rangle\left\langle\varphi_{\mathrm{core}}^{k}\right|\left.\overline{\varphi}_{\mathrm{val}}^{i}\right\rangle\right)$$
$$=\epsilon_{\mathrm{val}}^{i}\left(\left|\overline{\varphi}_{\mathrm{val}}^{i}\right\rangle-\sum_{k}\left|\varphi_{\mathrm{core}}^{k}\right\rangle\left\langle\varphi_{\mathrm{core}}^{k}\right|\left.\overline{\varphi}_{\mathrm{val}}^{i}\right\rangle\right) \tag{F.62}$$

The left hand of this equation can be transformed to

$$\hat{H}_{\mathrm{KS}}\left(\left|\overline{\varphi}_{\mathrm{val}}^{i}\right\rangle-\sum_{k}\left|\varphi_{\mathrm{core}}^{k}\right\rangle\left\langle\varphi_{\mathrm{core}}^{k}\right|\left.\overline{\varphi}_{\mathrm{val}}^{i}\right\rangle\right)$$
$$=\hat{H}_{\mathrm{KS}}\left|\overline{\varphi}_{\mathrm{val}}^{i}\right\rangle-\sum_{k}\epsilon_{\mathrm{core}}^{k}\left|\varphi_{\mathrm{core}}^{k}\right\rangle\left\langle\varphi_{\mathrm{core}}^{k}\right|\left.\overline{\varphi}_{\mathrm{val}}^{i}\right\rangle \tag{F.63}$$

From these equations, we get

$$\left(\hat{H}_{\mathrm{KS}}+\sum_{k}\left|\varphi_{\mathrm{core}}^{k}\right\rangle\left(\epsilon_{\mathrm{val}}^{i}-\epsilon_{\mathrm{core}}^{k}\right)\left\langle\varphi_{\mathrm{core}}^{k}\right|\right)\left|\overline{\varphi}_{\mathrm{val}}^{i}\right\rangle=\epsilon_{\mathrm{val}}^{i}\left|\overline{\varphi}_{\mathrm{val}}^{i}\right\rangle \tag{F.64}$$

This equation shows that the wave functions $\{\overline{\varphi}_{\mathrm{val}}^{i}\}$ are the eigenfunctions of the KS Hamiltonian combined with the nonlocal potential given by

$$\upsilon_{\mathrm{nloc}}\left(\boldsymbol{r},\boldsymbol{r}'\right)=\sum_{k}\left(\epsilon_{\mathrm{val}}^{i}-\epsilon_{\mathrm{core}}^{k}\right)\varphi_{\mathrm{core}}^{k}\left(\boldsymbol{r}\right)\varphi_{\mathrm{core}}^{k*}\left(\boldsymbol{r}'\right) \tag{F.65}$$

Importantly, the eigenvalue of $\overline{\varphi}_{\mathrm{val}}^{i}$ is the same as that of $\varphi_{\mathrm{val}}^{i}$. The wave functions are called pseudo wave functions since their spatial behaviors are different from those of the wave functions for the valence electrons. Since the difference $\epsilon_{\mathrm{val}}^{i}-\epsilon_{\mathrm{core}}^{k}$ in Eq. (F.65) is always positive by definition, the potential $\upsilon_{\mathrm{nloc}}\left(\boldsymbol{r},\boldsymbol{r}'\right)$ is also always positive. Thus, the valence electrons represented by $\{\overline{\varphi}_{\mathrm{val}}^{i}\}$ feel effectively the repulsive potential due to the core electrons.

It is important to note that the choice of the pseudo wave functions $\{\overline{\varphi}_{\mathrm{val}}^{i}\}$ is not *unique*. This can be understood by the following discussion. We construct a new wave function $\left|\widetilde{\varphi}_{\mathrm{val}}^{i}\right\rangle$ by adding $\left|\overline{\varphi}_{\mathrm{val}}^{i}\right\rangle$ a linear combination of the wave functions for the core electrons, thus,

$$\left|\widetilde{\varphi}_{\mathrm{val}}^{i}\right\rangle=\left|\overline{\varphi}_{\mathrm{val}}^{i}\right\rangle+\sum_{k'}c_{k'}\left|\varphi_{\mathrm{core}}^{k'}\right\rangle \tag{F.66}$$

Replacing the wave function $\left|\overline{\varphi}_{\mathrm{val}}^{i}\right\rangle$ in the right side of Eq. (F.61) by $\left|\widetilde{\varphi}_{\mathrm{val}}^{i}\right\rangle$, we obtain,

$$\left(\left|\overline{\varphi}^i_{\mathrm{val}}\right\rangle + \sum_{k'} c_{k'} \left|\varphi^{k'}_{\mathrm{core}}\right\rangle\right) - \sum_k \left|\varphi^k_{\mathrm{core}}\right\rangle\left\langle\varphi^k_{\mathrm{core}}\right| \left(\left|\overline{\varphi}^i_{\mathrm{val}}\right\rangle + \sum_{k'} c_{k'} \left|\varphi^{k'}_{\mathrm{core}}\right\rangle\right)$$
$$= \left|\overline{\varphi}^i_{\mathrm{val}}\right\rangle + \sum_{k'} c_{k'} \left|\varphi^{k'}_{\mathrm{core}}\right\rangle - \sum_k \left|\varphi^k_{\mathrm{core}}\right\rangle\left\langle\varphi^k_{\mathrm{core}}\middle| \overline{\varphi}^i_{\mathrm{val}}\right\rangle$$
$$- \sum_k \sum_{k'} c_{k'} \left|\varphi^k_{\mathrm{core}}\right\rangle\left\langle\varphi^k_{\mathrm{core}}\middle| \varphi^{k'}_{\mathrm{core}}\right\rangle$$
$$= \left|\overline{\varphi}^i_{\mathrm{val}}\right\rangle - \sum_k \left|\varphi^k_{\mathrm{core}}\right\rangle\left\langle\varphi^k_{\mathrm{core}}\middle| \overline{\varphi}^i_{\mathrm{val}}\right\rangle$$
$$= \left|\varphi^i_{\mathrm{val}}\right\rangle \tag{F.67}$$

It is thus shown that the set $\{\widetilde{\varphi}^i_{\mathrm{val}}\}$ can also be regarded as the pseudo wave functions. In fact, it is possible to prove that $\left|\widetilde{\varphi}^i_{\mathrm{val}}\right\rangle$ also satisfies Eq. (F.64). This property can be used to design the corresponding pseudopotential so that the resultant pseudo wave function to satisfies the desired properties. An important quantity to characterize the pseudo wave function is the cut off radius r_c. When the distance r of the electron from the nucleus is larger than r_c, that is $r \geq r_c$, then, it is desired that the radial behavior of the pseudo wave function $\overline{\varphi}^i_{\mathrm{val}}(r)$ is strictly the same as that of the corresponding *true* valence wave function $\varphi^i_{\mathrm{val}}(r)$. The normalization of $\overline{\varphi}^i_{\mathrm{val}}$ is also the essential requirement. The first and the second derivatives of $\overline{\varphi}^i_{\mathrm{val}}(r)$ with respect to r must be continuous at r_c to guarantee the smoothness of the corresponding pseudopotential.

Since the p orbital has a node at the atomic core, the repulsive potential due to the corresponding pseudopotential is smaller than that for the s orbital. Thus, it will be better to construct the pseudopotentials so that they are dependent on the angular-momentum of the wave function. For the details to construct the specific pseudopotentials, we refer readers to the literatures, e.g. Refs. [31, 32].

F.3.2 Real-Space Grid Implementation

In the real-space grid implementation of the pseudopotential on the 'Vmol' package, we employ the fully nonlocal form proposed by Kleinman and Bylander [32]. Explicitly, the KB pseudopotential for an atom A can be written in the form as

$$\upsilon^{\mathrm{A,KB}}_{\mathrm{ps}}\left(\boldsymbol{r} - \boldsymbol{R}_{\mathrm{A}}, \boldsymbol{r}' - \boldsymbol{R}_{\mathrm{A}}\right) = \upsilon^{\mathrm{A}}_{\mathrm{loc}}\left(r_{\mathrm{A}}\right) + \sum_l \sum_{m_l=-l}^{l} \frac{\left\langle \boldsymbol{r} \middle| \phi^{\mathrm{A}}_{l,m_l} \Delta V^{\mathrm{A}}_l\right\rangle\left\langle \phi^{\mathrm{A}}_{l,m_l} \Delta V^{\mathrm{A}}_l \middle| \boldsymbol{r}'\right\rangle}{\left\langle \phi^{\mathrm{A}}_{l,m_l}\right| \Delta V^{\mathrm{A}}_l \left|\phi^{\mathrm{A}}_{l,m_l}\right\rangle} \tag{F.68}$$

where the argument r_{A} in the local potential $\upsilon^{\mathrm{A}}_{\mathrm{loc}}$ is the distance $|\boldsymbol{r} - \boldsymbol{R}_{\mathrm{A}}|$ between the electron at $\boldsymbol{r}$ and the atomic core placed at $\boldsymbol{R}_{\mathrm{A}}$. l is the angular momentum quantum number and m_l is the corresponding magnetic quantum number. $\phi^{\mathrm{A}}_{l,m_l}$ is the atomic pseudo wave function for atom A. ΔV_l^{A} is the difference between the l-dependent pseudopotential υ_l^{A} and the local potential $\upsilon_{\mathrm{loc}}^{\mathrm{A}}$,

$$\Delta V_l^{\mathrm{A}}\left(r_{\mathrm{A}}\right)=\upsilon_l^{\mathrm{A}}\left(r_{\mathrm{A}}\right)-\upsilon_{\mathrm{loc}}^{\mathrm{A}}\left(r_{\mathrm{A}}\right) \tag{F.69}$$

Provided that $Z_{\mathrm{val}}^{\mathrm{A}}$ is the number of valence electrons, the potential $\upsilon_{\mathrm{loc}}^{\mathrm{A}}(r_{\mathrm{A}})$ has a long-range nature and asymptotically goes to $-Z_{\mathrm{val}}^{\mathrm{A}}/r_{\mathrm{A}}$ when r_{A} becomes large. Thus, the difference ΔV_l^{A} decays within a short range. The operation of the local pseudopotential $\upsilon_{\mathrm{loc}}^{\mathrm{A}}$ in Eq. (F.68) on a KS wave function $\varphi_i^{\mathrm{KS}}(\boldsymbol{r})$ is trivial. The operation of the nonlocal term $\upsilon_{\mathrm{nloc}}^{\mathrm{A,KB}}\left(\boldsymbol{r},\boldsymbol{r}'\right)$, that is the second term in the right side of Eq. (F.68), on $\varphi_i^{\mathrm{KS}}(\boldsymbol{r})$ can be expressed as

$$\int d\boldsymbol{r}'\upsilon_{\mathrm{nloc}}^{\mathrm{A,KB}}\left(\boldsymbol{r}-\boldsymbol{R}_{\mathrm{A}},\boldsymbol{r}'-\boldsymbol{R}_{\mathrm{A}}\right)\varphi_i^{\mathrm{KS}}\left(\boldsymbol{r}'\right)=\sum_l\sum_{m_l=-l}^{l}\left\langle\boldsymbol{r}\left|\phi_{l,m_l}^{\mathrm{A}}\Delta V_l^{\mathrm{A}}\right.\right\rangle G_{i,l,m_l}^{\mathrm{A}} \tag{F.70}$$

where the quantity G_{i,l,m_l}^{A} is defined as

$$G_{i,l,m_l}^{\mathrm{A}}=\frac{\left\langle\phi_{l,m_l}^{\mathrm{A}}\Delta V_l^{\mathrm{A}}\left|\varphi_i^{\mathrm{KS}}\right.\right\rangle}{\left\langle\phi_{l,m_l}^{\mathrm{A}}\right|\Delta V_l^{\mathrm{A}}\left|\phi_{l,m_l}^{\mathrm{A}}\right\rangle} \tag{F.71}$$

The major computational cost in the pseudopotential calculation arises from the integration of $\left\langle\phi_{l,m_l}^{\mathrm{A}}\Delta V_l^{\mathrm{A}}\left|\varphi_i^{\mathrm{KS}}\right.\right\rangle$. However, as noted above, since ΔV_l^{A} is a short-range potential ($\Delta V_l^{\mathrm{A}}(r)\approx 0, r>r_c$), the computational cost associated with the integration can be suppressed substantially.

In the integration in the numerator of Eq. (F.71), the KS wave function φ_i^{KS} is supposed to be sufficiently smooth. However, the pseudopotential shows in general a rapid behavior near the atomic core region. Therefore, it is better to introduce the denser grids [21] near the atomic cores to ensure the numerical accuracy of the integration. In our implementation of the 'Vmol' package, we employed the time-saving double grid method developed by Ono and Hirose [9, 33, 34]. The notable feature of their approach is that the values of the wave functions on the dense grids are to be evaluated by an interpolation of the values on the coarse grids, and only the values of the pseudopotential on the denses grids are explicitly computed. It was found that this treatment greatly reduces the computational cost. In fact, the integration over the dense grids can be reduced to the weighted integration over the coarse grids. Explicitly, the integration in the numerator in Eq. (F.71) along an axis with the dense grids can be equivalently transformed to the integration over the coarse grids, thus,

$$\sum_J\sum_{j=0}^{n-1}\upsilon_j^J\varphi_j^J h'=\sum_J w_J\Psi_J h \tag{F.72}$$

where j denotes the dense grids between the coarse grids J and $J+1$ along the axis, and w_J is the weight factor that incorporates the pseudopotentials υ_j^J and the wave functions φ_j^J. Ψ_J is the wave function on the coarse grid J. h' and h are the widths of the dense and coarse grids, respectively. In the above equation, it is assumed that $h' = h/n$. Thus, the additional memory allocations for the dense grids are not needed. Furthermore, the values of the weights w_J do not change during the SCF calculation. Therefore, they only need to be calculated once before the SCF calculation.

F.3.3 Hellmann-Feynman Force

To obtain the force acting on the nucleus, the Hellmann-Feynman (HF) force must be computed. As differed from the basis set based on the linear-combination of the atomic orbital (LCAO) method, the real-space grid does not varies when the position of the atom will be moved. Thus, there is no need to compute the Pulay forces [35] additionally. This is, of course, one of the advantages of the real-space grid method. The method to compute the HF force in the 'Vmol' package follows the way proposed in Ref. [22]. In the following, we show the outline of the method. We consider separately the energies due to the local and nonlocal parts of the KB pseudopotential. The interaction energy E_{ps}^{A} between the pseudopotential of atom A and the KS wave functions $\{\varphi_i^{\text{KS}}\}$ is decomposed as

$$E_{\text{ps}}^{\text{A}} = E_{\text{loc}}^{\text{A}} + E_{\text{nloc}}^{\text{A}} \tag{F.73}$$

The HF force associated with the energy $E_{\text{loc}}^{\text{A}}$ is given by

$$\boldsymbol{F}_{\text{loc}}^{\text{A}} = -\sum_i^{occ} \int d\boldsymbol{r}\, \varphi_i^{\text{KS}}(\boldsymbol{r})^* \frac{\partial}{\partial \boldsymbol{R}_{\text{A}}} v_{\text{loc}}^{\text{A}}(\boldsymbol{r} - \boldsymbol{R}_{\text{A}})\, \varphi_i^{\text{KS}}(\boldsymbol{r}) \tag{F.74}$$

What we want to do is to change the differentiation of the pseudopotential to that of the wave functions. To this end, we introduce a variable $\boldsymbol{x}$ defined as $\boldsymbol{x} = \boldsymbol{r} - \boldsymbol{R}_{\text{A}}$. Then, the above equation can be transformed to

$$\boldsymbol{F}_{\text{loc}}^{\text{A}} = -\sum_i^{occ} \int d\boldsymbol{x}\, \varphi_i^{\text{KS}}(\boldsymbol{x} + \boldsymbol{R}_{\text{A}})^* \frac{\partial}{\partial \boldsymbol{R}_{\text{A}}} v_{\text{loc}}^{\text{A}}(\boldsymbol{x})\, \varphi_i^{\text{KS}}(\boldsymbol{x} + \boldsymbol{R}_{\text{A}}) \tag{F.75}$$

Now we replace the argument $\boldsymbol{x} + \boldsymbol{R}_{\text{A}}$ with $\boldsymbol{r}$. Then, we obtain,

$$\begin{aligned}
\boldsymbol{F}_{\mathrm{loc}}^{\mathrm{A}} &= -\sum_{i}^{occ} \int d\boldsymbol{r}\, \varphi_i^{\mathrm{KS}}(\boldsymbol{r})^* \upsilon_{\mathrm{loc}}^{\mathrm{A}}(\boldsymbol{r} - \boldsymbol{R}_{\mathrm{A}}) \frac{\partial \boldsymbol{r}}{\partial \boldsymbol{R}_{\mathrm{A}}} \frac{\partial}{\partial \boldsymbol{r}} \varphi_i^{\mathrm{KS}}(\boldsymbol{r}) + c.c. \\
&= -\sum_{i}^{occ} \int d\boldsymbol{r}\, \varphi_i^{\mathrm{KS}}(\boldsymbol{r})^* \upsilon_{\mathrm{loc}}^{\mathrm{A}}(\boldsymbol{r} - \boldsymbol{R}_{\mathrm{A}}) \frac{\partial}{\partial \boldsymbol{r}} \varphi_i^{\mathrm{KS}}(\boldsymbol{r}) + c.c. \qquad \text{(F.76)} \\
&= -\int d\boldsymbol{r}\, \upsilon_{\mathrm{loc}}^{\mathrm{A}}(\boldsymbol{r} - \boldsymbol{R}_{\mathrm{A}}) \frac{\partial}{\partial \boldsymbol{r}} n(\boldsymbol{r}) \qquad \text{(F.77)}
\end{aligned}$$

where the relation

$$n(\boldsymbol{r}) = \sum_{i}^{occ} \varphi_i^{\mathrm{KS}}(\boldsymbol{r})^* \varphi_i^{\mathrm{KS}}(\boldsymbol{r}) \qquad \text{(F.78)}$$

is used.

For the energy $E_{\mathrm{nloc}}^{\mathrm{A}}$ due to the nonlocal part of the pseudopotential, the differentiation of the potential can also be replaced by that of the wave functions in the same way as the local potential. Actually, it reads,

$$\begin{aligned}
\boldsymbol{F}_{\mathrm{nloc}}^{\mathrm{A}} &= \int d\boldsymbol{r} d\boldsymbol{r}' \varphi_i^{\mathrm{KS}}(\boldsymbol{r})^* \upsilon_{\mathrm{nloc}}^{\mathrm{A,KB}}\left(\boldsymbol{r} - \boldsymbol{R}_{\mathrm{A}}, \boldsymbol{r}' - \boldsymbol{R}_{\mathrm{A}}\right) \frac{\partial}{\partial \boldsymbol{r}'} \varphi_i^{\mathrm{KS}}(\boldsymbol{r}') \\
&\quad + \int d\boldsymbol{r} d\boldsymbol{r}' \frac{\partial}{\partial \boldsymbol{r}} \varphi_i^{\mathrm{KS}}(\boldsymbol{r})^* \upsilon_{\mathrm{nloc}}^{\mathrm{A,KB}}\left(\boldsymbol{r} - \boldsymbol{R}_{\mathrm{A}}, \boldsymbol{r}' - \boldsymbol{R}_{\mathrm{A}}\right) \varphi_i^{\mathrm{KS}}(\boldsymbol{r}') \\
&= \int d\boldsymbol{r} d\boldsymbol{r}' \varphi_i^{\mathrm{KS}}(\boldsymbol{r})^* \upsilon_{\mathrm{nloc}}^{\mathrm{A,KB}}\left(\boldsymbol{r} - \boldsymbol{R}_{\mathrm{A}}, \boldsymbol{r}' - \boldsymbol{R}_{\mathrm{A}}\right) \frac{\partial}{\partial \boldsymbol{r}'} \varphi_i^{\mathrm{KS}}(\boldsymbol{r}') + c.c. \qquad \text{(F.79)}
\end{aligned}$$

Note that the nonlocal potential $\upsilon_{\mathrm{nloc}}^{\mathrm{A,KB}}$ is defined in Eq. (F.70), and explicitly written as

$$\upsilon_{\mathrm{nloc}}^{\mathrm{A,KB}}\left(\boldsymbol{r} - \boldsymbol{R}_{\mathrm{A}}, \boldsymbol{r}' - \boldsymbol{R}_{\mathrm{A}}\right) = \sum_{l} \sum_{m_l=-l}^{l} \frac{\langle \boldsymbol{r} \left| \phi_{l,m_l}^{\mathrm{A}} \Delta V_l^{\mathrm{A}} \right\rangle \langle \phi_{l,m_l}^{\mathrm{A}} \Delta V_l^{\mathrm{A}} \left| \boldsymbol{r}' \right\rangle}{\left\langle \phi_{l,m_l}^{\mathrm{A}} \right| \Delta V_l^{\mathrm{A}} \left| \phi_{l,m_l}^{\mathrm{A}} \right\rangle} \qquad \text{(F.80)}$$

Thus, corresponding to the energy $E_{\mathrm{ps}}^{\mathrm{A}}$ in Eq. (F.73), the total force $\boldsymbol{F}_{\mathrm{ps}}^{\mathrm{A}}$ acting on atom A is given by

$$\boldsymbol{F}_{\mathrm{ps}}^{\mathrm{A}} = \boldsymbol{F}_{\mathrm{loc}}^{\mathrm{A}} + \boldsymbol{F}_{\mathrm{nloc}}^{\mathrm{A}} \qquad \text{(F.81)}$$

where $\boldsymbol{F}_{\mathrm{loc}}^{\mathrm{A}}$ and $\boldsymbol{F}_{\mathrm{nloc}}^{\mathrm{A}}$ are given in Eqs. (F.77) and (F.79), respectively.

Appendix G
Density-Functional Theories for Electrons and for Solutions

It is quite helpful to discuss the relationship between the density-functional theories (DFTs) for electrons and for solutions. Actually, it is famous that the Hohenberg-Kohn (HK) theorem [36], introduced in Sect. 1.3, was extended to a system with a finite temperature by Mermin [37]. The Hohenberg-Kohn-Mermin (HKM) theorem serves as the basis of the DFT for solutions. Given that $\mathbf{x}$ represents the coordinate of a solvent molecule with respect to the solute of interest, the one-to-one correspondence between the set of the solute-solvent interaction potentials $\{u(\mathbf{x})\}$ and the set of solvent distributions $\{\rho(\mathbf{x})\}$ is guaranteed by the HKM theorem in parallel to the DFT for electrons. In 2000, Matubayasi et al. developed a theory of solutions [38] where the solvent distribution $\rho^e(\varepsilon)$ on the energy coordinate ε plays a role in the construction of the solvation free energy μ of the solute in solution. In 2018, H.T. extended the method to the DFT for electrons [39] as was shown in Chap. 3. The DFT for the energy electron density $n^e(\epsilon)$ serves as a novel theoretical framework to develop the static correlation and the kinetic energy functionals, as described in Chaps. 4 and 5, respectively. Thus, there has been a 'give-and-take'relationship between the two DFTs for electrons and for solutions. In the following, we provide the proof for the HKM theorem and derive the mean field for the solvent molecules around the solute of interest.

G.1 Hohenberg-Kohn-Mermin Theorem

Suppose that a solute molecule is embedded in a solvent in an infinite dilution. For the sake of simplicity, we consider here a closed system for the solution where the intermolecular potentials are assumed to be pairwise additive. Let $u(\mathbf{x})$ and $w(\mathbf{x})$ represent the different solute-solvent interaction potentials, and $\mathbf{x}$ specifies the solvent relative coordinate with respect to the solute of interest. Then, the solvation free energy $\Delta\mu$ of the solute with the potential $u(\mathbf{x})$ in the solution can be described as

H. Takahashi, Density-Functional Theory for Electrons,
https://doi.org/10.1007/978-981-95-1229-4

$$\exp\left(-\beta\Delta\mu\right) = \frac{\int d\mathbf{X} \exp\left[-\beta\left(\sum_i u\left(\mathbf{x}_i\right) + U\left(\mathbf{X}\right)\right)\right]}{\int d\mathbf{X} \exp\left[-\beta U\left(\mathbf{X}\right)\right]} \tag{G.1}$$

where $\mathbf{X}$ collectively represents the solvent coordinates $\mathbf{x}_1, \mathbf{x}_2, \cdots, \mathbf{x}_i, \cdots$, and U stands for the solvent-solvent interaction potential. β is the reciprocal of the temperature T multiplied by the Boltzmann constant k_B. Similarly, for the solute with the potential w, we obtain the solvation free energy $\Delta\mu'$

$$\exp\left(-\beta\Delta\mu'\right) = \frac{\int d\mathbf{X} \exp\left[-\beta\left(\sum_i w\left(\mathbf{x}_i\right) + U\left(\mathbf{X}\right)\right)\right]}{\int d\mathbf{X} \exp\left[-\beta U\left(\mathbf{X}\right)\right]} \tag{G.2}$$

Then, the free energy difference $\Delta\mu' - \Delta\mu$ can be written as

$$\begin{aligned}\Delta\mu' - \Delta\mu &= -k_B T \ln \frac{\int d\mathbf{X} \exp\left(-\beta \sum_i u\left(\mathbf{x}_i\right)\right) \exp\left[-\beta \sum_i \left(w\left(\mathbf{x}_i\right) - u\left(\mathbf{x}_i\right)\right)\right]}{\int d\mathbf{X} \exp\left(-\beta \sum_i u\left(\mathbf{x}_i\right)\right)} \\ &= -k_B T \ln \left\langle \exp\left(-\beta \sum_i \left(w\left(\mathbf{x}_i\right) - u\left(\mathbf{x}_i\right)\right)\right)\right\rangle_u \end{aligned} \tag{G.3}$$

where $\langle\cdots\rangle_u$ denotes the statistical average under the solute-solvent interaction u. Since $f(x) = -k_B T \ln x$ is a downward convex function, it is possible to apply Jensen's inequality to the right side of the second equality of Eq. (G.3), which leads to

$$-k_B T \ln \left\langle \exp\left(-\beta \sum_i \left(w\left(\mathbf{x}_i\right) - u\left(\mathbf{x}_i\right)\right)\right)\right\rangle_u < \left\langle \sum_i \left(w\left(\mathbf{x}_i\right) - u\left(\mathbf{x}_i\right)\right)\right\rangle_u \tag{G.4}$$

Using the solvent distribution function, the right side of the equation can be rewritten as

$$\left\langle \sum_i \left(w\left(\mathbf{x}_i\right) - u\left(\mathbf{x}_i\right)\right)\right\rangle_u = \int d\mathbf{x}\, \rho_u\left(\mathbf{x}\right)\left(w\left(\mathbf{x}\right) - u\left(\mathbf{x}\right)\right) \tag{G.5}$$

where the distribution $\rho_u(\mathbf{x})$ is explicitly given by

$$\rho_u\left(\mathbf{x}\right) = \frac{\int d\mathbf{X} \sum_i \delta\left(\mathbf{x} - \mathbf{x}_i\right) \exp\left[-\beta\left(\sum_i u\left(\mathbf{x}_i\right) + U\left(\mathbf{X}\right)\right)\right]}{\int d\mathbf{X} \exp\left[-\beta\left(\sum_i u\left(\mathbf{x}_i\right) + U\left(\mathbf{X}\right)\right)\right]} \tag{G.6}$$

Thus, we have the inequality,

$$-k_B T \ln \left\langle \exp\left(-\beta \sum_i \left(w\left(\mathbf{x}_i\right) - u\left(\mathbf{x}_i\right)\right)\right)\right\rangle_u < \int d\mathbf{x}\, \rho_u\left(\mathbf{x}\right)\left(w\left(\mathbf{x}\right) - u\left(\mathbf{x}\right)\right) \tag{G.7}$$

By exchanging the potentials u and w, we obtain the inequality

$$-k_BT\ln\left\langle\exp\left(-\beta\sum_i\left(u\left(\mathbf{x}_i\right)-w\left(\mathbf{x}_i\right)\right)\right)\right\rangle_w<\int d\mathbf{x}\,\rho_w\left(\mathbf{x}\right)\left(u\left(\mathbf{x}\right)-w\left(\mathbf{x}\right)\right) \tag{G.8}$$

Here, we assume that the distribution $\rho_u(\mathbf{x})$ is equal to $\rho_w(\mathbf{x})$ when the solute-solvent interaction potentials u and w differ by more than a constant shift. Then, Eq. (G.8) can be transformed to

$$-k_BT\ln\left\langle\exp\left(-\beta\sum_i\left(w\left(\mathbf{x}_i\right)-u\left(\mathbf{x}_i\right)\right)\right)\right\rangle_u>\int d\mathbf{x}\,\rho_u\left(\mathbf{x}\right)\left(w\left(\mathbf{x}\right)-u\left(\mathbf{x}\right)\right) \tag{G.9}$$

It is thus found that the assumption leads to the contradiction between Eqs. (G.7) and (G.9).

From the above discussion, it can be concluded that there exist one-to-one correspondence between the set of solute-solvent interaction potentials $u(\mathbf{x})$ and the resultant set of solvent distributions $\rho(\mathbf{x})$. This conclusion justifies the construction of the density-functional theory for solutions, where the solvent distribution plays a fundamental role in providing the properties of the solution of interest. The method to prove the one-to-one correspondence for the open system (grand canonical ensemble) was provided in Refs. [37, 40].

G.2 Mean Field in Solution

In the density-functional theory (DFT) for electrons, the creation of the mean field for the electrons plays a central role in determining the ground state electron density and its energy. In the DFT for solutions, the mean field is also introduced for the solvent molecules around the solute. In the following, we formulate the mean field ω and the solvation free energy $\Delta\mu$ of a solute in solution will be described in terms of ω. To this end, we introduce Kirkwood's charging formula [40]. Suppose that $\overline{u}(\mathbf{X})$ represents the interaction energy between the solute and *all* the solvent molecules of which coordinates are collectively denoted by $\mathbf{X}$. Then, the solvation free energy $\Delta\mu$ of the solute can be expressed as

$$\begin{aligned}\Delta\mu&=-k_BT\ln\frac{\int d\mathrm{X}\exp\left[-\beta\left(\overline{u}\left(\mathbf{X}\right)+U\left(\mathbf{X}\right)\right)\right]}{\int d\mathrm{X}\exp\left[-\beta U\left(\mathbf{X}\right)\right]}\\&=-k_BT\ln\left\langle\exp\left(-\beta\overline{u}\left(\mathbf{X}\right)\right)\right\rangle_0\end{aligned} \tag{G.10}$$

where $\langle\cdots\rangle_0$ indicates that the average is taken for the system with $\overline{u}=0$. Now we introduce coupling parameter λ that ranges from 0 to 1 to describe the gradual change of the potential $\overline{u}_\lambda$ from $\overline{u}_0=0$ to $\overline{u}_1=\overline{u}$. Then, Eq. (G.10) can be reformulated as

$$\Delta\mu = -k_B T \int_0^1 d\lambda \, \frac{d}{d\lambda} \ln \langle \exp\left(-\beta \overline{u}_\lambda(\mathbf{X})\right) \rangle_0$$
$$= \int_0^1 d\lambda \, \frac{\left\langle \frac{d}{d\lambda} \overline{u}_\lambda(\mathbf{X}) \exp\left(-\beta \overline{u}_\lambda(\mathbf{X})\right) \right\rangle_0}{\langle \exp\left(-\beta \overline{u}_\lambda(\mathbf{X})\right) \rangle_0}$$
$$= \int_0^1 d\lambda \left\langle \frac{d}{d\lambda} \overline{u}_\lambda(\mathbf{X}) \right\rangle_\lambda \tag{G.11}$$

For the above formulation to hold, the solute-solvent potential needs not necessarily be pairwise additive. When the potential $\overline{u}$ is assumed to be given by sum of the pairwise potentials u, that is,

$$\overline{u}(\mathbf{X}) = \sum_i u(\mathbf{x}_i), \tag{G.12}$$

the free energy $\Delta\mu$ can be described in terms of the distribution function $\rho_\lambda(\mathbf{x})$ of the solvent molecules, thus,

$$\Delta\mu = \int_0^1 d\lambda \left\langle \frac{d}{d\lambda} \overline{u}_\lambda(\mathbf{X}) \right\rangle_\lambda$$
$$= \int_0^1 d\lambda \left\langle \frac{d}{d\lambda} \sum_i u_\lambda(\mathbf{x}_i) \right\rangle_\lambda$$
$$= \int_0^1 d\lambda \, \frac{\int d\mathbf{X} \frac{d}{d\lambda} \sum_i u_\lambda(\mathbf{x}_i) \exp\left[-\beta\left(\sum_i u_\lambda(\mathbf{x}_i) + U(\mathbf{X})\right)\right]}{\int d\mathbf{X} \exp\left[-\beta\left(\sum_i u_\lambda(\mathbf{x}_i) + U(\mathbf{X})\right)\right]}$$
$$= \int_0^1 d\lambda \int d\mathbf{x} \frac{d}{d\lambda} u_\lambda(\mathbf{x}) \frac{\int d\mathbf{X} \sum_i \delta(\mathbf{x} - \mathbf{x}_i) \exp\left[-\beta\left(\sum_i u_\lambda(\mathbf{x}_i) + U(\mathbf{X})\right)\right]}{\int d\mathbf{X} \exp\left[-\beta\left(\sum_i u_\lambda(\mathbf{x}_i) + U(\mathbf{X})\right)\right]}$$
$$= \int_0^1 d\lambda \int d\mathbf{x} \, \frac{d}{d\lambda} u_\lambda(\mathbf{x}) \left\langle \sum_i \delta(\mathbf{x} - \mathbf{x}_i) \right\rangle_\lambda$$
$$= \int_0^1 d\lambda \int d\mathbf{x} \, \frac{d}{d\lambda} u_\lambda(\mathbf{x}) \, \rho_\lambda(\mathbf{x}) \tag{G.13}$$

where $\rho_\lambda(\mathbf{x})$ is the distribution of the solvent molecules under the pairwise potential u_λ. Equation (G.13) is referred to as Kirkwood's charging formula.

Now we introduce the mean field $\omega(\mathbf{x})$ referred to as the indirect part of the solute-solvent potential of mean force. The mean field ω_λ for the coupling parameter λ is defined as

$$\frac{\rho_\lambda(\mathbf{x})}{\rho_0(\mathbf{x})} = \exp\left(-\beta\left(u_\lambda(\mathbf{x}) + \omega_\lambda(\mathbf{x})\right)\right) \tag{G.14}$$

Provided that the solvent-solvent interactions are absent, the distribution of the solvent molecules around the solute is merely described with the Boltzmann distribution and the potential $\omega(\mathbf{x})$ vanishes in Eq. (G.14). Thus, the potential $\omega_\lambda(\mathbf{x})$ gives rise to the deviation of $\rho_\lambda(\mathbf{x})$ from the Boltzmann distribution. Thus, the potential $\omega_\lambda(\mathbf{x})$ is responsible for the distribution change created by the complex correlations among the solvent molecules. From Eq. (G.14), it is possible to rewritten the potential $u_\lambda(\mathbf{x})$ in terms of the mean field $\omega_\lambda(\mathbf{x})$, thus,

$$u_\lambda(\mathbf{x}) = -k_B T \ln\left(\frac{\rho_\lambda(\mathbf{x})}{\rho_0(\mathbf{x})}\right) - \omega_\lambda(\mathbf{x}) \tag{G.15}$$

Substituting this equation to Kirkwood's charging equation in Eq. (G.13), we get

$$\Delta\mu = -k_B T \int d\mathbf{x} \left[\rho_1(\mathbf{x}) - \rho_0(\mathbf{x}) + \beta\rho_1(\mathbf{x})\omega_1(\mathbf{x}) \right.$$
$$\left. -\beta \int_0^1 d\lambda\ \omega_\lambda(\mathbf{x}) \frac{d}{d\lambda}\rho_\lambda(\mathbf{x}) \right] \tag{G.16}$$

From the fact that the correspondence between the set of solute-solvent potentials $u(\mathbf{x})$ and the set of solvent distributions $\rho(\mathbf{x})$ is one-to-one, the mean field $\omega(\mathbf{x})$ can be regarded as a functional of the distribution $\rho(\mathbf{x})$ instead of the potential $u(\mathbf{x})$. Therefore, the construction of the approximate functional $\omega[\rho](\mathbf{x})$ is one of the major subjects in the solution theory. It is quite helpful to recognize that the exchange-correlation potential $v_{xc}[n](\boldsymbol{r})$ defined in Eq. (1.111) in the DFT for electrons is regarded as the counterpart of the potential $\omega[\rho](\mathbf{x})$ in the solution theory.

The review for the developments of the approximate functionals for $\omega[\rho](\mathbf{x})$ is beyond the scope of this book. However, it is worth noting that the development has its own difficulty. The major source of the difficulty comes from the degrees of freedom (DOF) in the variable $\mathbf{x}$ to fully specify the relative configuration of the solvent molecule with respect to the solute of interest. Even when the internal coordinates of the solute and solvent molecules are being fixed, the number of DOF becomes 6 to specify the position of the center of mass and the orientation of the solvent molecule. If the internal DOF are also taken into consideration, the number increases drastically according to the number of atoms in the solute or solvent molecule. In practice, it is very difficult to cope with the calculation using the functional in the numerical applications. An approximate solution to alleviate the difficulty is to introduce the interaction sites on the solute and solvent molecules and the $\rho(\mathbf{x})$ is replaced by a set of the distribution functions of the site-site pair distribution functions $\{\rho_{\mathrm{XA}}(r_{ia})\}$.

The notation 'X' in $\rho_{XA}(r_{ia})$ stands for the solute and 'A' is for the solvent molecule. r_{ia} is the distance between the $i$the site in solute X and the ath site in the solvent A. Then, the solvation free energy $\Delta\mu$ is given by the sum of the contributions from the site-site pairs. $\{\rho_{XA}(r_{ia})\}$ are obtained by solving a set of integral equations self-consistently using the approximations called closures. The correlations among the interaction sites are quite strong and the set $\{\rho_{XA}(r_{ia})\}$ obtained as the solution of the integral equations can not realizes the true distribution $\rho(\mathbf{x})$ in practice.

The method developed by Matubayasi et al, referred to as method of energy representation, offers a simple and beautiful route [38] to circumvent the problem of increasing number of dimensions of the spatial distribution functions $\rho(\mathbf{x})$. The major characteristic of the approach is to project the distribution function $\rho(\mathbf{x})$ onto the energy coordinate ε to form the distribution $\rho^e(\varepsilon)$ defined on the one-dimensional coordinate. Explicitly, using a defining potential $u_{\mathrm{def}}(\mathbf{x})$, $\rho^e(\varepsilon)$ is given as

$$\rho^e(\varepsilon) = \int d\mathbf{x}\, \rho(\mathbf{x})\, \delta(\varepsilon - u_{\mathrm{def}}(\mathbf{x})) \tag{G.17}$$

In Ref. [38], it is proved that there also exists one-to-one correspondence between the set of potentials $u^e(\varepsilon)$ and the set of the distributions $\rho^e(\varepsilon)$ associated with the defining potential in parallel to the proof for the Hohenberg-Kohn-Mermin theorem [37]. Then, it is possible to construct the DFT for the distribution $\rho^e(\varepsilon)$ on the energy coordinate. Accordingly, the solvation free energy $\Delta\mu$ can be represented in terms of $\rho^e(\varepsilon)$ in completely parallel to Eq. (G.16), thus,

$$\Delta\mu = -k_B T \int d\varepsilon \Big[\rho_1^e(\varepsilon) - \rho_0^e(\varepsilon) + \beta\rho_1^e(\varepsilon)\omega_1^e(\varepsilon) - \beta\int_0^1 d\lambda\, \omega_\lambda^e(\varepsilon;\lambda)\frac{d}{d\lambda}\rho_\lambda^e(\varepsilon)\Big] \tag{G.18}$$

It is also worth noting that the approach by Matubayasi exploits the molecular simulations to construct the distribution functions $\rho_1^e(\varepsilon)$ and $\rho_0^e(\varepsilon)$ for the interacting (solution) and the reference (pure solvent) systems, respectively [41]. Thus, the both ends of the coupling parameter, that is $\lambda = 0, 1$ of the system are described faithfully with the molecular simulations under a given force field, and only the coupling parameter integration in Eq. (G.18) is to be approximately evaluated using a density functional. It was shown in Ref. [42], that their method is capable of evaluating the solvation free energies with the accuracies comparable to those obtained by an elaborate and time-consuming approach. Their approach was further extended to a QM/MM hybrid system [13] where the solute molecule is described with a method of quantum mechanics and the solvent molecules are represented with a classical force field.

References

1. W. Yang, P.W. Ayers, Q. Wu, Phys. Rev. Lett. **92**(14), 146404(4) (2004)
2. L.H. Thomas, Proc. Camb. Phil. Soc. **23**, 541 (1927)
3. E. Fermi, Z. Phys. **48**, 73 (1928)
4. R.G. Parr, W. Yang, *Density-Functional Theory of Atoms and Molecules* (Oxford University Press, New York, 1989)
5. J. Lindhard, Kgl. Danske Videnskab. Selskab Mat.-fys. Medd. **28** (1954)
6. J.R. Chelikowsky, N. Troullier, Y. Saad, Phys. Rev. Lett. **72**(8), 1240 (1994)
7. J.R. Chelikowsky, N. Troullier, K. Wu, Y. Saad, Phys. Rev. B **50**(16), 11355 (1994)
8. H. Takahashi, T. Hori, T. Wakabayashi, T. Nitta, Chem. Lett. **3**, 222 (2000)
9. H. Takahashi, T. Hori, T. Wakabayashi, T. Nitta, J. Phys. Chem. A **105**(17), 4351 (2001)
10. H. Takahashi, T. Hori, H. Hashimoto, T. Nitta, J. Comp. Chem. **22**(12), 1252 (2001)
11. H. Takahashi, H. Hashimoto, T. Nitta, J. Chem. Phys. **119**(15), 7964 (2003)
12. T. Hori, H. Takahashi, T. Nitta, J. Chem. Phys. **119**(16), 8492 (2003)
13. H. Takahashi, N. Matubayasi, M. Nakahara, T. Nitta, J. Chem. Phys. **121**(9), 3989 (2004)
14. H. Takahashi, W. Satou, T. Hori, T. Nitta, J. Chem. Phys. **122**(4), 044504(1) (2005)
15. H. Takahashi, Y. Kawashima, T. Nitta, J. Chem. Phys. **123**(12), 124504(1) (2005)
16. H. Takahashi, A. Omi, A. Morita, N. Matubayasi, J. Chem. Phys. **136**(21), 214503 (2012)
17. H. Takahashi, S. Umino, Y. Miki, R. Ishizuka, S. Maeda, A. Morita, M. Suzuki, N. Matubayasi, J. Phys. Chem. B **121**, 2279 (2017)
18. H. Takahashi, H. Kambe, A. Morita, J. Chem. Phys. **148**(13), 134119(9) (2018)
19. H. Takahashi, D. Suzuoka, S. Sakuraba, A. Morita, J. Phys. Chem. B **123**, 7081 (2019)
20. H. Takahashi, S. Sakuraba, A. Morita, J. Chem. Inf. Model. **60**, 1376 (2020)
21. T.L. Beck, Rev. Mod. Phys. **72**(4), 1041 (2000)
22. K. Hirose, T. Ono, Y. Fujimoto, S. Tsukamoto, *First-Principles Calculations in Real-Space Formalism* (Imperial College Press, 2005)
23. F. Gygi, G. Galli, Phys. Rev. B **52**, R2229 (1995). https://doi.org/10.1103/PhysRevB.52.R2229
24. M. Frigo, S. Johnson, The Fastest Fourier Transform in the West. Tech. Rep. MIT-LCS-TR-728, Massachusetts Institute of Technology (1997)
25. M. Frigo, in *Proc. 1999 ACM SIGPLAN Conf. on Programming Language Design and Implementation*, vol. 34 (ACM, 1999), vol. 34, pp. 169–180
26. R.N. Barnett, U. Landman, J. Chem. Phys. **48**(4), 2081 (1993)
27. U. Schumann, R.A. Sweet, J. Comput. Phys. **75**(1), 123 (1988). https://doi.org/10.1016/0021-9991(88)90102-7
28. P.M. Ricker, Astrophys. J. Suppl. Ser. **176**(1), 293 (2008). https://doi.org/10.1086/526425
29. A.D. Becke, J. Chem. Phys. **88**(4), 2547 (1988)
30. W. Gropp, E. Lusk, A. Skjellum, *Using MPI* (The MPI Press, Cambridge, 1997)
31. G.B. Bachelet, D.R. Hamann, M. Schlüter, Phys. Rev. B **26**(8), 4199 (1982)
32. L. Kleinman, D.M. Bylander, Phys. Rev. Lett. **48**(20), 1425 (1982)
33. T. Ono, K. Hirose, Phys. Rev. Lett. **82**(25), 5016 (1999)
34. K. Hirose, T. Ono, Y. Fujimotmo, S. Tsukamoto, *First-Principles Calculations in Real-Space Formalism* (Imperial College Press, London, 2005)
35. P. Pulay, Mol. Phys. **17**(2), 197 (1969)
36. P. Hohenberg, W. Kohn, Phys. Rev. **136**(3B), B864 (1964)
37. N.D. Mermin, Phys. Rev. **137**(5A), A1441 (1965)
38. N. Matubayasi, M. Nakahara, J. Chem. Phys. **113**(15), 6070 (2000)
39. H. Takahashi, J. Phys. B: Atomic, Mol. Optic. Phys. **51**(5), 055102(11) (2018)
40. J.P. Hansen, I.R. Mcdonald, *Theory of Simple Liquids* (Academic Press, 2013)
41. N. Matubayasi, M. Nakahara, J. Chem. Phys. **117**(8), 3605 (2002)
42. Y. Karino, M.V. Fedorov, N. Matubayasi, Chem. Phys. Lett. **496**, 351 (2010)

MIX
Papier aus verantwortungsvollen Quellen
Paper from responsible sources
FSC® C105338

If you have any concerns about our products,
you can contact us on
ProductSafety@springernature.com

In case Publisher is established outside the EU,
the EU authorized representative is:
Springer Nature Customer Service Center GmbH
Europaplatz 3, 69115 Heidelberg, Germany

Printed by Libri Plureos GmbH
in Hamburg, Germany